NATURAL DISASTER RESEARCH, PREDICTION AND MITIGATION

FUKUSHIMA NUCLEAR ACCIDENT

GLOBAL IMPLICATIONS, LONG-TERM HEALTH EFFECTS AND ECOLOGICAL CONSEQUENCES

NATURAL DISASTER RESEARCH, PREDICTION AND MITIGATION

Additional books in this series can be found on Nova's website under the Series tab.

Additional e-books in this series can be found on Nova's website under the e-book tab.

NATURAL DISASTER RESEARCH, PREDICTION AND MITIGATION

FUKUSHIMA NUCLEAR ACCIDENT

GLOBAL IMPLICATIONS, LONG-TERM HEALTH EFFECTS AND ECOLOGICAL CONSEQUENCES

SHIZUYO SUTOU, PHD
EDITOR

nova publishers
New York

Library of Congress Cataloging-in-Publication Data

Fukushima nuclear accident : global implications, long-term health effects, and ecological consequences / editor, Shizuyo Sutou, Ph.D. (Functional Genomics, School of Pharmacy, Shujitsu University, Nishigawara, Naka-ku, Okayama-shi, Okayama, Japan).
pages cm -- (Natural disaster research, prediction, and mitigation)
Includes bibliographical references and index.
ISBN 978-1-63463-748-0 (hardcover)
1. Fukushima Nuclear Disaster, Japan, 2011--Environmental aspects. 2. Fukushima Nuclear Disaster, Japan, 2011--Health aspects. 3. Radioactive pollution--Japan--Fukushima-ken. 4. Environmental disasters--Japan--Fukushima-ken. I. Sutou, Shizuyo.
TD196.R3F86 2015
363.17'9940952117--dc23

2014047565

Published by Nova Science Publishers, Inc. † New York

CONTENTS

PREFACE

It is close to four years since the magnitude 9.0 Great East Japan Earthquake (GEJE) occurred on March 11, 2011. It claimed the lives of more than 20,000 people. Evacuees are still 263,392 as of February 13, 2014. Of them, 132,500 are from Fukushima and more than half the families (55%) live the lifestyle of an evacuee. About half of the families (49%) are separated, with 33% being in 2 places, 12% in 3, 3% in 4, and 0.5% in ≧5 places. Approximately half of the evacuees (45%) want to return to their former homes. Refugee life is so difficult that many families have been affected by suicide, divorce, separation of family members, migration and settlement to other places, etc. Indeed, 67% of the people experienced mental or physical disorders. No one was killed by radiation itself, but 129 had committed suicide until July, 2014; 56 of these were from Fukushima (Chapter 16 by Sutou). The major effects of the Fukushima tragedy arose from the fear of low-dose radiation. Therefore, it is important to ascertain whether such fear is justified. Towards this goal, the current comprehensive knowledge of radiation physics, chemistry, and biology is presented to help understand the impact of the Fukushima accident as well as that of Chernobyl accident comparatively (Chapter 1 by Suzuki and Yamashita).

Fear of low-dose radiation comes from the concept that even the smallest amount of radiation dose is hazardous, as implied by the linear no-threshold (LNT) model used by the International Commission on Radiological Protection (ICRP) and other advisory bodies for radiation safety purposes. The LNT model is mainly based on analyses of cancers among atomic bomb survivors in Hiroshima and Nagasaki. A dose of instantaneous exposure to radiation (at high dose rate, as from atomic bombs) would result in much higher biological responses as compared to the same dose from long-term irradiation (at low dose rate, like in Fukushima). The dose-rate effectiveness factor (high dose-rate/low dose-rate) should be ~16, whereas the factor recommended by advisory bodies is ~2 (Chapter 6 by Tanooka). Re-examination of the leukemia data in atomic bomb survivors published in 1958 clearly shows that the concept of zero threshold dose is not valid (Chapter 2 by Cuttler). The latest dose-response data for solid cancer mortality is not linear, because of less than expected cancers in the low-dose range (Chapter 3 by Doss). The biphasic dose response where low doses exert beneficial effects while high doses exert deleterious effects seems to be an adaptive response attained through the evolutional history of life on the earth over billions of years. A large amount of data accumulated over the past 30 years supports the concept of adaptive response, which some prefer to call hormesis (Chapter 5 by Sugimoto). The Keap1/Nrf2 pathway has been shown to be responsible for the adaptive response or hormesis (Chapter 7 by Maruyama and Itoh). A very important result obtained from intensive analyses of atomic bomb survivors is that no genetic effects have been detected in their offspring (Chapter 4 by Kodama). Indeed, when the trans-generational effects of chronic low dose-rate internal radiation exposure were investigated by giving drinking water containing ^{137}CsCl (100 Bq/ml) to mice, no hazardous effects have been seen even up to the tenth generation. Rather, the presence of radiation hormesis was indicated as shown by the suppression of tumor growth (Chapter 13 by Nakajima). In 2012, high rates of abnormalities of butterflies captured in the areas surrounding Fukushima have been sensationally reported as genetic hazards, but Sutou (Chapter 14) and Obinata (Chapter 15) argue that these were not due to genetic effects, but demonstrably due to decomposition of some proteins or small RNA molecules.

There are several areas with high background levels of radiation in the world. Epidemiological studies evaluating the solid cancer risk in these areas suggest that the cancer risk is minimal, if any (Chapter 10 by Akiba). Experimentally, life-time exposure of β-rays to the back skin of mice three times a week induced skin cancers for the groups having 0.75-11.8 Gy/dose, but at 0.5 Gy/dose, the cancers did not occur at all until the death of the last mouse, with the lifespan being the same as the control group. The author deliberately refers to this non-carcinogenic dose (0.5 Gy) as the threshold-like dose (Chapter 9 by Ootsuyama). Another study of lifetime exposure of γ-rays to mice showed that despite the significant over-all shortening of lifespans in the 21 mGy/day groups, there was no significant increase in the incidence of neoplastic and non-neoplastic lesions (Chapter 8 by Tanaka et al.). These natural and experimental findings strongly support the existence of thresholds.

In addition to topics mentioned above, image analyses of leaves contaminated with ^{137}Cs collected in Chernobyl and Fukushima were conducted (Chapter 11 by Nakajima). This method provides an easy and quick method of biological monitoring of contamination, and complements the usual physical measurements. Sutou (Chapter 12) reports on the volunteer activities of monitoring the radio-contamination in returnees from evacuation zones in Fukushima. Although these activities are just the tip of an iceberg, this record offers a glimpse of what Japanese people did for the refugees at the time of the historical accident.

As described above, this book denounces the LNT model and maintains that low dose radiation is not hazardous, but possibly even beneficial. This book is compiled primarily to get rid of the spell of the LNT model and release Fukushima people from the undue torture from the fear of low-dose radiation. Also, there are many people that use radiation worldwide such as workers at nuclear power plants, radiologists, radiological technologists including X-ray operators, radiation researchers, etc. These people would feel relieved to learn that the alleged risk of low-dose radiation is illusionary and negligible, and the low-dose radiation may even be beneficial. This would also be applicable to the public in general who receive radiological examinations. Japanese public receive 4 mSv radiation dose per year from medical procedures on the average and enjoy significant benefit of longevity, because it is not harmful. In addition to the people directly associated with radiation, policy makers of nuclear energy and radiation who are working for governmental and/or regulatory agencies are encouraged to read this book. They claim that they are protecting the general public from hazardous effects of radiation and often set severe regulatory limits such as 1 mSv/y. Though this dose limit was chosen as a conservative measure, the harm from the resultant fear of low-dose radiation is much higher than the benefits when we consider of the tremendous harm to the Fukushima population. The same holds true for some civil activists and journalists who emphasize the dangers of low-dose radiation raising the fear of low-dose radiation. With the increase in scientific knowledge, it is the time to shed new light on the outdated LNT model now. Useful strategies (Chapter 1, by Suzuki and Yamashita) and recommendations (Chapter 2, by Cuttler) are presented.

Shizuyo Sutou
(Editor-in-chief)
Hiroshi Tanooka
Mohan Doss

October 31, 2014

In: Fukushima Nuclear Accident
Editor: Shizuyo Sutou

ISBN: 978-1-63463-748-0

Chapter 1

PERSPECTIVE: HEALTH-RISK IMPLICATIONS OF THE FUKUSHIMA NUCLEAR POWER PLANT ACCIDENT

Keiji Suzuki** and Shunichi Yamashita
Department of Radiation Medical Sciences, Atomic Bomb Disease Institute, Nagasaki University, Sakamoto, Nagasaki, Japan

ABSTRACT

After the Fukushima Daiichi Nuclear Power Plant (NPP) accident, health effects from radiation exposure have been of deep concern. Currently, the linear no-threshold (LNT) model is applied for the purpose of radiation protection, however, our knowledge on human cancer risk from doses of 100 mSv or less is quite limited. Furthermore, although a dose-and-dose-rate effectiveness factor (DDREF) has been applied for radiation protection, the risks of cancer from chronic low dose radiation exposure are insufficiently understood.

Such uncertainties result in an establishment of prolonged "radiation phobia" in the public, which should be overcome. Since the elucidation of the mechanism of radiation-induced carcinogenesis is indispensable to evaluate radiation effects from low-dose and low dose-rate radiation exposure, we intend to review previous studies providing the mechanistic insights into the effects of radiation at cellular and molecular levels, and discuss possible roles of radiation in radiation-induced carcinogenesis, especially for thyroid cancers.

In this chapter, current scientific knowledge of radiation physics, radiation chemistry, and radiation biology is overviewed in order to comprehend the radiation health effects in Fukushima. Experiences after the Fukushima accident and on-going projects aiming at recovery from the disaster will be discussed.

Keywords: Ionizing radiation - low-dose - epidemiology – thyroid cancer - A-bomb survivors – Chernobyl accident – Fukushima accident

---

* Correspondence to: TEL: +81-95-819-7116, FAX: +81-95-819-7117, E-mail: kzsuzuki@nagasaki-u.ac.jp

Introduction

Soon after the discovery of X-rays by Röntgen in 1895, it was recognized that radiation exposure had a potential to cause acute tissue damage. Later, it was known that cancer, particularly leukemia and skin cancer, could be induced by radiation exposure. Since then, accumulated evidences have demonstrated that radiation is capable of inducing cancer in many of the tissues and organs. Subsequently, general needs of protecting people, especially occupational workers from harmful radiation exposure have lead to the establishment of international organization such as ICRP.

In contrast to the acute deterministic effects, the knowledge about the dose-response relationship for the risk of late effects, such as cancer, was quite limited until the risk of leukemia and solid cancer was estimated based primarily on the atomic bomb (A-bomb) survivors in Hiroshima and Nagasaki in 1945 [1]. During the 1980s the data from the follow-up of A-bomb survivors led to a revision of the earlier risk estimates [2, 3], and provided the fundamental principle of the LNT model. It assumes that the cancer risk will increase linearly dependent upon the dose. However, since the risk estimates have been obtained from the epidemiological studies of A-bomb survivors, who received moderate to high doses at extremely high-dose rates, the estimates may not be applicable for populations receiving lower doses at low-dose rates. Thus, a reduction factor of 2, which is called a dose and dose-rate effectiveness factor (DDREF), has been proposed for chronic exposures at low-dose rates [4], while another report proposed a DDREF value of 1.0 – 1.5 [5].

Further information from a number of epidemiological studies on cancer induction by exposure to external and internally incorporated radioactive nuclides has indicated that caution is needed in interpreting the dose-response relationships obtained by direct extrapolation from epidemiological studies in A-bomb survivors, particularly at lower doses [6]. Furthermore, recent results with regard to radiation-induced adaptive response, radiation-induced genomic instability, and bystander effects have indicated that simple extrapolation from high-dose effects may not be justified. Thus, the comprehensive understanding of the mechanism of radiation-induced carcinogenesis is indispensable to evaluate radiation health effects from low dose and low dose-rate radiation exposure. We aimed at overviewing previous studies, which could provide the mechanistic insights into the effects of radiation at cellular and molecular levels. Moreover, molecular studies, which intend to identify molecular changes associated with radiation-induced childhood thyroid cancer after the Chernobyl accident, are compiled in order to discuss possible roles of radiation in radiation-induced carcinogenesis. Finally, experiences after the Fukushima accident will be introduced in relation to the establishment of resilience to radiation for rebirth from the disaster.

1. Physical Aspects of Ionizing Radiation

1.1. Definition of Low-Dose Radiation

Based upon the dose response for mortality from solid cancers among A-bomb survivors, United Nations Scientific Committee on the Effects of Atomic Radiation (UNSCEAR) 1993 report considered that a low dose could be less than 200 mGy [7]. Recently, it has been reported that statistically significant risk elevation was not observed at doses of 100 mSv or less [8], so that low dose could be defined to be 100 mSv or less. The Biological Effects of Ionizing Radiation (BEIR) VII report of the US National Academy of Sciences defined low dose as doses up to about 100 mSv [5]. Accordingly, low doses are defined as those less than 100 mSv.

1.2. Units of Radiation Exposure

The quantity used to determine the amount of ionizing radiation is called the absorbed dose, which is defined as the energy absorbed per unit mass. The unit of absorbed dose is the gray (Gy), and 1 Gy equals to 1 joule of energy absorbed per kilogram of matter. It is well recognized that different types of radiation produce different biological effects. For example, alpha particles and neutrons are more effective than electrons and photons. High linear-energy-transfer (LET) radiations, such as heavy-ion particles, have profound effects. Therefore, the equivalent dose, which is the product of absorbed dose and radiation weighting factor, has been introduced. The unit of equivalent dose is the sievert (Sv). Finally, effective dose is used to limit health risks from radiation exposure. Effective dose is the sum of all of the weighted equivalent doses in all tissues and organs exposed. Since different tissues show very different radiation sensitivities, tissue-weighting factors are used to calculate weighted equivalent doses. Thus, if the effective dose is used for radiation exposure, radiation health effects are the same between external and internal exposure. The unit of effective dose is again the sievert (Sv).

1.3. Direct and Indirect Effects

Absorption of radiation energy into DNA directly induces structural alterations of DNA, which is called direct effects. Alternatively, interaction of radiation with water molecules surrounding DNA molecules produces water-derived free radicals that indirectly cause structural alterations of DNA. This action is called indirect effect. Structural changes in DNA molecules are collectively called DNA damage, and it is estimated that approximately 60 - 70% of DNA damage induced by low-LET radiation is resulted from indirect effects, while about 30 – 40% of the damage is caused by direct absorption of radiation energy to DNA [9,10]. Similar biochemical process of indirect effects was well recognized as the physiological source of the endogenous DNA damage [11, 12].

1.4. Induction of DNA Damage by Radiation

Absorption of radiation energy into DNA gives rise to a variety of DNA damage. Those include base damage, apyrimidinic/apurinic (AP) sites, single-strand breaks (SSBs), double-strand breaks (DSBs) and crosslinks [13-15]. As presented in Table 1.1, the number of each type of DNA damage was calculated theoretically, and some of them have been confirmed experimentally. The most prevalent DNA damage is base damage, followed by AP sites and SSBs. While the yields of DSBs are the lowest, they are the most detrimental damage. It has been estimated that the yield of DSB per cell per Gy is approximately 40 [9]. Recently, various surrogate markers for DSBs have been developed. For example, phosphorylated histone H2AX, a member of histone H2A, is stimulated by ataxia-telangiectasia mutated (ATM), which is activated by DSB induction. Phosphorylated histone H2AX forms discrete foci within the chromatin regions surrounding DSB, so that the number of foci can be considered as the number of DSBs. By using such surrogate markers, the number of DSB per cell has been determined (Figure 1.1), and approximately 40 foci per cell are reported to be induced by 1 Gy of low-LET radiation in cultured human cells [16-18].

Table 1.1. Number of DNA damage induced by 100 mGy of γ-rays

DNA damage	No. damage per cell
Base damage	~100
A-base sites	~30
Single-strand breaks	~100
Double-strand breaks	~4

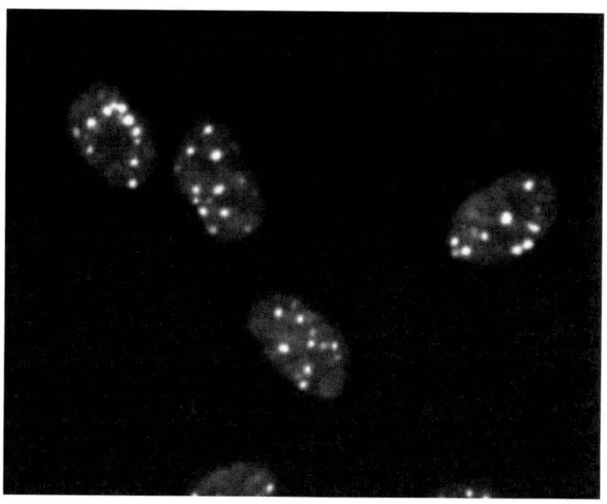

Figure 1.1. Induction of foci of DNA damage response factor (53BP1). Normal human cells were exposed to 0.25 Gy of X-rays. After 1 hour, cells were fixed with 4% formalin and analyzed by immunofluorescence using anti-53BP1 antibody. Pale grey indicates the nucleus, while white dots show 53BP1 foci.

Induction of DSBs by low-dose radiation also quantified by foci counting, and over the few mGy to 1000 mGy dose range the induction shows linear dose-response [16,17]. More recently, *in vivo* formation of DSBs was examined in lymphocytes obtained from individuals undergoing computed tomography (CT) examinations. It was found that the number of DSBs increased linearly dependent on the dose-length product (DLP) [19], which reflect radiation doses.

1.5. Endogenous DNA Damage

It is well established that free radicals created along the low-LET radiation tracks are the source of indirect effects. While ionizing radiation is definitely an efficient inducer of free radicals, there are also the endogenous (intracellular) free radicals, collectively called reactive oxygen species (ROS), which are generated from mitochondrial oxidative metabolism and other reactions in cells [11]. Therefore, similar to the indirect effect of ionizing radiation, it has been shown that DNA damage is also induced by endogenous ROS. According to the previous estimation, the average generation rate is approximately 10^9 ROS per cell per day [12], which results in 10^6 oxidative DNA damage, 10^5 SSBs, and, therefore, spontaneous DSBs are estimated to be approximately 0.1 per cell per day [11]. It should be noted that this DSB level seems to be equivalent to the estimated number of DNA damage in cells receiving approximately 1000 mGy per year which is equivalent to 0.1 mGy/h. Experimentally, levels of spontaneous DSBs have been quantified in several previous studies. For example, the studies using cultured normal human cells reported that the number of DSBs in the control cells is approximately 0.1 DSB per cell. Detection of DSBs in mouse tissues also demonstrated that spontaneous DSB levels were between 0.1 and 2.8 per cell. Thus, it can be concluded that spontaneous DSBs are commonly induced in cells of tissues and organs even without radiation exposure.

2. Cellular Response to Radiation

2.1. DNA Damage Repair

DNA damage repair is a fundamental system inherited by living organisms. Accordingly, most of the repair systems found in prokaryotes existed in mammalian cells [20]. Multiple mechanisms are involved in repairing various types of DNA damage. For example, oxidative DNA damage, such as base damage, AP sites and SSBs, is efficiently repaired through the base excision repair and single-strand break repair pathways [21-24]. The initial step is governed by DNA glycosylase, which excises modified base. Then, resultant AP sites are cleaved by AP endonuclease, giving rise to SSBs. At least two independent types of systems are involved in processing SSBs. The nucleotides gaps are filled with DNA polymerases, and DNA termini are rejoined by DNA ligases. Thus, while the latter steps use the common mechanism, different types of enzymes respond to various DNA damage structure. It is well know that oxidative base damage, such as 8-oxoguanine, causes mis-match base pairing during DNA replication and eventually induces mutation [25]. Thus, cells have systems, by which damage base or damaged base-containing nucleotides are eliminated from the nucleotide pool [26].

Since DNA double-strand breaks result in disruption of higher-order structure of chromatin, these lead to gross alterations of the genome structure. Thus, multiple DNA repair pathways are also involved in repairing DSBs [27-29]. Nonhomologous end joining (NHEJ) is the major repair pathway for DSBs through the cell cycle. Classical NHEJ does not use any sequence homology, therefore, it does not need DNA end processing. However, alternative-NHEJ and homologous recombination (HR) are dependent on DNA end resection. HR is functional only after the sister chromatid is provided through DNA replication. In principle, HR is an error-free pathway, whereas NHEJ, particularly alternative-NHEJ is error-prone [30, 31].

While DSBs are spontaneously generated during DNA replication, or produced by specific nucleases during V(D)J recombination, class switch recombination and meiotic recombination, in addition to those induced by endogenous ROS, repair of DNA double-strand breaks has been an efficient process, and reparable DSBs are generally eliminated within 24 hours after radiation exposure. In fact, it has been shown that DSBs induced by X-ray doses up to 200 mGy were completely repaired in proliferating human cells after 24 hours [16]. Thus, while the initial induction of DSBs shows linear dose-response relationship, DNA damage caused by low-dose radiation has little chance to persist in cells. This is in sharp contrast to those induced by high dose radiation, which often result in residual DSBs [16, 17].

Repair DSBs was also examined following chronic low-dose-rate irradiation. *In vitro* experiments showed that normal human diploid fibroblasts exposed to γ-rays at a dose rate of 18 mGy/h did not accumulate DSBs nor phosphorylation of p53 [32]. According to the previous estimation, endogenous DSBs are formed from single-strand DNA lesions, including SSBs, AP sites and oxidative base damage, in replicating cells at a rate equivalent to that of DSBs induced by radiation at a dose rate of 282 mGy/h [33]. Therefore, at around this dose rate, human cells are expected to repair DSBs efficiently and faithfully. This assumption is in good agreement with the data showing that DSBs induced at a dose rate of 238 mGy/h were repaired with no error [34]. Although increased levels of residual DSBs were observed with 102 mGy/h in confluent non-dividing cells [35], there was a similar paper reporting that DSBs induced in quiescent normal human fibroblasts by very low-dose radiation, such as 1 mGy of X-rays, remained unrepaired for many days, but it was rapidly repaired if the cells were allowed to proliferate [16]. Thus, it is obvious that low level DSBs are efficiently repaired with high fidelity especially in proliferating human cells.

2.2. DNA Damage Response

While DNA repair pathways efficiently amend DSBs, a certain fraction of the initial breaks possibly remains unrepaired depending upon the dose. Such lesions could be complex lesions or clustered damage, or DSBs induced in heterochromatic regions [36, 37]. If cells with residual DSBs are replicated, the stability of the genome is threatened. Therefore, cells have evolved a system called DNA damage checkpoint, which protects the integrity of the genome [27, 29]. The central players of DNA damage checkpoint pathway are ATM and p53 proteins (Figure 1.2). Once DSBs are sensed by ATM, it is activated as a protein kinase and catalyzes phosphorylation of downstream factors, such as p53. p53 protein is a well known tumor suppressor and regulates transcription of various genes, whose products regulate cell death or irreversible growth arrest [27, 29]. Other downstream factors include DNA repair proteins, DNA damage checkpoint mediators, chromatin modifiers, and histones.

Recently, it has been shown that DNA damage signal is amplified through formation of multiple protein complexes [38]. The initial event is ATM-dependent phosphorylation of a histone H2AX, a member of histone H2A and comprises nucleosome cores, which plays a critical role in anchoring the protein complex of DNA damage checkpoint factors. As soon as the histone H2AX at serine 139 is phosphorylated, its phosphorylation recruits MDC1 protein that harbors BRCT motif recognizing phosphorylated serine/threonine. MDC1 is autophosphorylated and brings MRE11/RAD50/NBS1 complex, which are the phosphorylation targets of ATM. MDC1 by itself is also phosphorylated by ATM, resulting in a recruitment of BRCT-containing RNF8 protein. RNF8 forms heterodimer with UBC13, which catalyzes ubiquitination of proteins. As a result, RNF163/UBC13, another ubiquitination enzyme, is recruited and ubiquitylates histone H2A at lysine 13/15. Exposure of dimethylation of histone H4 and ubiquitination of histone H2A recruits 53BP1, which stimulated a certain pathway of DNA repair and facilitate further DNA damage response.

Thus, multiple proteins are sequentially recruited to the chromatin containing phosphorylated histone H2AX. Further studies have demonstrated that phosphorylation of histone H2AX expands over several mega-base chromatin regions surrounding the initial DSB site [39]. Therefore, thousands of protein complexes are visualized as discrete foci under fluorescence microscope, indicating that the foci could be used as the surrogate marker for DSBs. In fact, earlier studies have already shown that the foci of phosphorylated histone H2AX, also known as γ-H2AX foci, were induced dose-dependently. Moreover, the number of foci is well correlated with the actual number of DSBs, so that the foci are now widely used to quantify the number of DSBs [39].

Several previous studies have also demonstrated substantial change in the foci size. The initial tiny foci, which are detectable immediately after radiation exposure, become large in size, if the residual foci persist for a long time. In fact, one focus persisted for 24 hours after irradiation became more than 2 μm in diameter, so that over a thousand molecules are estimated to have accumulated. Thus, DNA damage signals are remarkably amplified at the site of chromatin neighboring residual DSBs.

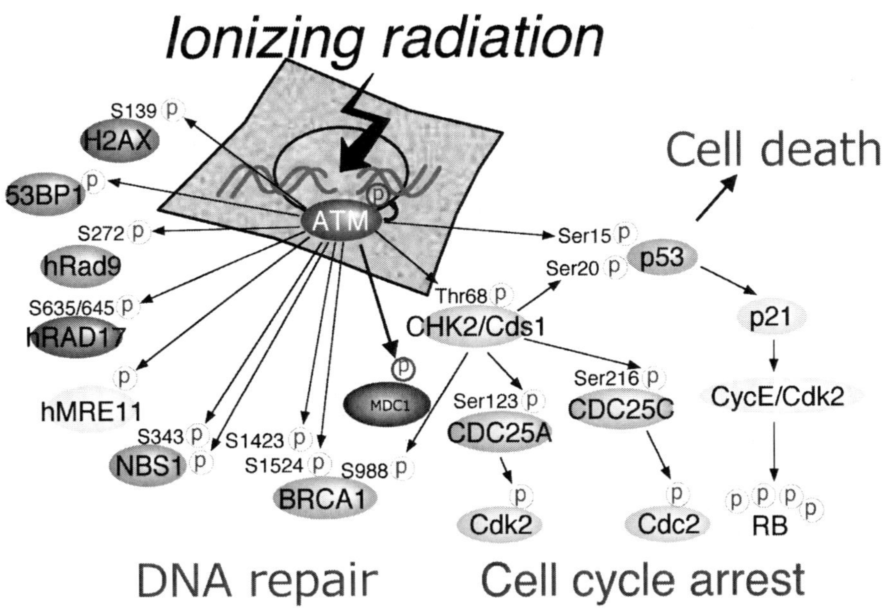

Figure 1.2. Schematic illustration of DNA damage signaling pathway. ATM-dependent phosphorylation (P) transduces DNA damage signals to the downstream factors, which execute DNA damage response, including enhanced DNA repair, regulation of cell cycle progression, and induction of cell death.

Amplification of DNA damage signal plays a crucial role when the amount of DNA damage is small. Amplification of ATM-dependent DNA damage signals is essential for executing cell-cycle arrest, particularly G_1 arrest, as AT cells defective in ATM function fail to initiate G_1 arrest [38]. Although recent study reported that the G_1 checkpoint was inefficiently maintained [40], it should be mentioned that cells with DNA damage terminate cell proliferation at G_1 phase within the next few cell cycle [41]. Thus, cells have evolved a sophisticated system, by which even a small number of DSBs can be detected efficiently.

2.3. Non-Targeted Effects

2.3.1. Radiation-Induced Genomic Instability

It is well established that radiation effects are manifested in the targeted cells, which absorb radiation energy directly. However, accumulating evidences have shown that non-targeted effects, which are expressed in cells that have not been directly exposed to radiation, are also induced. Manifestation of non-targeted radiation effects was first described in the studies investigating the descendants of targeted cells surviving from lethal effect of radiation exposure. Subsequently, several reports confirmed that delayed cell death was induced when the surviving cells propagate in culture. For example, colonies formed by surviving cells frequently contained dead cells. So far, various phenotypes of delayed effects were confirmed in surviving cells, such as delayed chromosome aberrations, delayed micronuclei formation and delayed mutations [42-46]. It was hypothesized that a sort of instability is transmitted through the progeny of surviving cells, so that the phenomenon was called as radiation-induced genomic instability.

While most of the studies have been performed in cultured cells, radiation-induced genomic instability was also identified in mouse models as trans-generational effects [46]. For example, male mice were exposed to radiation and mated with unirradiated female mice. Then, the F_1 progeny descended from irradiated spermatozoa were analyzed. The results showed that mutations in minisatellite sequences, which are composed of a stretch of short tandem repeats, were increased in the paternal allele as well as in the maternal allele [47]. Thus, it turned out to be clear that indirect radiation effects exist not only *in vitro* but also *in vivo* [48-50], and they greatly contribute to the overall consequences of radiation exposure.

Since the incidence of delayed cell death was in excess of the frequency of gene mutation at similar doses, mutation in a single gene or gene family was unlikely to be responsible for delayed cell death. It can be hypothesized that radiation signature is assumed to be memorized in the genome of the progeny of surviving cells, which resulted in a manifestation of delayed phenotypes. As restriction enzymes were able to induce radiation-induced genomic instability, DNA double-strand breaks appeared to be the nature of radiation signature [45, 51]. We have proposed that mis-rejoining of DNA breaks created potentially unstable chromosome regions, which could be transmitted as radiation signature, and they might cause unwanted DNA breakage in the progeny of surviving cells [45]. Delayed DNA damage induction was also postulated to be created by elevated oxidative level in surviving progenitors, since oxidative stress was often observed in the hematopoietic cells derived from bone marrow receiving radiation exposure [44, 52]. Furthermore, isolation of unstable clones after radiation exposure suggested an alternative pathway, which was an involvement of soluble protein factor inducing delayed cell death [53, 54]. So-called death-inducing factors were identified by medium transfer from unstable clones to unirradiated cells. These observations indicate that endogenous DNA damage as well as exogenous signals through secretion of soluble factors could be the mechanism of indirect radiation effects. Although the induction of gene transcriptions and proteins in targeted cells was reported previously [47-57], the new paradigm has disclosed that similar inducible phenotypes are detectable in non-targeted cells.

2.3.2. Adaptive Response

Adaptive response was first described in human lymphocytes, in which a low "priming" radiation dose preceding a larger "challenging" dose protected cells from excess chromosomal aberrations [58, 59]. Amelioration of cell survival by the challenging dose was demonstrated by the priming dose. Decreased mutation frequency as well as suppression of neoplastic transformation was confirmed as a result of adaptive response [60]. In general, doses of 5 – 100 mGy have been used for priming, and more than several hours, mostly 3 – 6 hours, are required for adequate induction of adaptive response. The occurrence of adaptive response was also reported *in vivo*, and the induction of leukemia and lymphoma as well as the development of heritable mutations were reduced by priming dose [61, 62].

The mechanism of adaptive response has not been fully understood yet, but some studies have suggested an involvement of p53 and DNA repair functions [63]. Although there is no clear evidence that factors associated with DSB repair are up-regulated after the priming dose, induction of base damage repair and activation of DNA damage response were reported. While adaptive response has been identified in many of the systems, the response is not always detectable. For example, adaptive response for chromosomal aberrations has been shown to vary with individuals, and some donors did not show any adaptive response [64]. Therefore, it is difficult to evaluate the potential significance of the adaptive response for the estimation of cancer risk.

2.3.3. Bystander Response

The radiation-induced bystander effect was initially demonstrated in cells exposed to low dose α-particles [65]. While 1% of the cellular nuclei were traversed by a α-particle, increased chromosome aberrations were observed in over 30% of cells, indicating that cells that were not directly targeted by α-particle manifested radiation effects. These non-targeted cells are those in close proximity to the targeted cells, therefore, they are referred to as bystander cells [66-68]. Subsequently, microbeam technology, which allows deposition of radiation energy to specific cells at defined location, was applied to demonstrate bystander effects [69, 70]. A variety of biological effects have been confirmed in bystander cells after microbeam irradiation, including cell death, chromosome aberration, and mutation.

Until now, two types of bystander mechanisms have been proposed [69-71]. In a confluent culture, where cells are attached each other, gap-junctional intercellular communication (GJIC) plays a role in transmitting bystander signals. Previous study in normal human fibroblasts showed that chemical inhibition of GJIC prevented bystander signaling [72], whose effects were not observed in tumor cells lost GJIC. The second mechanism involves the release of soluble factors from the targeted cells. While bystander factors have not been thoroughly described yet, signal transduction pathways activated in bystander cells indicate major roles of cytokines, including IL-6 and IL-8, transforming growth factor (TGF)-β1, tumor necrosis factor (TNF)-α, reactive oxygen species (ROS), reactive nitrogen species (RNS), and lipid peroxidation products [66-68,73-78]. Thus, it is not surprising that many of the aspects of bystander signaling have some connection to inflammatory response. In fact, cyclooxygenase (COX) 2-dependent signaling pathway, which is a central player in cellular inflammatory response, also plays a crucial role in bystander response [79, 80].

Bystander effect has been confirmed in various types of cells, including fibroblasts, lymphocytes, keratinocytes, and endothelial cells [81-83], however, there are also reports that failed to detect bystander effects in various cell types [84, 85]. Lack of evidence for radiation-induced bystander effects was also reported after microbeam irradiation [86-89]. In cultured stem cells, contradictory results were obtained as well [90-93]. Therefore, attention should be paid to the confounding factors, such like cell types, cell lines, culture conditions, endpoints, etc., that influence bystander response.

Recent studies have shown that bystander response is also observed in reconstituted, three-dimensional tissues [94-101]. In a skin-like model, damaged cells were detectable in unirradiated cells up to 1 mm away from the targeted region [96]. Bystander effect was also found in lung reconstituted model [96]. *In vivo* evidence of abscopal bystander effects are rather limited, but it was reported that partial irradiation of rat lung induced DNA damage in the other unirradiated lung [102]. Another study showed that irradiation of lower abdomen of mice gave rise to COX-2 induction in out-of-field lung tissues [103], in which TGFβ signaling is involved [104]. Thus, accumulating evidences have suggested the potential clinical relevance of bystander effects in tissue homeostasis.

3. Radiation Exposure and Carcinogenesis

3.1. Epidemiological Study

3.1.1. A-Bomb Survivors

The most informative epidemiological study, called the Life Span Study (LSS), has been conducted in the atomic bomb survivors in Hiroshima and Nagasaki. It is based upon large numbers of persons with various whole-body doses [105]. Excess relative risk (ERR), which is a relative increase in cancer risk in the study population due to the radiation at given doses, has shown the close relationship between the radiation dose and the risk of cancer induction/mortality. The latest report on LSS mortality stated that the ERR estimate for solid cancers was 0.47 per Gy. The dose-response relationship at low doses below 1 Gy was described by both a linear and a curvilinear function [106]. In the dose range 0 - 150 mSv the excess risk of solid cancer was assumed to be linear, however, there is no statistically significant risk elevation at doses below 100 mSv. In relation to the Fukushima issue, the strong link between radiation exposure and thyroid cancer was also provided by studies of the A-bomb survivors, but for higher doses [107]. The dose-response relationship was assumed to be linear, and gender-averaged ERR estimate was 0.57 per Gy [106]. The most important modifier of thyroid cancer risk is age at exposure, and elevated risk is no loner detectable among survivors exposed after the age of 30.

The A-bomb survivors received higher external doses over short period, and the LSS cohort has played a critical role in obtaining the basic coefficients of the risk estimation. In addition, the LSS cohort has provided the unique opportunity to evaluate the scientific validity of the LNT model. So far, the dose-response relationship supposed to fit to the LNT model in principle at higher doses, however, the dose-response relationship below 100 mGy tend to fluctuate, which limits statistical significance of the increase of cancer at lower doses. While the LNT model has been used for evaluating the cancer risk from radiation exposure, even the most celebrated epidemiological study could not uncover the uncertainties of the radiation effects below 100 mSv, which requires understandings of the molecular mechanisms of radiation-induced carcinogenesis.

3.1.2. Chernobyl Accident and Childhood Thyroid Cancer

The Chernobyl nuclear power plant accident (Chernobyl NPP) on April 26, 1986 released large amount of radioactive materials, which resulted in radiation exposures of the residents living in the affected regions [108-110]. In particular, fallout of radioactive iodine caused internal exposure through ingestion of contaminated foods and inhalation, which causes childhood thyroid cancer as one of the main health effects of the accident [111]. Among children and adolescents under 18 years in 1986, 6848 cases of thyroid cancer were reported between 1991 and 2005 [109]. The epidemiological studies showed a very strong dose-response relationship, and the risk appeared to increase linearly with dose up to 1.5 to 2 Gy, although statistically significant increase in risk was not observed below 200 mGy [112]. The estimated excess relative risk of thyroid cancer among children younger than 15 years at the time of the accident was 5.6 per Gy. Recent analyses of thyroid cancer prevalence in the Belarus and Ukrainian cohorts showed a linear dose-response below 5 Gy with an excess risk of 2.15 and 1.91 per Gy, respectively. In both cases, no statistically significant increase in risk was observed below 100 mGy [113-116].

3.2. Mechanism of Radiation-Induced Childhood Thyroid Carcinogenesis

It is generally accepted that cancer has arisen as a result of accumulation of oncogenic mutations. Mathematical considerations estimate that solid cancers demonstrate age-dependent increase in the incidence roughly by the fifth power of age. This could be the most appropriate explanation of the reason why adulthood cancers make an appearance late in life.

After the Chernobyl accident, early childhood thyroid cancer cases showed significantly higher prevalence of rearrangements between the *rearranged during transfection* (*RET*) gene and the *PTC3* gene (*RET/PTC3* rearrangement) [117-119]. Subsequently, the *RET/PTC1* rearrangement was also identified, which is the most prevailing genetic alteration in childhood thyroid cancers after the Chernobyl accident overall [120,121]. The RET/PTC rearrangements were created thorough the fusions of the *RET* proto-oncogene with several partner genes, which collectively been designated as the *PTC* genes (Table 1.2) [122]. The *RET* gene encodes a transmembrane receptor tyrosine kinase. The fusion partner proteins are commonly expressed in thyroid follicular cells and possess coiled-coil domains that enable homodimerization of the fusion RET/PTC proteins. As a result, RET/PTC proteins constitutively activate the mitogen activated protein kinase (MAPK) pathway without any ligand binding [120, 123-126].

RET/PTC1 and *RET/PTC3* rearrangements are created through the intrachromosomal inversion within chromosome 10, which requires at least two independent DNA double strand breakages. Since radiation exposure does cause DSBs, it had been hypothesized that radiation exposure from the internal ^{131}I caused DSBs resulting in the oncogenic genome rearrangements after illegitimate recombination [120]. The possibility was supported by the experimental studies, in which radiation-induced *RET/PTC* rearrangements were confirmed in X-irradiated primary thyroid tissues transplanted into SCID mice, although the doses used exceeded 50 Gy [127].

Later, several independent studies following the earlier reports have evaluated the prevalence of *RET/PTC* rearrangements in childhood thyroid cancer after the Chernobyl accident, and the compiled data demonstrated that *RET/PTC* rearrangements were detectable to the same extent both in childhood thyroid cancers after the Chernobyl accident and sporadic childhood thyroid cancers [128-130]. Subsequent studies demonstrated that the frequency of thyroid cancer with *RET/PTC* rearrangements decreases with age in sporadic cases, while those with the *BRAF* mutation becomes greater. It is well established that these two genetic changes are mutually exclusive. Individuals born before the accident are now twenty-eight years or older, and recent report has suggested that the frequency of thyroid cancer harboring *BRAF* mutation tends to increase in the affected group [131]. Thus, it seems likely that molecular changes in thyroid cancer after Chernobyl accident mirror those occurring spontaneously.

Table 1.2. Oncogenic RET/PTC rearrangements in childhood thyroid cancers related to the Chernobyl accident

Oncogenes	Rearrangement partners
RET/PTC1	*CCDC6* (also *H4*)
RET/PTC2	*PRKAR1A*
RET/PTC3	*NCOA4* (also *Ele*)
RET/PTC4	*NCOA4* (also *Ele*)
RET/PTC5	*GOLGA5* (also *RFG5*)
RET/PTC6	*TRIM24*
RET/PTC7	*TRIM33* (also *RFG7*)
RET/PTC8	*KTN1*
RET/PTC9	*RFG9* (also *MBD1*)

Then, what would be the role of radiation in thyroid carcinogenesis? Possible involvement of tissue disturbance in thyroid carcinogenesis has been discussed in the observations, in which chronic autoimmune thyroiditis, such as Hashimoto's thyroiditis, is sometimes accompanied by cancer [132]. Although the link is still debated, it seems likely that PTCs may develop if the cells with oncogenic mutations preexisted in the region with Hashimoto's thyroiditis. It should be noted that proliferative response was observed in Hashimoto's thyroiditis [133], and therefore, the disturbance of tissue homeostasis by chronic inflammation could create a condition for the cells harboring spontaneous RET/PTC rearrangement to undergo cell proliferation [134].

The disturbance of thyroid tissue has brought about radiation-induced cell death, and recently, it has been recognized that ionizing radiation induces senescence-like cell death in thyroid follicular cells [135]. Several recent studies have indicated that senescence-like cell death promotes secretion of inflammatory cytokines [136], so that it is temping to speculate that radiation-induced tissue disruption could result in inflammatory circumstances that promote the initial stage of thyroid carcinogenesis. More recently, it has been shown that some inflammatory cytokines stimulate reprogramming of differentiated somatic cells. In fact, we have demonstrated that differentiated thyroid follicular cells are able to be reprogrammed to immature cells. Thus, taking all these information into consideration, it is plausible to propose that a role of radiation in childhood thyroid cancers after the Chernobyl accident could be an introduction of tissue disturbance by inducing thyroid follicular cell death as well as introducing secretory phenotype of dead cells, which may contribute to provide proliferative microenvironment to thyroid follicular cells harboring spontaneous *RET/PTC* rearrangements.

4. From Radiation Biology to Radiation Protection

4.1. Difficult and Painful Experiences in Fukushima

On the basis of the progress of radiation effects research, the experts and authorities can contribute to the crisis and post-crisis communications in response to radiation emergency situations including the nuclear power reactor accident. Especially the scientific reviews and updated analyses of radiation effects research by the UNSCEAR, and a cooperative work with the ICRP, the IAEA and the WHO, are significantly important to realize the radiation protection countermeasure. However, the facts occurred in Fukushima after March 11, 2011 have been far beyond our understanding of radiation protection criteria, even far from our capabilities to handle the confusion and chaos of the unprecedented NPP accident using such conventional criteria. Compared to the Chernobyl NPP accident, the doses received internally and externally were quite low and very limited after the Fukushima NPP accident because of early and successful intervention such as prompt evacuation, indoor sheltering and food restriction/safe food chain supply [137-139]. However, since the LNT model, originally used for radiation protection, has been misused to estimate individual cancer risk, even tiny and small dose exposures have been assumed to cause increased occurrence of cancers. Especially, unnecessary fear for childhood thyroid cancers has been established, even though the levels of thyroid dose are quite different from those estimated after the Chernobyl NPP accident.

It is now apparent that there have been no acute radiation injuries among nuclear workers in Fukushima and the radiation-induced physical health consequences to general public, including evacuees, are likely much lower than those arising from the Chernobyl NPP accident [138]. However, the social, psychological and secondary physical, and even economic impacts of Fukushima NPP accident are expected to be considerable [140]. Continued monitoring and characterization of the level of radioactivity in the environment and food in Fukushima are vital for obtaining informed consent for decisions on living in the areas already radiocontaminated and returning back to the evacuated areas once re-entry is permitted. Until now, the WHO reports in 2012 [141] and 2013 [142] have made the risk assessment theoretically based on preliminary dose assessment, but the UNSCEAR 2014 report [138] provided a more realistic risk assessment using much more concrete data obtained from Fukushima.

How did the recent progress and knowledge of radiation biology contribute to radiation protection and risk communication just after the Fukushima NPP accident? Although it is too early to extract the essence and lessons learned from the Fukushima, the uncertainty of low-dose and low-dose radiation effects make it difficult to communicate the risk to the public as it has been following the Chernobyl accident. This is exactly a pitfall of understanding of the relationship between low dose radiation and health risk; diverse public perception of uncertain radiation risk. The concept of risk at first should be defined when we discuss about radiation risk, which mainly means as equal to health risk at the time of nuclear accident. What is health risk induced by low dose radiation exposure? This time it is a strong fear and anxiety about the late occurrence of malignancy, especially a concern of childhood thyroid cancer in Fukushima [143]. In general epidemiological approach is mainly targeting group risk. However, the obstacle we ran into this time was how to assess each individual's risk---an incredibly difficult task. There is a stochastic effect of cancer risk on a basis of LNT model from radiation protection criteria, but a risk of cancer risk also exists merely by virtue of being alive, and there is no situation in which cancer risk would be zero. So we need a common standard to measure these risks on a basis of the results of radiation biology and cancer biology. Apparently, one of the cancer causes is radiation exposure, so that we should look at radiation dose level received. We also need to pay more attention to dose-rate, since the effect must be different between the exposures in a second, a minute, an hour, a year, and a lifetime. In addition, while a health risk of low-dose radiation is a late occurrence of cancer, a major principle is that radiation exposure is not the sole contributing risk. We cannot ignore the past and future history of various cancer risks. It is really important to understand how to barter these past risks with future risks and reduce the risk of predictable cancers by such a trade-off, conducting comprehensive health and mental care.

4.2. Strategies How to Establish Resilience for Recovery from the Disaster

Maternal concern in Fukushima is the most serious and important consideration, especially for children's health and future. During the middle of confusion in May 2011, therefore, a health examination committee was established in Fukushima, targeting evacuees, children and pregnant women. Since then, the official plans of the Fukushima Health Management Survey has been implemented, which includes a basic survey for the estimation of the external dose that were received during the first four months after the accident, and four more detailed surveys which includes thyroid ultrasound examination for children, comprehensive health check-up and mental health and life-style survey for evacuees and survey for pregnant women and nursing mothers. The aims of the survey program are to take care of the health of all the residents of Fukushima prefecture for a long time.

The original study protocol was reported in 2012 [143] and the early results of comprehensive health management also presented at the National Council on Radiation Protection and Measurements annual meeting in March 11, 2013 [140]. The past three years' results have been periodically updated [144]. According to those data, health effects due to radiation exposure are highly unlikely under the current situations and radiation levels in Fukushima. However, the risk of childhood thyroid cancer is the major issue to be paid careful attentions because of an increased number of asymptomatic cancers detected in general by a newly introduction of sophisticated ultrasound mass screening [140,145]. Furthermore the intermediate results have suggested that the rates of obesity, glucose metabolism dysfunction, hyperlipidemia and liver dysfunction after the accident were elevated compared with those before the accident which seems to be closely related to changes of life-style, diet, exercise and other personal habits caused by forced evacuation [140]. Changes in mental and psychological status are also reported [140, 146,147] and so mental health care in Fukushima is essential for a long time; this is exactly similar to the Chernobyl accident.

The comprehensive health management project based on the Fukushima Health Management Survey is one of the challenging programs in Fukushima to overcome the unnecessary radiophobia and even groundless rumors about low-dose radiation health risk [140,148]. It is also necessary to communicate smoothly and honestly with the public on the evidences and results of the projects. Beyond the radiation protection criteria and health risk in Fukushima, the individual risk perception should be seriously considered because a large number of evacuees still remain in interdeterminate and uncertain situation facing the difficult issues such as decision of returning or not, aging, life-style related diseases and psychosocial problems in addition to restoration of infrastructure of daily life. It is, therefore, critically important that best activities and practices related to recovery and resilience in Fukushima should be encouraged, supported and implemented at local and regional levels. Since psychosocial well-being of individuals and communities is the core element of resilience, local individuals, health professionals and authorities are uniquely positioned to identify and provide insight into what would provide the best resolution for their specific needs.

Conclusion

While epidemiological studies have demonstrated the dose-dependent relationships for the risk of radiation-induced cancer, statistically significant increase is hardly described for radiation doses below 100 mSv. The LNT model has been applied for assessing the risks resulting from exposures externally or internally to ionizing radiation, however, uncertainties still exist in determining the shape of the dose-response relationship especially at low doses. Thus, understandings of the mechanisms of radiation carcinogenesis based upon the updated knowledge of radiation biology, stem cell biology and molecular biology are essential for further insights into health effects of low dose radiation. Current models for radiation carcinogenesis have paid too much attention to the stochastic process of energy deposition in cells, however, accumulating evidences have implicated that the nature of tissue and stem cell reaction, i.e. tissue turnover, stem cells competition, and reprogramming of the progenitor cells, need to be taken into consideration.

These information and knowledge should greatly improve our assessment of the likely form of the dose-response at low dose and low dose rate radiation exposures. With such scientific knowledge it is believed that basic research is a beneficial and strong tool of radiation risk communication. However, during the recovery phase the co-existence of low dose and chronic low dose-rate radiation exposure has brought more realistic and painful difficulties in Fukushima because of the gap between radiation protection criteria and real heal risk. Therefore, a scientific approach which establish more appropriate risk to establish radiation protection criteria would be helpful to overcome individual risk perception as well as community levels. This is also essential to establish resilience to radiation, which should surely accelerate recovery from the disaster.

References

[1] United Nations (1972) Ionizing radiation: Levels and effects. Volume I: Levels, volume II: Effects. United Nations Scientific Committee on the Effects of Atomic Radiation. Report to the general assembly, with annexes. United Nations, New York.

[2] United Nations (1977) Sources and effects of ionizing radiation. United Nations Scientific Committee on the Effects of Atomic Radiation. Report to the general assembly, with annexes. United Nations, New York.

[3] United Nations (1988) Sources, effects and risks of ionizing radiation. United Nations Scientific Committee on the Effects of Atomic Radiation. Report to the general assembly, with annexes. United Nations, New York.

[4] International Commission on Radiological Protection (1991) Recommendations of the International Commission on Radiological Protection. *Annals of the ICRP* 21 (1-3). ICRP Publication 60. Pergamon Press, Oxford.

[5] Committee to assess health risks from exposure to low levels of ionizing radiation (2006) Health risks from exposure to low levels of ionizing radiation: BEIR VII - Phase 2. National Academy of Sciences, National Research Council. National Academy Press, Washington.

[6] United Nations (2000) Sources and effects of ionizing radiation. Volume I: Sources, volume II: Effects. United Nations Scientific Committee on the Effects of Atomic Radiation. Report to the general assembly, with annexes. United Nations, New York.

[7] United Nations (1993) Sources and effects of ionizing radiation. United Nations Scientific Committee on the Effects of Atomic Radiation. Report to the general assembly, with annexes. United Nations, New York.

[8] United Nations (2011) Summary of low-dose radiation effects on health. United Nations Scientific Committee on the Effects of Atomic Radiation. Report to the general assembly. United Nations, New York.

[9] Ward JF (1988) DNA damage produced by ionizing radiation in mammalian cells: Identities, mechanisms of formation, and repairability. *Progress in Nucleic Acid Research and Molecular Biology* 35: 95-125.

[10] Hall EJ, Giaccia AJ (2006) Radiobiology for the radiologist, 6th Ed., Philadelphia, Baltimore, New York: Lippincott Williams & Wilkins.

[11] Pollycove M, Feinendegen LE (2003) Radiation-induced versus endogenous DNA damage: Possible effect of inducible protective responses in mitigating endogenous damage. *Human Experimental Toxicology* 22: 290-306.

[12] Beckman KD, Ames BN (1998) The free radical theory of aging matures. *Physiology Review* 78: 547-581.

[13] Frankenberg-Schwager M (1990) Induction, repair and biological relevance of radiation-induced DNA lesions in eukaryotic cells. *Radiation Environmental Biophysics* 29: 273-292.

[14] Lindahl T (1993) Instability and decay of the primary structure of DNA. *Nature* 362: 709-715.

[15] Ward JF (1995) Radiation mutagenesis: The initial DNA lesions responsible. *Radiation Research* 142: 362-368.

[16] Rothkamm K, Lobrich M (2003) Evidence for a lack of DNA double-strand break repair in human cells exposed to very low x-ray doses. *Proceedings of the National Academy of Sciences of the United States of America* 100: 5057-5062.

[17] Suzuki K, Okada H, Yamauchi M, Oka Y, Kodama S, et al. (2006) Qualitative and quantitative analysis of phosphorylated ATM foci induced by low-dose ionizing radiation. *Radiation Research* 165: 499-504.

[18] Markova E, Schultz N, Belyaev IY (2007) Kinetics and dose-response of residual 53BP1/γ-H2AX foci: colocalization, relationship with DSB repair and clonogenic survival. *International Journal of Radiation Biology* 83: 319-329.

[19] Lobrich M, Rief N, Kuhne M, Heckmann M, Fleckenstein J, et al. *In vivo* formation and repair of DNA double-strand breaks after computed tomography examinations. *Proceedings of the National Academy of Sciences of the United States of America* 102: 8984-8989.

[20] Bootsma D, Westerveld A, Hoeijmakers JH (1988) DNA repair in human cells: from genetic complementation to isolation of genes. *Cancer Surveys* 7: 303-315.

[21] Barnes DE, Lindahl T (2004) Repair and genetic consequences of endogenous DNA base damage in mammalian cells. *Annual Review of Genetics* 38: 445-476.

[22] Fortini P, Dogliotti E (2007) Base damage and single-strand break repair: mechanisms and functional significance of short- and long-patch repair subpathways. *DNA Repair* 6: 398-409.

[23] Robertson AB, Klungland A, Rognes T, Leiros I (2009) Base excision repair: the long and short of it. *Cell and Molecular Life Science* 66: 981-993.

[24] Cadet J, Douki T, Ravanat J-L (2011) Measurement of oxidatively generated base damage in cellular DNA. *Mutation Research* 711: 3-12.

[25] Nakabeppu Y, Sakumi K, Sakamoto K, Tsuchimoto D, Tsuzuki T, et al. (2006) Mutagenesis and carcinogenesis caused by the oxidation of nucleic acids. *Biological Chemistry* 387: 373-379.

[26] Caldecott KW (2003) DNA single-strand break repair and spinocerebellar ataxia. *Cell* 112: 7-10.

[27] Ciccia A, Elledge SJ (2010) The DNA damage response: Making it sate to paly with knives. *Molecular Cell* 40: 179-204.

[28] Kasparek TR, Humphery TC (2011) DNA double-strand break repair pathways, chromosomal rearrangements and cancer. *Seminar in Cell and Developmental Biology* 22: 886-897.

[29] Polo SE, Jackson SP (2011) Dynamics of DNA damage response proteins at DNA breaks: 1 focus on protein modifications. *Genes and Developments* 25: 409-433.

[30] Jeggo PA (2009) Risks from low dose/dose rate radiation: what an understanding of DNA damage response mechanisms can tell us. *Heath Physics* 97: 416-425.

[31] Mullenders L, Atkinson M, Paretzke H, Sabatier L, Bouffler S. Assessing cancer risks of low-dose radiation. *Nature Review of Cancer* 9: 596-604.

[32] Ishizaki K, Hayashi Y, Nakamura H, Yasui Y, Komatsu K, et al. (2004) No induction of p53 phosphorylation and few focus formation of phosphorylated H2AX suggest efficient repair of DNA damage during chronic low-dose-rate irradiation in human cells. *Journal of Radiation Research* 45: 521-525.

[33] Vilenchik MM, Knudson AG (2003) Endogenous DNA double-strand breaks: production, fidelity of repair, and induction of cancer. *Proceedings of the National Academy of Sciences of the United States of America* 100: 12871-12876.

[34] Rothkamm K, Kuhne M, Jeggo PA, Lobrich M (2001) Radiation-induced genomic rearrangements formed by nonhomologous end-joining of DNA double-strand breaks. *Cancer Research* 61: 3886-3893.

[35] Kato TA, Nagasawa H, Weil MM, Little JB, Bedford JS (2006) Levels of □-H2AX foci after low-dose-rate irradiation reveals a DNA DSB rejoining defect in cells from human ATM heterozygotes in two AT families and in another apparently normal individual. *Radiation Research* 166: 443-453.

[36] Eccles LJ, O'Neill P, Lomax ME (2011) Delayed repair of radiation induced clustered DNA damage: friend or foe? *Mutation Research* 711: 134-141.

[37] Goodarzi AA, Jeggo P, Lobrich M (2010) The influence of heterochromatin on DNA double strand break repair: Getting the strong, silent type to relax. *DNA Repair* 9: 1273-1282.

[38] Yamauchi M, Oka Y, Yamamoto M, Niimura K, Uchida M, et al. (2008) Growth of persistent foci of DNA damage checkpoint factors is essential for amplification of G1 checkpoint signaling. *DNA Repair* 7: 405-417.

[39] Bonner WM, Redon CE, Dickey JS, Nakamura AJ, Sedelnikova OA, et al. (2008) GammaH2AX and cancer. *Nature Review of Cancer* 8: 957-967.

[40] Deckbar D, Jeggo PA, Löbrich M (2011) Understanding the limitations of radiation-induced cell cycle checkpoints. *Critical Review of Biochemical and Molecular Biology* 46: 271-283.

[41] Oka Y, Yamauchi M, Suzuki M, Yamashita S, Suzuki K (2011) Persistence and Dynamics of DNA Damage Signal Amplification Determined by Microcolony Formation and Live-cell Imaging. *Journal of Radiation Research* 52: 766-774.

[42] Morgan WF, Day JP, Kaplan MI, McGhee EM, Limoli CL (1996) Genomic instability induced by ionizing radiation. *Radiation Research* 146: 247-258.

[43] Wright EG (1998) Radiation-induced genomic instability in haemopoietic cells. *International Journal of Radiation Biology* 74: 681-687.

[44] Little JB (2003) Genomic instability and bystander effects: a historical perspective. *Oncogene* 22: 6978-6987.

[45] Suzuki K, Ojima M, Kodama S, Watanabe M (2003) Radiation-induced DNA damage and delayed induced genomic instability. *Oncogene* 22: 6988-6993.

[46] Niwa O (2006) Radiation induced dynamic mutations and transgenerational effects. *Journal of Radiation Research* 47: B25-B30.

[47] Niwa O, Kominami R (2001) Untargeted mutation of the maternally derived mouse hypervariable minisatellite allele in F1 mice born to irradiated spermatozoa. *Proceedings of the National Academy of Sciences of the United States of America* 98: 1705-1710.

[48] Natarajan AT (2006) Induced transgenerational genetic effects in rodents and humans. *Journal of Radiation Research* 47: B39-B43.

[49] Streffer C (2006) Transgenerational transmission of radiation damage: genomic instability and congenital malformation. *Journal of Radiation Research* 47: B19-B24.

[50] Nomura T (2008) Transgenerational effects from exposure to environmental toxic substances. *Mutation Research* 659: 185-193.

[51] Limoli CL, Corcoran JJ, Milligan JR, Ward JF, Morgan WF (1999) Critical target and dose and dose-rate responses for the induction of chromosomal instability by ionizing radiation. *Radiation Research* 151: 677-685.

[52] Lorimore SA, Coates PJ, Scobie GE, Milne G, Wright EG (2001) Inflammatory-type responses after exposure to ionizing radiation in vivo: a mechanism for radiation-induced bystander effects? *Oncogene* 20: 7085-7095.

[53] Nagar S, Smith LE, Morgan WF (2003) Characterization of a novel epigenetic effect of ionizing radiation: the death-inducing effect. *Cancer Research* 63: 324-328.

[54] Nagar S, Smith LE, Morgan WF (2003) Mechanisms of cell death associated with death-inducing factors from genomically unstable cell lines. *Mutagenesis* 18: 549-560.

[55] Boothman DA, Bouvard I, Hughes EN (1989) Identification and characterization of X-ray-induced proteins in human cells. *Cancer Research* 49: 2871-2878.

[56] Boothman DA, Wang M, Lee SW (1991) Induction of tissue-type plasminogen activator by ionizing radiation in human malignant melanoma cells. *Cancer Research* 51: 5587-5595.

[57] Boothman DA, Meyers M, Fukunaga N, Lee SW (1993) Isolation of x-ray-inducible transcripts from radioresistant human melanoma cells. *Proceedings of the National Academy of Sciences of the United States of America* 90: 7200-7204.

[58] Olivieri G, Bodycote, Wolff S (1984) Adaptive response of human lymphocytes to low concentrations of radioactive thymidine. *Science* 223: 594-597.

[59] Shaley JD, Afzal V, Wolff S (1987) Characterization of the adaptive response to ionizing radiation induced by low doses of X rays to human lymphocytes. *Radiation Research* 111: 511-517.

[60] Ikushima T (1989) Radio-adaptive response: characterization of a cytogenetic repair induced by low-level ionizing radiation in cultured Chinese hamster cells. *Mutation Research* 227: 241-246.

[61] Wojcik A, Tuschl H (1990) Indications of an adaptive response in C57BL mice pre-exposed in vivo to low doses of ionizing radiation. *Mutation Research* 243: 67-73.

[62] Rigaud O, Moustacchi E (1996) Radioadaptation for gene mutation and the possible molecular mechanisms of the adaptive response. *Mutation Research* 358: 127-134.

[63] Sasaki MS, Ejima Y, Tachibana A, Yamada T, Ishizaki K, et al. (2002) DNA damage response pathway in radioadaptive response. *Mutation Research* 504: 101-118.

[64] Sorensen KJ, Attix CM, Christian AT, Wyrobek AJ, Tucker JD (2002) Adaptive response induction and variation in human lymphoblastoid cell lines. *Mutation Research* 519: 15-24.

[65] Nagasawa H, Little JB (1992) Induction of sister chromatid exchanges by extremely low doses of alpha-particles. *Cancer Research* 52: 6394-6396.

[66] Morgan WF (2003) Non-targeted and delayed effects of exposure to ionizing radiation: I. Radiation-induced genomic instability and bystander effects *in vitro*. *Radiation Research* 159: 567-580.

[67] Hei TK, Zhou H, Chai Y, Ponnaiya B, Ivanov VN (2011) Radiation induced non-targeted response: mechanism and potential clinical implications. *Current Molecular Pharmacology* 4: 96-105.
[68] Prise KM, O'Sullivan JM (2009) Radiation-induced bystander signalling in cancer therapy. *Nature Review of Cancer* 9: 351-360.
[69] Prise KM, Schettino G, Vojnovic B, Belyakov O, Shao C (2009) Microbeam studies of the bystander response. *Journal of Radiation Research* 50: A1-A6.
[70] Hei TK, Ballas LK, Brenner DJ, Geard CR (2009) Advances in radiobiological studies using a microbeam. *Journal of Radiation Research* 50: A7-A12.
[71] Prise KM, Schettino G (2011) Microbeams in radiation biology: review and critical comparison. *Radiation Protection Dosimetry* 143: 335-339.
[72] Zhou H, Suzuki M, Randers-Pehrson G, Vannais D, Chen G, et al. (2001) Radiation risk to low fluences of alpha particles may be greater than we thought. *Proceedings of the National Academy of Sciences of the United States of America* 98: 14410-14415.
[73] Lehnert BE, Goodwin EH, Deshpande A (1997) Extracellular factor(s) following exposure to alpha particles can cause sister chromatid exchanges in normal human cells. *Cancer Research* 57: 2164-2171.
[74] Iyer R, Lehnert BE, Svensson R (2000) Factors underlying the cell growth-related bystander responses to alpha particles. *Cancer Research* 60: 1290-1298.
[75] Matsumoto H, Hayashi S, Hatashita M, Ohnishi K, Shioura H, et al. (2001) Induction of radioresistance by a nitric oxide-mediated bystander effect. *Radiation Research* 155: 387-396.
[76] Azzam EI, De Toledo SM, Spitz DR, Little JB (2002) Oxidative metabolism modulates signal transduction and micronucleus formation in bystander cells from alpha-particle-irradiated normal human fibroblast cultures. *Cancer Research* 62: 5436-5442.
[77] Shao C, Stewart V, Folkard M, Michael BD, Prise KM (2003) Nitric oxide-mediated signaling in the bystander response of individually targeted glioma cells. *Cancer Research* 63: 8437-8442.
[78] Buonanno M, de Toledo SM, Pain D, Azzam EI (2011) Long-term consequences of radiation-induced bystander effects depend on radiation quality and dose and correlated with oxidative stress. *Radiation Research* 175: 405-415.
[79] Zhou H, Ivanov VN, Gillespie J, Geard CR, Amundson SA, et al. (2005) Mechanism of radiation-induced bystander effect: role of the cyclooxygenase-2 signaling pathway. *Proceedings of the National Academy of Sciences of the United States of America* 102: 14641-14646.
[80] Zhou H, Ivanov VN, Lien YC, Davidson M, Hei TK (2008) Mitochondrial function and nuclear factor-kappaB-mediated signaling in radiation-induced bystander effects. *Cancer Research* 68: 2233-2240.
[81] Prise KM, Belyakov OV, Folkard M, Michael BD (1998) Studies of bystander effects in human fibroblasts using a charged particle microbeam. *International Journal of Radiation Biology* 74: 793-798.
[82] Shao C, Aoki M, Furusawa Y (2004) Bystander effect in lymphoma cells vicinal to irradiated neoplastic epithelial cells: nitric oxide is involved. *Journal of Radiation Research* 45: 97-103.

[83] Gaugler MH, Neunlist M, Bonnaud S, Aubert P, Benderitter M, et al. (2007) Intestinal epithelial cell dysfunction is mediated by an endothelial-specific radiation-induced bystander effect. *Radiation Research* 167: 185-193.

[84] Groesser T, Cooper B, Rydberg B (2008) Lack of bystander effects from high-LET radiation for early cytogenetic end points. *Radiation Research* 170: 794-802.

[85] Sowa MB, Goetz W, Baulch JE, Pyles DN, Dziegielewski J, et al. (2010) Lack of evidence for low-LET radiation induced bystander response in normal human fibroblasts and colon carcinoma cells. *International Journal of Radiation Biology* 86: 102-113.

[86] Fournier C, Barberet P, Pouthier T, Ritter S, Fischer B, et al. (2009) No evidence for DNA and early cytogenetic damage in bystander cells after heavy-ion microirradiation at two facilities. *Radiation Research* 171: 530-540.

[87] Morgan WF, Goetz W, Sowa MB (2009) No bystander effect after irradiation of mammalian cells with a variable energy electron microbeam. *Journal of Radiation Research* 50: A91-A95.

[88] Durante M. Bystander response measured at the GSI microbeam (2009*) Journal of Radiation Research* 50: A117-A121.

[89] Cherubini R, De Nadal V, Gerardi S, Guryev D (2011) Lack of hyper-radiosensitivity and induced radioresistance and of bystander effect in V79 cells after proton irradiation of different energies. *Radiation Protection Dosimetry* 143: 315-319.

[90] Bowler DA, Moore SR, Macdonald DA, Smyth SH, Clapham P, et al. (2006) Bystander-mediated genomic instability after high LET radiation in murine primary haemopoietic stem cells. *Mutation Research* 597: 50-61.

[91] Sokolov MV, Neumann RD (2010) Radiation-induced bystander effects in cultured human stem cells. *PLoS One* 5: e14195.

[92] Ermakov AV, Kon'kova MS, Kostiuk SV, Smirnova TD, Kameneva LV, et al. (2010) Bystander effect development in human mesenchymal stem cells after exposure to adaptive dose of X-radiation. *Radiation Biology Radioecology* 50: 42-51.

[93] Ozeki A, Suzuki K, Suzuki M, Ozawa H, Yamashita S (2012) Acceleration of astrocytic differentiation in neural stem cells surviving X-irradiation. *Neuroreport* 23: 290-293.

[94] Prise KM, Belyakov OV, Newman HC, Patel S, Schettino G, et al. (2002) Non-targeted effects of radiation: bystander responses in cell and tissue models. *Radiation Protection Dosimetry* 99: 223-226.

[95] Belyakov OV, Folkard M, Mothersill C, Prise KM, Michael BD (2003) A proliferation-dependent bystander effect in primary porcine and human urothelial explants in response to targeted irradiation. *British Journal of Cancer* 88: 767-774.

[96] Belyakov OV, Mitchell SA, Parikh D, Randers-Pehrson G, Marino SA, et al. (2005) Biological effects in unirradiated human tissue induced by radiation damage up to 1 mm away. *Proceedings of the National Academy of Sciences of the United State of America* 102: 14203-14208.

[97] Belyakov OV, Folkard M, Mothersill C, Prise KM, Michael BD (2006). Bystander-induced differentiation: a major response to targeted irradiation of a urothelial explant model. *Mutation Research* 597: 43-49.

[98] Pinto M, Azzam EI, Howell RW (2006) Bystander responses in three-dimensional cultures containing radiolabelled and unlabelled human cells. *Radiation Protection Dosimetry* 122: 252-255.

[99] Sedelnikova OA, Nakamura A, Kovalchuk O, Koturbash I, Mitchell SA, et al. (2007) DNA double-strand breaks form in bystander cells after microbeam irradiation of three-dimensional human tissue models. *Cancer Research* 67: 4295-4302.

[100] Shuryak I, Sachs RK, Brenner DJ (2007) Biophysical models of radiation bystander effects: 1. Spatial effects in three-dimensional tissues. *Radiation Research* 168: 741-749.

[101] Leonard BE (2009) The range of the bystander effect signal in three-dimensional tissue and estimation of the range in human lung tissue at low radon levels. *Radiation Research* 171: 374-378.

[102] Khan MA, Hill RP, Van Dyk J (1998) Partial volume rat lung irradiation: an evaluation of early DNA damage. *International Journal of Radiation Oncology Biology Physics* 40: 467-476.

[103] Chai Y, Calaf GM, Zhou H, Ghandhi SA, Elliston CD, et al. (2013) Radiation induced COX-2 expression and mutagenesis at non-targeted lung tissues of gpt delta transgenic mice. *British Journal of Cancer* 108: 91-98.

[104] Chai Y, Lam RK, Calaf GM, Zhou H, Amundson S, et al. (2013) Radiation-induced non-targeted response in vivo: role of the TGFβ-TGFBR1-COX-2 signalling pathway. *British Journal of Cancer* 108: 1106-1112.

[105] Preston DL, Pierce DA, Shimizu Y, Cullings HM, Fujita S, et al. (2004) Effect of recent changes in atomic bomb survivor dosimetry on cancer mortality risk estimates. *Radiation Research* 162: 377-389.

[106] Ozasa K, Shimizu Y, Suyama A, Kasagi F, Soda M, et al. (2012) Studies of the mortality of atomic bomb survivors, report 14, 1950-2003: an overview of cancer and noncancer diseases. *Radiation Research* 177: 229-243.

[107] Wood JW, Tamagaki H, Neriishi S, Sato T, Sheldon WF, et al. (1969) Thyroid carcinoma in atomic bomb survivors Hiroshima and Nagasaki. *American Journal of Epidemiology* 89: 4-14.

[108] United Nations Chernobyl Forum (2006) Health effects of the Chernobyl accident and special health care programmes. In: Bennett B, Repacholi M, Carr Z, editors. World Health Organization report of the UN Chernobyl Forum expert group "Health" (EGH). Geneva: World Health Organization.

[109] United Nations (2011) Sources and effects of ionizing radiation. United Nations Scientific Committee on the Effects of Atomic Radiation, 2008 Report to the general assembly, with annexes C, D and E. United Nations, New York.

[110] Saenko V, Ivanov V, Tsyb A, Bogdanova T, Tronko M, et al. (2011) The Chernobyl accident and its consequences. *Clinical Oncology* 23: 234-243.

[111] Demidchik YE, Saenko VA, Yamashita S (2007) Childhood thyroid cancer in Belarus, Russia, and Ukraine after Chernobyl and at present. *Arquivos Brasileiros de Endocrinologia & Metabologia* 51: 748-762

[112] Cardis E, Kesminiene A, Ivanov V, Malakhova I, Shibata Y, et al. (2005) Risk of thyroid cancer after exposure to 131I in childhood. *Journal of National Cancer Institute* 97: 724-732.

[113] Tronko MD, Howe GR, Bogdanova TI, Bouville AC, Epstein OV, et al. (2006) A cohort study of thyroid cancer and other thyroid diseases after the chornobyl accident: thyroid cancer in Ukraine detected during first screening. *Journal of National Cancer Institute* 98: 897-903.
[114] Zablotska LB, Ron E, Rozhko AV, Hatch M, Polyanskaya ON, et al. (2011) Thyroid cancer risk in Belarus among children and adolescents exposed to radioiodine after the Chornobyl accident. *British Journal of Cancer* 104: 181-187.
[115] Jacob P, Kenigsberg Y, Zvonova I, Goulko G, Buglova E, et al. (1999) Childhood exposure due to the Chernobyl accident and thyroid cancer risk in contaminated areas of Belarus and Russia. *British Journal of Cancer* 80: 1461-1469.
[116] Likhtarov I, Kovgan L, Vavilov S, Chepurny M, Ron E, et al. (2006) Post-Chernobyl thyroid cancers in Ukraine. Report 2: risk analysis. *Radiation Research* 166: 375-386.
[117] Fugazzola L, Pilotti S, Pinchera A, Vorontsova TV, Mondellini P, et al. (1995) Oncogenic rearrangements of the *RET* proto-oncogene in papillary thyroid carcinomas from children exposed to the Chernobyl nuclear accident. *Cancer Research* 55: 5617-5620.
[118] Klugbauer S, Lengfelder E, Demidchik EP, Rabes HM (1995) High prevalence of RET rearrangement in thyroid tumors of children from Belarus after the Chernobyl reactor accident. *Oncogene* 11: 2459-2467.
[119] Nikiforov YE, Rowland JM, Bove KE, Monforte-Munoz H, Fagin JA (1997) Distinct pattern of *ret* oncogene rearrangements in morphological variants of radiation-induced and sporadic thyroid papillary carcinomas in children. *Cancer Research* 57: 1690-1694.
[120] Nikiforov YE, Nikiforova MN (2011) Molecular genetics and diagnosis of thyroid cancer. *Nature Review of Endocrinology* 7: 569-580.
[121] Santoro M, Carlomagno F (2013) Oncogenic rearrangements driving ionizing radiation-associated human cancer. *Journal of Clinical Investigation* 11: 4566-4568.
[122] Shaw AT, Hsu PP, Awad MM, Engelman JA (2013) Tyrosine kinase gene rearrangements in epithelial malignancies. *Nature Review of Cancer* 13: 772-787.
[123] Kodama Y, Asai N, Kawai K, Jijiwa M, Murakumo Y, et al. (2005) The RET proto-oncogene: a molecular therapeutic target in thyroid cancer. *Cancer Science* 96: 143-148.
[124] Wang X (2013) Structural studies of GDNF family ligands with their receptors-Insights into ligand recognition and activation of receptor tyrosine kinase RET. *Biochimica et Biophysica Acta* 1834: 2205-2212.
[125] Santoro M, Carlomagno F (2013) Central role of RET in thyroid cancer. *Cold Spring Harbor Perspective Biology* 5: 1-17.
[126] Mulligan LM (2014) RET revisited: expanding the oncogenic portfolio. *Nature Review of Cancer* 14: 173-186.
[127] Mizuno T, Iwamoto KS, Kyoizumi S, Nagamura H, Shinohara T, et al. (2000) Preferential induction of RET/PTC rearrangement by X-ray irradiation. *Oncogene* 19: 438-443.
[128] Fenton CL, Lukes Y, Nicholson D, Dinauer CA, Francis GL, et al. (2000) The ret/ptc mutations are common in sporadic papillary thyroid carcinoma of children and young adults. *Journal of Clinical Endocrinology Metabolism* 85: 1170-1175.
[129] Xing M (2005) BRAF mutation in thyroid cancer. *Endocrinology Related Cancer* 12: 245-262.

[130] Romei C, Elisei R (2012) RET/PTC translocations and clinico-pathological features in human papillary thyroid carcinoma. *Frontiers in Endocrinology* 3: Article 54, 1-8.
[131] Tronko M, Bogdanova T, Voskoboynyl L, Zurnadzhy L, Shpak V, et al. (2010) Radiation induced thyroid cancer: Fundamental and applied aspects. *Experimental Oncology* 32: 200-204.
[132] Muzza M, Degl'Innocenti D, Colombo C, Perrino M, Ravasi E, et al. (2010) The tight relationship between papillary thyroid cancer, autoimmunity and inflammation: clinical and molecular studies. *Clinical Endocrinology* 72: 702-708.
[133] Okayasu I, Saegusa M, Fujiwara M, Hara Y, Rose NR (1995) Enhanced cellular proliferative activity and cell death in chronic thyroiditis and thyroid papillary carcinoma. *Journal of Cancer Research Clinical Oncology* 121: 746-752.
[134] Guarino V, Castellone MD, Avilla E, Melillo RM (2010) Thyroid cancer and inflammation. *Molecular and Cellular Endocrinology* 321: 94-102.
[135] Podtchenko A, Namba H, Saenko VA, Ohtsuru A, Starenki D, et al. (2005) Radiation-induced senescence-like terminal growth arrest in thyroid cells. *Thyroid* 15: 306-313.
[136] Suzuki K, Yamashita S (2014) Radiation-induced bystander response: Mechanism and clinical implications. *Advances in Wound Care* 3: 16-24.
[137] Nagataki S (2012) Thyroid consequences of the Fukushima nuclear reactor accident. *European Thyroid Journal* 1: 148-158.
[138] United Nations (2013) Sources, effects and risks of ionizing radiation. United Nations Scientific Committee on the Effects of Atomic Radiation. Report to the general assembly, with annex A. Levels and effects of radiation exposure due to the nuclear accident after the 2011 great east-Japan earthquake and tsunami. United Nations, New York.
[139] Nagataki S, Takamura N, Kamiya K, Akashi M (2013) Measurement of individual radiation doses in residents living around the Fukushima nuclear power plant. *Radiation Research* 180: 439-447.
[140] Yamashita S (2014) Tenth Warren K. Sinclair keynote address-The Fukushima nuclear power plant accident and comprehensive health risk management. *Health Physics* 106: 166-180.
[141] World Health Organization (2012) Preliminary dose estimation from the nuclear accident after the 2011 Great East Japan earthquake and tsunami. WHO press, World Health Organization, Geneva.
[142] World Health Organization (2013) Health risk assessment from the nuclear accident after the 2011 Great East Japan earthquake and tsunami, based on a preliminary dose estimation. WHO press, World Health Organization, Geneva.
[143] Yamashita S, Suzuki S (2013) Risk of thyroid cancer after the Fukushima nuclear power plant accident. *Respiratory Investigation* 51: 128-133.
[144] Homepage of Radiation Medical Science Center for the Fukushima Health Management Survey, Fukushima Medical University; http://www.fmu.ac.jp/radiationhealth/results/
[145] Hayashida N, Imaizumi M, Shimura H, Okubo N, Asari Y, et al. (2013) Thyroid ultrasound findings in children from three Japanese prefectures: Aomori, Yamanashi and Nagasaki. *PLoS One* 8: e83220.

[146] Matsumoto J, Kunii Y, Wada A, Mashiko H, Yabe H, et al. (2014) Mental disorders that exacerbated due to the Fukushima disaster, a complex radioactive contamination disaster. *Psychiatry and Clinical Neuroscience* 68: 182-187.

[147] Yabe H, Suzuki Y, Mashiko H, Nakayama Y, Hisata M, et al. (2014) Psychological distress after the Great East Japan earthquake and Fukushima Daiichi nuclear power plant accident: results of a mental health and lifestyle survey through the Fukushima health management survey in FY2011 and FY2012. *Fukushima Journal of Medical Science* 60: 57-67.

[148] Yasumura S, Hosoya M, Yamashita S, Kamiya K, Abe M, et al. (2012) Study protocol for the Fukushima Health Management survey. *Journal of Epidemiology* 22: 375-383.

In: Fukushima Nuclear Accident
Editor: Shizuyo Sutou
ISBN: 978-1-63463-748-0

Chapter 2

NUCLEAR ENERGY AND THE LNT HYPOTHESIS OF RADIATION CARCINOGENESIS

Jerry M. Cuttler*
Cuttler & Associates Inc., Ontario, Canada

ABSTRACT

This chapter outlines the origin of the linear no-threshold (LNT) hypothesis for assessing the risk of cancer following exposure to ionizing radiation. Heavy reliance is placed on information from the careful research by Edward Calabrese. Facts on early and modern discoveries of the real health effects of x-rays and nuclear radiation are presented to demonstrate that a low dose or a low dose rate actually induces very important beneficial effects. These health benefits have been ignored by radiation protection organizations and nuclear regulators. Instead, they focus on risks of hypothetical harmful effects. The fears, uncertainties and doubts they have been raising link low radiation to a risk of excess cancer mortality, which has not been observed. This ongoing cancer scare blocks social acceptance of the benefits of an enormous supply of nuclear energy and very significant medical applications of low radiation to diagnose and treat serious illnesses. The issues and concerns are briefly outlined and shown to be invalid. The Scientific Method should be relearned and employed to discredit the LNT hypothesis by comparing its predictions against more than a century of observations. Seven recommendations are provided for nuclear science professionals, medical practitioners and regulators. Included is the development and fulfillment of a public communication program to dispel the myth that links low radiation to an increased risk of cancer. Also included are the most important lessons from the Chernobyl and Fukushima accidents that have not been learned. Humanity risks losing many life-saving health benefits if it continues to ignore the evidence.

Keywords: Nuclear energy, Fukushima Daiichi Nuclear Power Plant (FDNPP), linear no-threshold (LNT), Chernobyl, Fukushima, ICRP, NAS, x-rays, CT-scans, AHARS, ALARA, radiation hormesis

---

* Correspondence to: Jerry M. Cuttler, Tel: +1 416 837 8865, E-mail: jerrycuttler@rogers.com

INTRODUCTION

Nuclear technologies already provide humanity and the environment with the enormous benefit of a large supply of non-polluting electrical energy. The benefit from nuclear energy could be increased substantially by employing it to desalinate sea water. However, nuclear technologies are being abandoned because of the social concerns and fears that have been created about the health effects of nuclear radiation, specifically about the risk of excess cancer deaths from any exposure, no matter how small. This is apparent from public and government reactions to the consequences of the Fukushima NPP accident and the concerns of many physicians and patients regarding potential risks of x-rays and nuclear radiations used in medical imaging. The real effects of ionizing radiation are very different from current and past beliefs. High, intense radiation is indeed harmful, but a low dose or a low dose rate actually produces remarkable, positive effects in all organisms. How can this be?

The reason is not difficult to understand. Every living thing adapts to its environment, which includes natural background radiation. When a perturbation occurs—for instance, an increase in the radiation level or a high short-term exposure—the activities of the protection systems increase, both the immediately-acting ones and the delayed adaptive defences, some of which persist for days, weeks and much longer. The stimulated protection systems act, not only on the additional damage that was or is being produced by the radiation increase, but also on the much more extensive damage that is occurring continuously due to the natural endogenous processes (e.g., oxygen in metabolism, thermal effects) and the changes being caused by the external disturbances, such as injuries, infections and ingestion of chemicals. The overall response to the radiation increase results in an improvement in health, including a reduction in the risk of cancer.

Powerful medicinal properties of x-rays and radium were observed soon after their discovery [1, 2], and medical practitioners began to cure many illnesses using these radiations in imaging and therapies. However, the early radiologists suffered from burns and a higher incidence of neoplasm mortality than did their peers. Protection advice began to appear in 1913. Soon after, the British Roentgen Society issued a warning and then recommendations in 1921. [3] A 1981 study of the British radiologists, covering the period 1897-1954, revealed that those who entered the profession after 1920 had a lower incidence of cancer mortality and a lower mortality from all causes. [4] In 1925, the International Congress of Radiology was formed, evolving into the International Commission of Radiation Units and Measurements (ICRU) with the subsequent creation of the International Commission of Radiological Protection (ICRP). In 1928, the ICRP issued recommendations for radiologists, emphasizing that they should avoid unnecessary exposures. Its 1934 meeting recommended a dose limit, a tolerance dose of 0.2 roentgens[†] per day, implying the concept of a safe threshold. [3] This limit, about 700 mGy[*] per year, had been recommended in the 1924 meeting of the American Roentgen Ray Society. The ICRP became active again after WWII and began addressing the concerns of potential harm being raised by geneticists at the time. Its 1950 report recommended that the maximum permissible dose be limited to 0.5 roentgen per week.

---

† An exposure of 1 R (roentgen) means that the radiation dose to ordinary tissue is about 9.3 mGy. [40]

* The Gray (Gy) is the System International unit of absorbed radiation dose. 1 Gy = 1 joule/kg.

While physicians were treating patients with radiation, biologists were studying its effects on all other organisms. They discovered that low doses produced beneficial effects, while high doses caused harm. A 1915 study showed high x radiation rendered mice susceptible to transplanted tumors, but the damaged defences were regenerated. However, a small dose to the animals stimulated very strong protection, a complete immunity, against their spontaneous cancers. [5]

Meanwhile, geneticists were studying potential causes of genetic change. Several of them reported that germinal changes could be induced by x or radium rays. Hermann Muller was the first to demonstrate inducible heritable changes by an environmental agent, i.e., x-rays. His 1927 paper, which led to his Nobel Prize, indicated that he was a eugenicist who wanted to produce mutations "to order" and create artificial races. [6] He produced true gene (sex-linked recessive lethal) mutations in the germ cells of fruit flies, by a "heavy treatment" of x-rays—a dose rate of 8000 R/h and irradiation times of 12, 24, 36 and 48 minutes. [7] At the highest dose tested, the mutation rate was 150 times the rate in untreated germ cells. The rate varied as the square root of the dose. Experiments by others at high dose rates indicated a linear dose-response relationship. A linear no-threshold (LNT) model supported the physics-based "target theory" concept that had been established prior to Muller's discovery of inducible mutations. [8]

Radiation target theory as applied to mutations was formulated by the detailed interactions and collaborations of leading radiation geneticists and theoretical physicists during the mid-1930s. They established a conceptual framework for gene structure, target theory for the induction of mutations via ionizing radiation, the single-hit mechanism hypothesis to account for the shape of the LNT dose response, and the application of this dose-response model for what is modern cancer risk assessment.

This theory saw mutations as a purely physical action following an "all or none" law. The energy of ionizing radiation was assumed essentially to be transformed into a genetic effect, which is in contrast to normal physiology that invariably deals with large numbers of molecules of many kinds in the evolution of genetically determined body changes, and where the elimination of a single molecule would not result in observable effects. [8]

Muller, who collaborated in the establishment of target theory, was so committed to the LNT model of radiation-induced mutations that he ignored the scientific evidence of a threshold in an experiment that he reviewed prior to receiving his Nobel Prize in December 1946. Muller, not understanding the complexity of generating a body change from an alteration of genes, misled the world during his acceptance speech in which he stated that there is "no escape from the conclusion that there is no threshold." [9-11]

Scientists had questioned whether the dose-response would still be linear with a low dose-rate, so Ernest Caspari measured the fruit fly mutation rate for a dose of 52.5 R delivered in 21 days. This dose was 30 times lower than the lowest dose that Muller used, and the 0.10 R/h dose rate was 80 thousand times lower than used by Muller. Caspari measured a threshold dose-response, which contradicted the LNT model. The control cultures and the experimental cultures numbered 56,252, and 51,963 respectively.

The first two statements in the summary of his paper are:

1. The rate of lethal sex-linked mutations in Drosophila exposed to gamma-rays of 2.5 r units per day through 21 days (total 52.5 r) was determined.

2. In a total material of 108,215 chromosomes tested, no significant difference between experimentals and controls was found.

The supervisor Curt Stern, a long-time supporter of the linear dose-response model, deleted the threshold dose-response conclusion and published Caspari's paper with a negative discussion to explain why these data should not be accepted and utilized. This marginalized paper [12] did not receive an independent peer review. [10, 13, 14] The dropping of atomic bombs in 1945 to end the war in Japan was followed by development, testing and large-scale production of nuclear weapons. Fears of nuclear warfare drove many scientists to strong political actions against the nuclear arms race.

One of their strategies was to create and promote extreme social fear of low-level radiation from radioactive "fallout." [15] With sixty years of prior knowledge and experience in the use x-rays and radium to diagnose and treat a wide variety of illnesses, without a significant increase in cancer mortality, how was it possible for scientists to create such a strong fear of radiation in the 1950s? And how could this phobia persist over the subsequent sixty years, and continue even today, as important scientific advances were being made in biology and genetics and effects of radiation on cells, tissues and organisms? The Stern and Muller deceptions contributed to the 1956 decision of the U.S. National Academy of Sciences to recommend a linear dose-response policy for assessing risks to the genome from ionizing radiation, replacing the threshold dose-response model. [16]

This recommendation initiated a series of advisory and regulatory actions in essentially all countries to adopt linearity and apply it to somatic effects, that is, cancer risk assessment, for ionizing radiation and later for chemical carcinogens. [13]

The ICRP began to produce many new recommendations that were based on 1) the concept of the LNT hypothesis, 2) the notion of "stochastic" health effects, and 3) the high-dose cancer mortality statistics of the atomic bomb survivors.

The author perceives that the radiation protection organizations and the nuclear regulators have made errors of omission by ignoring information on the following very important discoveries:

1. DNA molecules are not stable—single and double-strand breaks occur spontaneously at very high rates due to natural, internal causes. [17]
2. Organisms have very powerful protection systems, immediate and adaptive, that deal with potential and actual cell damage and with all other injuries and health risks. [18]
3. Many of these protection systems (more than 200 genes) are up-regulated by radiation. A low acute dose or a low chronic dose rate produces beneficial health effects that outweigh the harmful effects. The benefits include a reduced risk of cancer. [19, 20]
4. A variety of very important medical treatments were carried out, starting in 1896, using low doses of radiation to cure serious illnesses without causing cancers. [21, 22] Humanity risks losing many life-saving health benefits if it continues to ignore this evidence. [23]
5. The 1981 study of the British radiologists [4] demonstrated that radiation protection based on the tolerance dose concept, introduced in 1921, is effective and more than adequate.

The author makes the following recommendations:

1. Nuclear science, medical and other professional societies should organize events to learn and discuss the facts that were discovered about biological effects of radiation. They should question risk concepts that are based on primitive target theory.
2. Professionals should urge the nuclear regulators and the radiation protection organizations to examine all the data, both beneficial and harmful, and to comply with the requirements of The Scientific Method in setting safe limits for acute and chronic radiation exposures.
3. Regulators should develop and implement a public communication program to dispel the myth that links radiation to an excess risk of cancer. They should stop regulating harmless radiation sources, such as radon in homes. The radon levels in many homes in Ramsar, Iran are much higher than in most homes elsewhere. No significant detrimental effect, such as an increased incidence of cancer mortality, has been observed in Ramsar. [84]
4. For radiation protection, the dose-response relationship should be based on scientific facts. [83] Calculating risk of excess cancers using the invalid LNT methodology should end.
5. In medical diagnostics, physicians should not be intimidated from prescribing x-rays and CT-scans by unjustifiable concerns about cancer risks that are calculated using the LNT hypothesis. A low dose of radiation up-regulates adaptive protection systems that induce beneficial health effects, which include a lower incidence of cancer mortality.
6. The radiation level for precautionary measures in the event of a radiological release should be set as high as relatively safe, based on the thresholds for biological harm. Reference [15] recommends the threshold dose rate of 700 mGy per year for a continuous exposure and the threshold dose of 500 mGy for an acute exposure.
7. Nuclear safety analysts should learn the three most important lessons from the Chernobyl and Fukushima accidents that they have not yet learned:
   - Severe accidents release radioactive materials that result mainly in low radiation levels.
   - Long-term evacuation of residents would not be appropriate because the low radiation levels would not be harmful. The dose-rate limit should be based on the threshold for harm.
   - Precautionary measures should be avoided because the fear induced by these actions causes severe psychological stress that results in many premature deaths.

NUCLEAR ENERGY RADIATION CONCERNS

Energy from fission has an important advantage over energy from burning coal, oil or methane because the thermal energy released when a uranium nucleus divides in two is about 170 MeV, which is 40 million times the energy released (4.1 eV) when a carbon atom reacts chemically with an oxygen molecule to form carbon dioxide. The amount of fuel that must be brought to a coal-burning plant and the amount of waste produced are therefore enormous compared with the relatively small amount of fuel that is brought and stored at a nuclear energy plant of the same size. And the amount of radioactive material to be managed will decrease by a factor of about 200 when reusable fuel is recycled in breeder reactors to increase uranium utilization from about 0.5% to about 95%. [24] And when conversion of thorium to fissile material is implemented, the long-term supply of energy will be assured.

When campaigning against nuclear energy, activists raise concerns about small releases of radioactive materials, "the problem of the nuclear waste" and the "severe" consequences of an accident, like the Chernobyl or the Fukushima accidents, on the population and the environment.

The basis for all of these concerns was created when the U.S. National Academy of Science (NAS) adopted the linear no-threshold (LNT) hypothesis in 1956 to assess the excess risk of cancer from ionizing (nuclear or x-ray) radiation. [16] The dropping of atomic bombs in 1945 to end the war in Japan was followed by development, testing and large-scale production of nuclear weapons. Fears of nuclear warfare drove many scientists to strong political actions against the nuclear arms race. One of their strategies was to create and promote extreme social fear of low-level radiation from radioactive "fallout." This was a motive for the NAS recommendation of a linear dose-response policy for assessment of risks to genetic material (DNA molecules) from radiation, replacing the threshold dose-response model. This initiated a series of advisory and regulatory actions in essentially all countries to adopt linearity and apply it for predicting somatic effects (cancer risk assessment) for radiation and later for chemical carcinogens. [25, 8, 13] Adoption of the LNT hypothesis created a new cancer scare because it links any radiation exposure, no matter how small, to an increased risk of cancer.

Figure 2.1 shows the typical radiation exposure to a physician. [26] Recommendations issued by the International Commission on Radiological Protection (ICRP) in 1934 to protect practitioners were based on the threshold dose concept. The limit was a "tolerance dose" of 0.2 roentgen per day. The 1981 study of British radiologists, covering the period 1897-1954, revealed that their cancer mortality and their mortality from all causes decreased below the comparable mortalities of their peers, after this concept was introduced in 1921. The threshold dose concept is therefore a satisfactory basis for protecting against delayed effects. [27]

In 1956, for the first time, the ICRP adopted recommendations for public exposures that included a concern about "stochastic effects." The major problem was believed to be hereditary harm. The ICRP began to produce many new recommendations. Since 1956, about 150 publications have been issued. [28] They are based on 1) the LNT hypothesis, 2) the notion of stochastic health effects, and 3) the high-dose cancer mortality statistics of the atomic bomb survivors.

These recommendations are used by all countries to protect humans and the environment against the LNT predicted hazards of ionizing radiations, which are generally low level (i.e., low dose-rate). The general practice is to ignore any evidence of beneficial effects. A U.S. Environmental Protection Agency staff report states, "as the purpose of a risk assessment is to identify risk (harm, adverse effects, etc.), effects that appear to be adaptive, non-adverse or beneficial may not be mentioned." [29]

An issue related to the radiation scare is the requirement for special, socially-acceptable facilities for permanent disposal of radioactive materials that are in waste storage areas. Another issue is insurance to cover liabilities for the potential costs of cleaning a contaminated area after an accident. The recent experiences in Japan have identified the very significant costs of caring for the residents near a damaged nuclear facility who are very fearful of the radiation. Suppliers of equipment for nuclear projects are also concerned about potential liabilities. Adopting a proper threshold dose-response would eliminate the concern about adverse health risks for exposures below the limit. This would allow potential liabilities to be estimated and would enable a proper determination of those employees and residents who have been harmed by radiation exposures. Radioactive materials, sealed in robust containers, could be stored long-term almost anywhere.

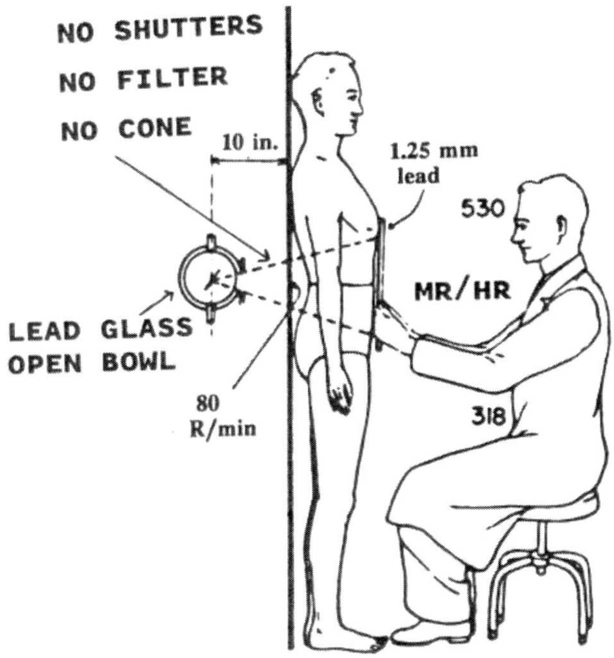

Figure 2.1. Radiation exposure to patient and physician from a non-protective fluoroscope in use circa. 1930. [26] Figure 1.

The limit for acceptable radiation exposures should be set as high as relatively safe (AHARS), as advocated by Wade Allison. [30] The concept of *as low as reasonably achievable* (ALARA) is very harmful. It reinforces the public perception that low radiation is very dangerous, and it requires that extreme emergency precautionary measures be carried out that cause unnecessary suffering and result in many premature deaths. [27] As required by the international standards to address the concerns about low-level radiation, the authorities in the Fukushima prefecture evacuated many homes, nursing homes and hospitals for more than three years. The human and economic consequences were severe. [31-34] Much more harm was caused by the permanent emergency evacuations following the Chernobyl accident. [35, 36] Evacuees continue to suffer and die prematurely from post-traumatic stress disorder. [37, 38]

The nuclear regulators insist on the requirement to learn lessons from nuclear accidents, but they fail to learn the most important lesson. Many people die from emergency precautions, but no one is, or would be, harmed by the low radiation levels. In an accident, the radiation dose rate in the area surrounding a properly designed facility will generally be below the threshold for harmful effects. The most important precautions that nuclear regulators should take would be to disregard the fear mongering and instead study the very important discoveries of the past century, as mentioned earlier.

The literature from 1895 to the 1920s contains many significant articles on the medicinal properties of X-rays and radium. [1, 2, 39] Dose thresholds to prevent late-appearing harmful effects in the radiologists were determined from this information. Scientists and nuclear regulators should become aware of this history and the subsequent evidence that has been documented by the medical practitioners. [40] The essence of this information should be summarized and communicated to the public in plain language. Understanding the true health effects is the best defence against a panic reaction to the release of radioactive material during a nuclear accident or by a terrorist radiological dispersion device.

In diagnostic imaging, patients are put at risk by the precautions that advocate reducing low-dose radiation exposures for the purpose of ALARA. [41] The fundamental principle that physicians must follow, "First, do no harm" is violated. The risks of serious harm arise from: 1) the increased likelihood of misdiagnosing serious illnesses from poor quality images that will result from a dose reduction policy, 2) physicians not ordering diagnostic studies when indicated, and 3) patients withholding consent for such imaging studies. [42]

LNT Hypothesis of Radiation Carcinogenesis

What is a hypothesis? A dictionary states: It is a proposition tentatively assumed in order to draw out its logical or empirical consequences and so test its accord with facts that are known or may be determined. It appears to be a condition of the most genuinely scientific hypothesis that it be … of such a nature as to be either proved or disproved by comparison with observed facts.[1]

What is this LNT hypothesis? This idea, adopted by the U.S. NAS in 1956, assumes there is an excess risk of cancer death that is proportional to the radiation dose, over the full range from a very high dose to *zero dose*. As shown in Table 9 of Ozasa et al. [43], the Life Span Study (LSS) of 86,611 Hiroshima-Nagasaki atomic bomb survivors identified that there are several hundred excess cancer deaths among several thousand survivors who received high radiation doses above 1 Gy or 100 rad. When the number of excess cancer deaths in the high dose range is plotted as the ordinate on a graph, with radiation dose as the abscissa, and a straight line is drawn from the origin to fit the high-dose data, then the LNT hypothesis predicts that the incidence of excess cancer death is given by the ordinate on this line, at any dose in the low dose range.

Figure 2.2, adapted from reference [44], is a representation of this idea. Radiation dose is on a log scale that spans the range of interest. The natural cancer mortality is shown as a negative health effect with an incidence of 25% of the population. The straight-line fit to the Hiroshima-Nagasaki excess cancer mortality data, in the range above 100 rad (1 Gy), appears as a solid curved on this log scale. The LNT model is shown as the dashed linear extrapolation of excess cancer mortality from the solid line to 0 dose, approaching the natural incidence of fatal cancer asymptotically. The hatched area shown is the range of the

[1] http://unabridged.merriam-webster.com/

beneficial health effects observed for an acute dose in the range from 50 to 0.1 rad. The decrease in the incidence of cancer is caused by radiation-induced up-regulation of adaptive protection systems, the amount of which depends on individual sensitivity to the stimulation. The LNT hypothesis prediction of a small excess cancer risk contradicts the evidence of a lower cancer risk in the dose range below 50 rad.

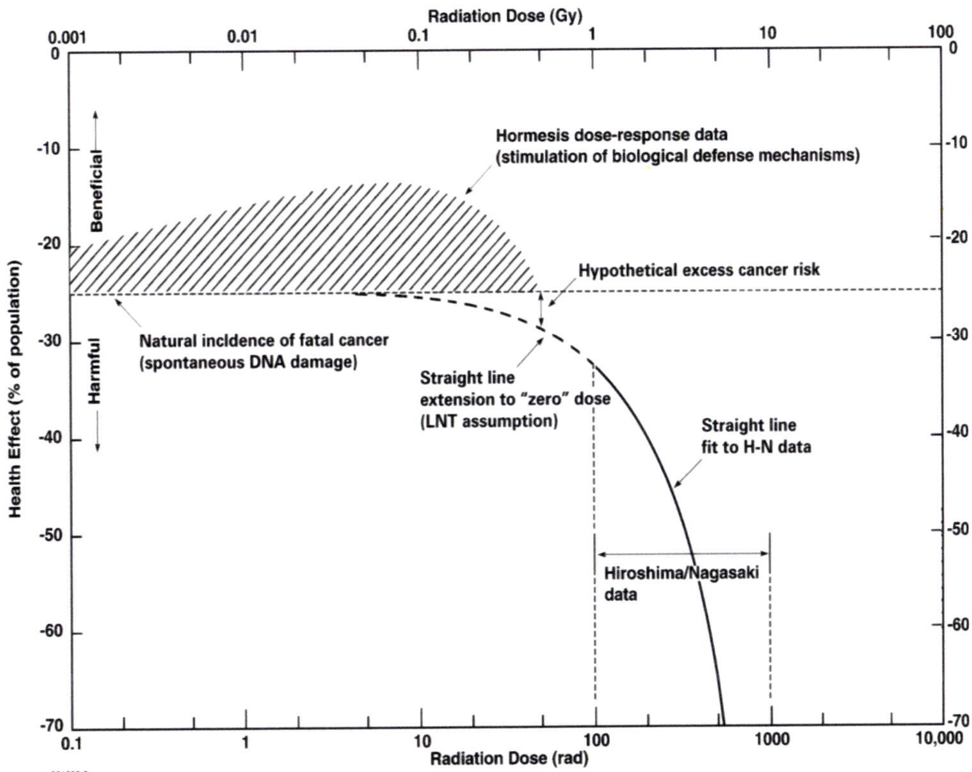

Figure 2.2. Comparison of LNT prediction of excess cancer risk (after an acute radiation dose) against observed dose-response data that show a reduction of cancer risk [44].

Studies are carried out to demonstrate that the LSS data, below 100 rad, fit the LNT hypothesis; however, they are unable to correct for the many non-radiation confounding factors that affect the cancer mortality of the survivors and the controls over their life spans. The causes of cancer are just not well understood. Furthermore, Socol and Dobrzyński have demonstrated that the LSS data have insufficient statistical power to fit any model. [45] The Scientific Method requires that any hypothesis be subjected to tests that may prove it wrong. [46] If its predictions cannot be shown to fit the observed data with adequate statistical certainty, the hypothesis must be rejected or modified. The scientific and regulatory radiation protection organizations have known about such uncertainties for decades. They continue using the LNT hypothesis because its predictions of excess cancer deaths are "conservative." However, they ignore the very carefully performed Nuclear Shipyard Worker Study, which demonstrated significant beneficial health effects, including a lower mortality from all cancers combined in the exposed cohort. [47]

The UNSCEAR 1958 report, Annex G page 165, contains the data shown in Table 2.1 on the occurrence of leukemia among the Hiroshima atomic bomb survivors. [48] The contribution from naturally-occurring leukemia was reduced by waiting for 4.4 years of latency time for radiation-induced leukemia to manifest itself and then measuring over the subsequent 8-year period when the incidence of this cancer was expected to peak. The acute dose to the control population, in Zone E, was about zero. Reference [45] points out that there were no significant lifestyle differences between these control and exposed populations.

Table 2.1. Leukemia incidence of Hiroshima survivors, 1950-57 [48] Annex G. Table VII

Zone	Distance from hypocentre (metres)	Dose (rem)	Persons exposed	L (Cases of leukemia)	$\sqrt{L}$	N* (total cases per 10^6)
A	under 1,000	1,300	1,241	15	3.9	12,087 ± 3,143
B	1,000-1,499	500	8,810	33	5.7	3,746 ± 647
C	1,500-1,999	50**	20,113	8	2.8	398 ± 139
D	2,000-2,999	2	32,692	3	1.7	92 ± 52
E	over 3,000	0	32,963	9	3.0	273 ± 91

* The standard error is taken as N times ($\sqrt{L}$ / L).

** It has been noted that almost all cases of leukemia in this zone occurred in patients who had severe radiation complaints, indicating that their doses were greater than 50 rem.

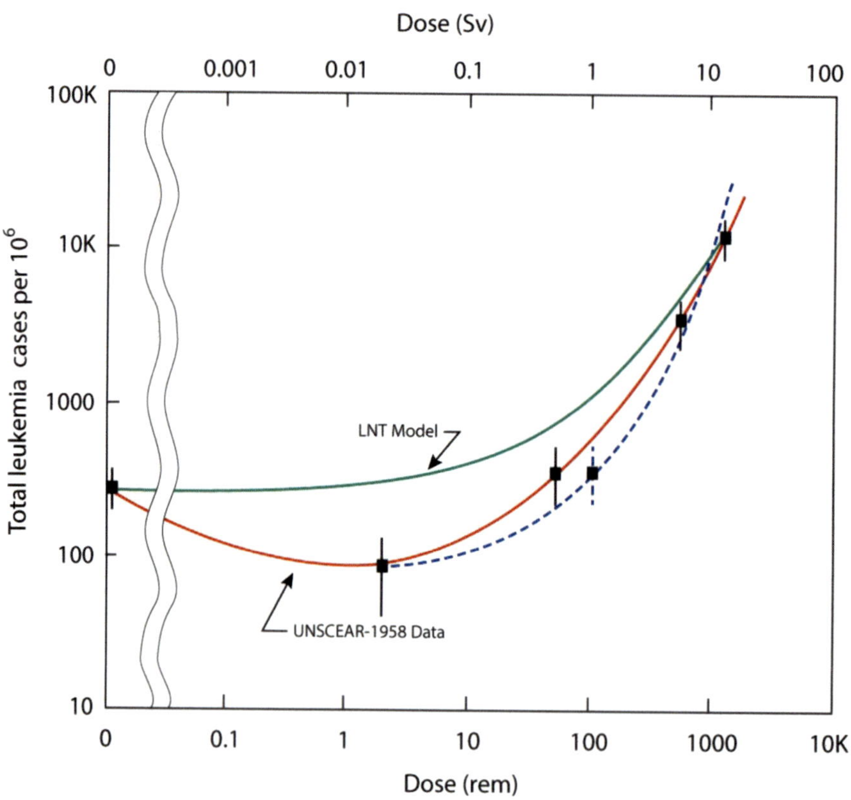

Figure 2.3. Leukemia incidence of the Hiroshima atomic bomb survivors [15] The dashed blue line through 100 rem dose addresses the UNSCEAR footnote on page 165 for the 50 rem dose in Zone C "that almost all cases of leukemia in this zone occurred in patients who had severe radiation complaints, indicating that their doses were greater than 50 rem."

These very important dose-response data were not presented in graphical form in the UNSCEAR report. They are shown in Figure 2.3, along with the response predicted by the LNT hypothesis, over the range from 1,300 to 0 rem (13 to 0 Sv). The line through 100 rem takes into account the observation concerning the patients from Zone C who received doses greater than 50 rem [15]. Clearly, these data contradict the linear dose-response predicted by the LNT hypothesis. On the contrary, they fit a hormetic *J*-shaped dose-response. Bone marrow cells are very sensitive to radiation, and these data indicate that there is no evidence of a significant excess risk of this cancer below an acute dose of about 50 rem or 0.5 Sv. This fact alone, this evidence of a threshold, is sufficient to reject the LNT hypothesis.

The linear concept of "stochastic risk" is consistent with the idea of adding the small radiation dose received by each person in a large population, living in a low radiation area, to evaluate the "collective dose." Radiation protection physicists multiply this aggregate dose by a fatal cancer risk factor to predict the number of cancer deaths that will occur in this population due to the low radiation exposure they received. Renowned radiation biologist Ronald Mitchel has pointed out that a fundamental principle of radiation protection, the assumption of a linear dose-response and dose additivity, is incorrect. [49]

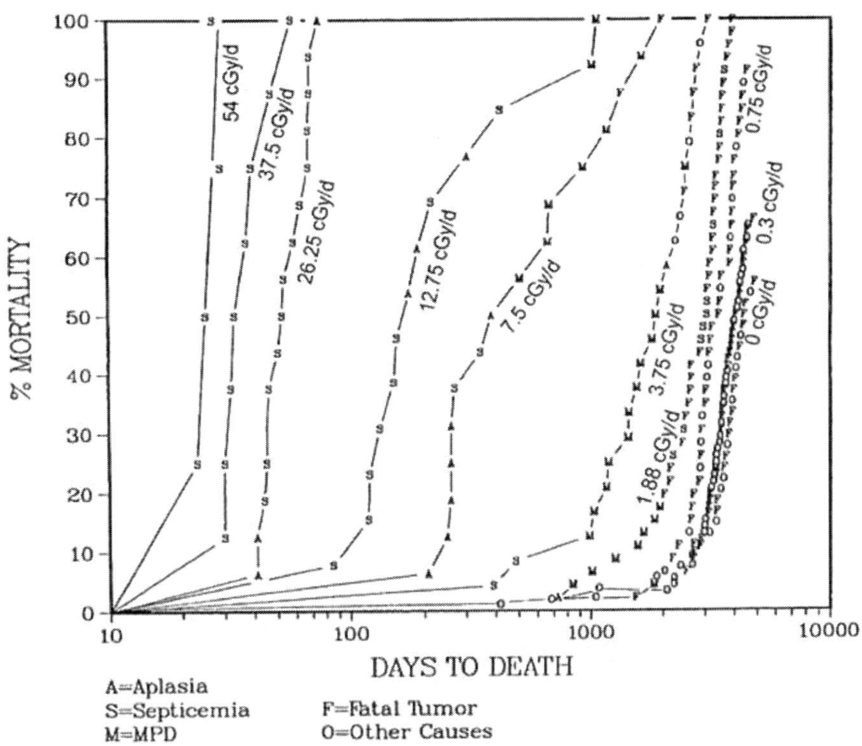

Figure 2.4. Mortality curves of groups of dogs that were subjected to whole-body chronic gamma irradiation at different dose rates, from [50] Figure 3.

The above leukemia data of the Hiroshima survivors pertain to an acute exposure. What are the health effects of a continuous exposure at a dose-rate that is significantly higher than the average background level? Many studies have been carried out since the 1960s on chronic exposures of mammals to low dose-rates of ionizing radiation. The review article by Fliedner et al [50] is particularly interesting because it covers a variety of studies on humans and animals. Of special interest is the carefully controlled study on the health effects of ten groups of dogs that were subjected to whole-body cobalt-60 radiation for their entire lives. The control group of dogs received normal background radiation, while the other 9 groups each received a fixed dose-rate, over a very wide range. The radiation affected primarily the dog's blood system, which adapted to this severe stress, until the dog died. The mortality data appear in Figure 2.4, including the cause of death. It is very interesting to note that the health of the 92 dogs exposed to the lowest dose-rate, 3 mGy per day or 1,100 mGy per year, was very similar to the health of the control dogs, and the causes of death in both groups were dominated by fatal tumors at about the same incidence.

Table 2.2. Determination of normalized lifespan versus dose rate

Dose Rate (cGy/day)	Dose Rate (mGy/year)	Lifespan – days (50% mortality)	Lifespan (normalized)
background	2.4×10^0	4300	1.00
0.3	1.1×10^3	4100	0.95
0.75	2.7×10^3	3300	0.77
1.88	6.9×10^3	3000	0.70
3.75	1.4×10^4	1900	0.44
7.5	2.7×10^4	410	0.095
12.75	4.7×10^4	160	0.037
26.25	9.6×10^4	52	0.012
37.5	1.4×10^5	32	0.0074
54	2.0×10^5	24	0.0056

The normalized lifespan of each group, at the 50% mortality level, is shown in Table 2.2 and in Figure 2.5. The data indicate that a threshold exits for increased mortality or shortened lifespan at a dose-rate of about 700 mGy per year, which is about 300 times the average background level. It is about the same as the radiologists' 0.2 roentgen per day tolerance dose. The LNT hypothesis, on the contrary, does not predict a dose-rate threshold. It predicts a risk of excess cancer mortality that increases linearly as the radiation level rises above background, leading to a shorten life span. Since the measured data contradict this prediction, the LNT hypothesis must be rejected.

Why is the LNT hypothesis incorrect? First of all, there is no statistically significant data that supports the use of this hypothesis to predict cancer risk in the low dose range, which is why it is still a hypothesis 58 years after it was adopted. Secondly, there are enormous amounts of data on the biological effects of: 1) a low radiation dose, 2) repeated low radiation doses, and 3) low radiation levels, which contradict the predictions of the LNT hypothesis by revealing rather high threshold levels before the onset of harmful late effects, such as lifespan reduction or a higher cancer incidence, are observed. [20, 15] These data have been accumulating since 1896, for more than 118 years.

Compliance with the requirements of the Scientific Method [46] should have resulted in the LNT hypothesis being rejected, rather than being adopted in 1956. Instead, it is employed to calculate hypothetical risks, creating uncertainty and great fear about potential cancer risks from low radiation. The common belief that the LNT hypothesis is needed for radiation protection purposes because it is "conservative" in over-predicting the number of fatal cancers is wrong. [51] The extreme precautions, such as emergency evacuations, cause many premature deaths from both physical and psychological stress.

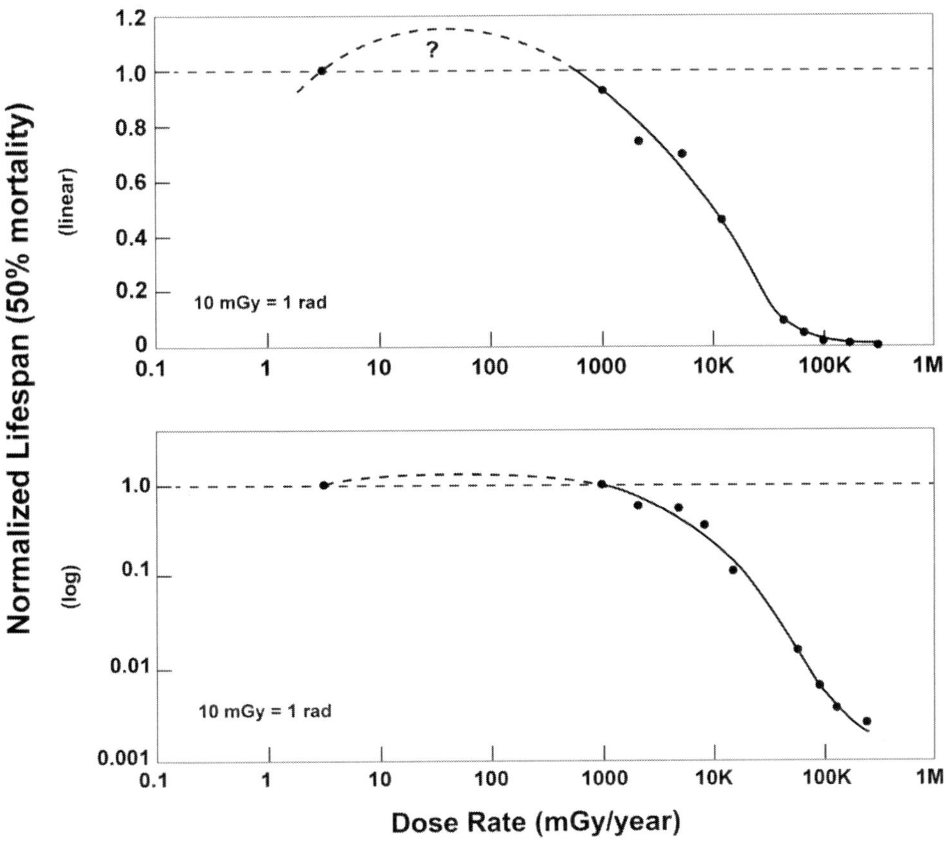

Figure 2.5. Lifespan of dogs versus cobalt-60 dose rate [20].

How then does ionizing radiation produce health effects? Feinendegen et al. point out that all living organisms possess very powerful immediate protection systems and adaptive ones that repair or remove cell, tissue and organ damage, and restore organism health. [19] Radiation is one of the stressors that modulate the protection systems; high radiation impairs protection, increasing the risk of cancer cells developing and progressing into tumors, while low radiation up-regulates many protection systems (> 200 genes) that act over extended periods of time to produce very important positive health effects, including a *lower* incidence of cancer. This is the mechanism for the significant net beneficial effects of low doses up to about 200 mSv or 20 rem. At even higher doses, additional protective mechanisms against cancer development operate.

Origin of Radiophobia

Scientists discovered in 1939 that a uranium nucleus often splits (fissions) when it absorbs a neutron. They realized that mutual electrostatic repulsion between the large fragments would release an enormous amount of heat. Coincident emission of several neutrons suggested the possibility of a chain reaction and the military application of this fission reaction. The nuclear explosions over Hiroshima and Nagasaki were followed by the intensive development, testing and large-scale production of weapons in the U.S., and proliferation to several other countries. Fear of nuclear warfare and feelings of remorse motivated some of the scientists involved in the original development to initiate strong antinuclear political actions against the nuclear arms race. They were joined by many other concerned scientists. One of their strategies was to create and promote extreme social fear of low-level radiation from the radioactive dust or "fallout" that is deposited downwind of a weapons test and around the world. Figure 2.6 [52] is the alarming telegram that Nobel Prize laureate Linus Pauling sent to President Kennedy in 1962 warning him of serious genetic harm to unborn children. With 60 years of knowledge and experience in the use x-rays and radium in research, diagnosis and the safe treatment of a wide variety of illnesses, how was it possible for scientists to create such a strong fear of radiation in the 1950s? How can it be that this phobia has persisted over the subsequent 60 years, and continues even today, as great scientific advances were being made in biology, genetics and effects of radiation on cells and organisms?

Immediately after x rays and nuclear radiation were discovered in 1895/6, thousands of scientists began intensive investigations into the characteristics, effects and potential applications of these ionizing radiations. A very powerful, stimulating effect on the protection systems of organisms was observed after the application of a low radiation dose. [1, 5] The mechanism of action was not understood; however, it resulted in remarkably rapid, positive health effects. Medical practitioners began diagnosing fractures and diseases with X-ray imaging and also treating many illnesses employing X- and radium radiations. They discovered that high acute radiation doses produced external and internal burns, which did not heal completely; scar tissue remained. High local doses destroyed cancers, but the surrounding tissues recovered. Exploiting the beneficial effects, many different kinds of low-dose radiation treatment were discovered and accepted by the medical community against a wide variety of very serious illnesses, including infections, inflammations and cancer. [2, 21, 22, 53, 54] No statistically significant increases in the incidence of cancers in these patients were noted, after decades of follow-up.

However, safety measures were needed for radiologists and medical practitioners who received significant doses after repeated exposures, Figure 2.1. British radiologists developed neoplasms after ten or more years of exposure. There was a statistically significant excess of deaths from cancers of the pancreas, lung and skin, and from leukemia. A study of 319 British radiologists who entered the profession between 1897 and the end of 1920 indicated they had a 75% higher death rate from cancer than that of the medical practitioners, and 44% higher rate than Social Class 1. [4]

The British X-ray and Radium Protection Committee was formed in 1921 and issued recommendations for a tolerance dose limit. In the study group of radiologists who entered the profession after 1920, those who died had about the same cancer death rate as the medical practitioners and a 21% *lower* rate than Social Class 1. It is interesting to note that their mortality from 'all causes' was *lower* by about 12% than the mortality from 'all causes' of the medical practitioners and of Social Class 1, whereas the mortality from 'all causes' of the early radiologists (those who entered before 1921) was about the same as the mortality from 'all causes' of the medical practitioners and of Social Class 1. This study demonstrated that the radiation protection recommendations issued in 1921 were effective in eliminating the risk of late effects of radiation, such as increased cancer mortality.

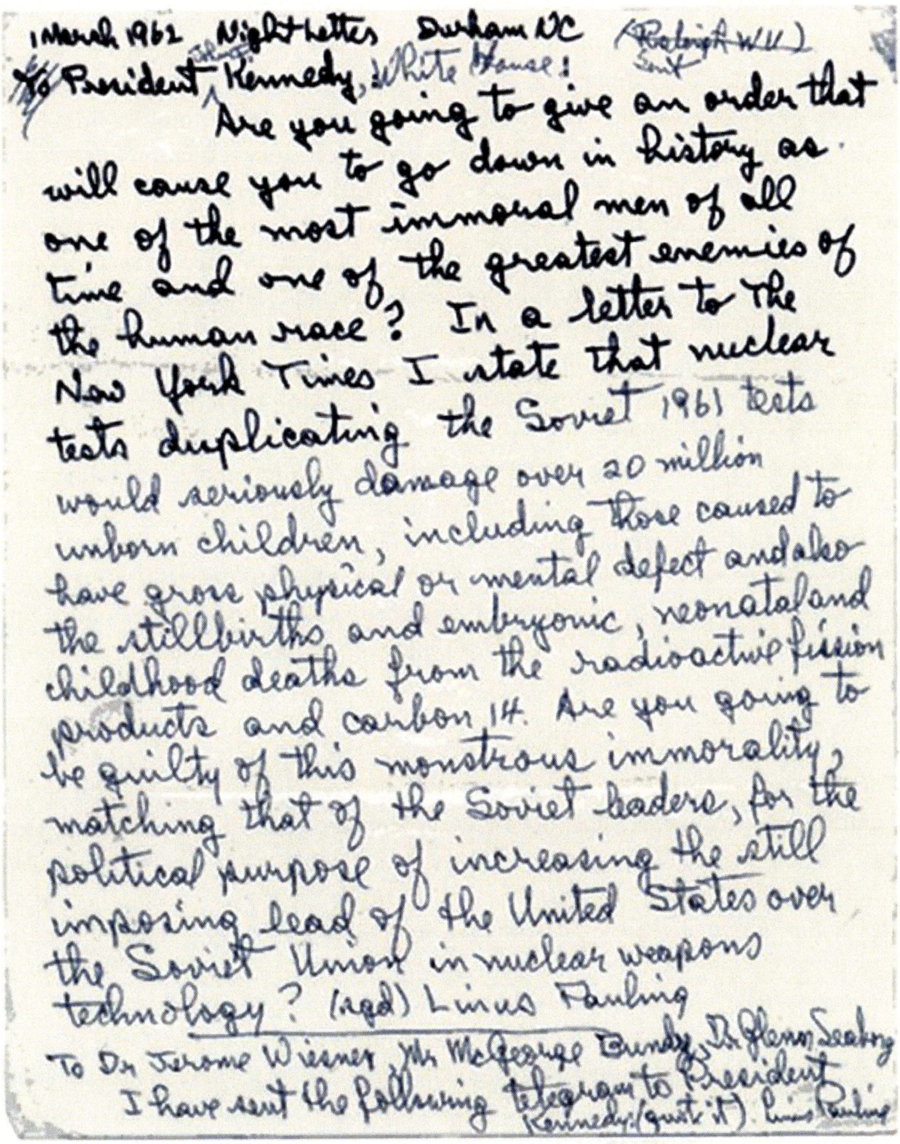
1 March 1962 Night Letter Durham NC (Raleigh WU)

To President Kennedy, White House:

Are you going to give an order that will cause you to go down in history as one of the most immoral men of all time and one of the greatest enemies of the human race? In a letter to the New York Times I state that nuclear tests duplicating the Soviet 1961 tests would seriously damage over 20 million unborn children, including those caused to have gross physical or mental defect and also the stillbirths and embryonic, neonatal and childhood deaths from the radioactive fission products and carbon 14. Are you going to be guilty of this monstrous immorality, matching that of the Soviet leaders, for the political purpose of increasing the still imposing lead of the United States over the Soviet Union in nuclear weapons technology? (sgd) Linus Pauling

To Dr Jerome Wiesner, Mr McGeorge Bundy, Dr Glenn Seaborg

I have sent the following telegram to President Kennedy: (quote it). Linus Pauling

Figure 2.6. Professor Linus Pauling's telegram[2] to President J. F. Kennedy in 1962 [52].

[2] The Ava Helen & Linus Pauling Papers, Courtesy of Oregon State University

After winning the Nobel Prize in Chemistry in 1953, Pauling became science's most prominent activist against nuclear weapons, a movement that led to the 1963 ban on above-ground testing and Pauling's Nobel Peace prize.

In 1924, Mutscheller proposed the concept of a tolerance dose. [55] He advocated using a value of 1% of the average erythema dose (600 roentgen) in 30 days, or 0.2 roentgen per day, to determine the amount of lead shielding needed. He argued that there was no safe dose, but safety had to be weighed against the costs to achieve it. The International Committee on X-Ray and Radium Protection was established in 1928 by the second International Congress of Radiology for the purpose of advising physicians on radiation safety measures in a non-regulatory framework. [25] It later became the International Commission on Radiological Protection (ICRP). The U.S. Bureau of Standards selected Lauriston Taylor to represent the U.S. After the 1928 congress, Taylor established the American X-ray and Radium Protection Committee. It later became the National Committee for Radiation Protection and Measurement (NCRPM). The tolerance dose was officially adopted in the 1934 ICRP standard, based on Mutscheller's concept, and it remained the foundation for radiation protection. The underlying assumption was that the biological effect was proportional to ionization. Any harm that occurs from exposures below the tolerance level was acceptable, and not that such effects were nonexistent. Nonetheless, the use of this specific limit was frequently interpreted to mean that there was a threshold below which no harm occurred. The committees did not disband after establishing this standard but continued to respond to questions from the public and government.

Universal fear of radiation began in 1956 when the NAS recommended a linearity dose-response policy for assessing risks to genetic material (DNA molecules) from radiation that replaced the threshold dose-response model. [16] This formal recommendation initiated a series of advisory and regulatory actions in essentially all countries to adopt linearity and apply it to somatic, late-appearing effects, i.e., cancer risk assessment, for radiation and later for chemical carcinogens. [8, 13, 25] The LNT hypothesis links any radiation exposure, no matter how small, to an increased risk of cancer.

In 1980, 52 years after the meeting that established the ICRP's first radiation protection standard and after establishing the NCRPM, Lauriston Taylor gave a very remarkable speech at the 5th Congress of the International Radiation Protection Association—the title, Some Non-Scientific Influences on Radiation Protection Standards and Practice. [56, 57] He began as follows. *"In the practical application of the principles of the achievement of protection against harmful radiation effects, our greatest obstacles today do not include a lack of knowledge about the biomedical effects of ionizing radiation. Today, we know about all we need to know to adequately protect ourselves from ionizing radiation. Let me repeat that. Today, we know about all we need to know to adequately protect ourselves against ionizing radiation. Therefore, I find myself charged to ask: What is the problem and why is there one? I suspect that most of us here today know in a general sort of way where the problem lies and that basically it is not a scientific one. Rather, it is a philosophical problem with all the ramifications implied in the term. Or perhaps it may be a political problem, that is, one requiring prudence and sagacity in devising and pursuing measures adapted to promote the public welfare, or perhaps the problem may not be as much protecting ourselves against radiation as protecting us against ourselves. In any case, I sometimes think that today we are – in many areas – tormenting ourselves through our obsession with health."*

Discussing biological effects, he stated, “Collectively there exists a vast array of facts and general knowledge about ionizing radiation effects on animal and man. It cannot be disputed that the depth and extent of this knowledge are unmatched by any of the myriads of other toxic agents known to man. It is because of this knowledge, portions of which have become known to the public, that the public has come to expect sharp, clear, definitive, and undisputed answers to any questions involving radiation. This is an understandable, if somewhat irrational, position. However, it leads to the difficulty that when there may be some indication of a lack of knowledge by, or disagreement among, scientists, the public feels that somehow they have been let down or led astray by the scientific community. A good example of this is the current so-called ‘controversy’ within the protection community centering around the effects of radiation delivered in low doses at low dose rates. Were it not for a few congressional committees, more interested in headlines than facts, aided and abetted by a willing press and a few publicity-seeking ‘scientists’, it is likely that the question would drone on in the normal scientific meetings and committees at a proper pace, commensurate with its importance. It's not that it is unimportant, but its priority should be low compared with so many other insults that man faces”.

He praised the “fantastically fine” radiation safety records that scientists have accomplished, and continued, “No one has been identifiably injured by radiation while working within the first numerical standards set by the NCRP and the ICRP in 1934. The theories about people being injured have still not led to the demonstration of injury and, if considered as facts by some, must only be looked upon as figments of the imagination.”

On morality, he stated, “An equally mischievous use of the numbers game is that of calculating the numbers of people who will die as a result of having been subject to routine diagnostic x-ray procedures. An example of such calculations is those made before a Congressional Committee in 1967 which were based on the literal application of the linear, non-threshold, dose-effect relationships, treated as a fact rather than a theory. By this procedure he calculates 30,000 deaths per year resulting from x-ray diagnosis. Of course, there has been no statistical or other verification of this calculation. Unfortunately, the technique has been picked up by others. These are deeply immoral uses of our scientific heritage.”

More recently, renowned Swedish scientist Gunnar Walinder, a colleague of Rolf Sievert, stated: “The LNT hypothesis is a primitive, unscientific idea that cannot be justified by current scientific understanding. As practiced by the modern radiation protection community, the LNT hypothesis is one of the greatest scientific scandals of our time.” His book, titled: Has Radiation Protection Become a Health Hazard? [58], includes the following summary of his conclusions (p 137).

1. I have found no rationale of distinguishing cancer risk assessments from those applied to other hazards. Generally speaking, cancer is not more nor less a stochastic event than the so called deterministic effects.
2. On grounds of principle, we cannot acquire any knowledge about the multi-iterative cancer development in a mutually adaptive human body in cases where the acting agent is non-dominant. This fundamental ignorance applies to carcinogenesis as well to its opposite (hormesis).
3. The current pretensions to knowledge about low-dose radiogenic cancer should be rejected. In particular, we have to realize that the dose-response relationship at very

low radiation doses (non-dominant doses) cannot be comprehended through or based on simple mathematical expressions and, least of all, on an equation of the first degree. The LNT hypothesis is deeply unscientific.

4. More often than not, countermeasures against the possible effects of very low radiation doses will imply more harm than benefit. For example, people who were evacuated to Kiev from low-contaminated, rural areas in the Ukraine, would probably run a greater cancer risk than what would have been the case had they remained in their native districts. (The age-adjusted cancer rate in big cities - like Kiev - is about 20% higher than that in rural areas.)
5. Sooner or later we will be forced to abandon the current radiation protection doctrines, not only with regard to cancer prognoses but also within the administrative work. The main reason for this conclusion is the 'categorical imperative': we must not let radiation protection become a health hazard.

In 2008, Bernard Cohen provided convincing reasons for rejecting the LNT theory. [86]

Origin of the LNT Hypothesis

It started in the 1920s with repeated reports that germinal changes could be induced by X-rays or radium rays. Hermann Muller was a geneticist and a eugenicist. In his seminal paper in Science, he laments that the study of gene mutations, which are the basis of organic evolution, "has heretofore been very seriously hampered by the extreme infrequency of their occurrence under ordinary conditions, and by the general unsuccessfulness of attempts to modify decidedly, and in a sure and detectable way, this sluggish 'natural' mutation rate. Modification of the innate nature of organisms, for more directly utilitarian purposes, has of course been subject to these same restrictions, and the practical breeder has hence been compelled to remain content with the mere making of recombinations of the material already at hand, providentially supplemented, on rare and isolated occasions, by an unexpected mutational windfall. To these circumstances are due the wide-spread desire on the part of biologists to gain some measure of control over the hereditary changes within the genes." [6]

Earlier experiments by medical physicists to produce mutations using radiation yielded data that was not reproducible. Muller thought the idea was promising on theoretical grounds, so he carried out a series of experiments on the fruit fly, *Drosophila melanogaster*, in an attempt to provide critical data. With his eight years of previous experience measuring the mutation rate in fruit flies, he was able to produce decisive mutational effects. He found that "treatment of the sperm with relatively heavy doses of X-rays induces the occurrence of true 'gene mutations' in a high proportion of the treated germ cells."... "Comparison of the mutation rates showed that heavy treatment had caused a rise of about fifteen thousand per cent in the mutation rate over that in the untreated germ cells." Regarding the types of mutations produced, it was found that the lethals (recessive for the lethal effect)[3] greatly outnumbered the non-lethals producing a visible morphological abnormality. The lethals were used as an index of mutation rate. He concluded that use of X-rays affords the opportunity of

[3] EPA 40 CFR 798.5275 Sex-linked recessive lethal test, available at: http://www.gpo.gov/fdsys/pkg/CFR-2012-title40-vol33/pdf/CFR-2012-title40-vol33-sec798-5275.pdf

creating a series of artificial races in the chosen organisms, allowing enough mutations to be produced 'to order' to furnish respectable genetic maps, in the selected species. "Similarly, for the practical breeder, it is hoped that the method will ultimately prove useful. The time is not ripe to discuss here such possibilities with reference to the human species."

Muller's experimental data appears in his paper at the 5th International Congress of Genetics in Berlin. The experimental conditions are given, i.e., a Coolidge x-ray tube with a tungsten target at a distance of 16 cm from the fruit-fly capsule, a 1 mm aluminum shield, 50 kV peak voltage and a 5 mA beam current. [7] However, the dose-rate is not mentioned. For these conditions, the Rad Pro Calculator gives a dose-rate of 7860 R/h or 131 R/min (1.1 Sv/min). [59] The irradiation times were used: 12, 24, 36 and 48 minutes. The doses for these four exposure times are: 1572, 3144, 4716 and 6288 R (or 13.2, 26.4, 39.6 and 52.8 Sv). If mutations result directly from hits by the X-ray photons, there should be an exact proportionality between point mutations and dosage. However, Muller's data indicated a square root relationship.

Oliver carried out a similar study in Muller's laboratory, covering the wider range 3.5, 7.0, 14, 28 and 56 minutes, and he observed a generally linear response for mutation rate vs. dose. [60, 61] Other geneticists performed experiments and found different dose-response shapes. Despite limited data and lack of overall consistent findings, Muller concluded that mutation frequency "is exactly proportional to the energy of the dose absorbed. There is, then, no trace of a critical or threshold dosage beneath which the treatment is too dilute to work." [62] Calabrese points to a letter Muller wrote to MIT professor Robley Evans in 1949. [25] In his letter, Muller stated that "many of the quantities are only very roughly known even for *Drosophila*, and we are admittedly extrapolating too far in applying this to man, but it is all we can do in our present state of ignorance and we must meanwhile remain on the safe side." Such a comment strongly suggests that Muller was guided more by a precautionary public health philosophy rather than the science with respect to the extrapolation of his findings for various types of extrapolations including across species and from high to low dose.

This was established when Muller deliberately misled the world during his Nobel Prize acceptance speech in which he stated that there was "no escape from the conclusion that there is no threshold." [9, 10, 11]

Prior to delivering his speech, he had reviewed Ernest Caspari's very carefully performed research on fruit flies. The data clearly demonstrated a threshold dose-response, which contradicted the linear dose-response model. Scientists had questioned whether the dose-response would be still be linear when a much lower dose-rate is used, so Caspari studied the same type of fruit fly that Muller had used; however, he employed a comparatively very low dose rate, 52.5 R (0.46 Sv) delivered in 21 days. The control cultures numbered 56,252, and the 51,963 experimental cultures received 52.5 R at a dose rate of about 0.104 R/h (1.53×10^{-5} Sv/min). The dose was 30 times lower than the lowest dose Muller used (1572 R). The dose-rate was about 80 thousand times lower. The first two statements in the summary of the Caspari and Stern paper [12] are:

1. The rate of lethal sex-linked mutations in Drosophila exposed to gamma-rays of 2.5 r units per day through 21 days (total 52.5 r) was determined.
2. In a total material of 108, 215 chromosomes tested, no significant difference between experimentals and controls was found.

This very important paper with critical evidence was not given any prominence in the scientific world because Caspari's supervisor, Curt Stern, believed that ionizing radiation affected the mutation rate in a linear dose-response manner. He deleted Caspari's threshold conclusion and added a discussion on why these data should not be accepted and used until it was determined why these low-dose, chronic irradiation findings differed from those of the high-dose, acute irradiation studies that supported linearity. The paper appeared in Genetics, without apparent independent peer review, a journal that Stern was editor of. [10, 13, 14]

After the discovery of X-ray-induced mutations, a proposal arose that cosmic/terrestrial radiation-induced mutations are the main mechanism for induction of heritable traits, providing the driving force for evolution. Muller evaluated his spontaneous and ionizing-induced mutation data in fruit flies and argued that background radiation had a negligible impact on spontaneous mutation rate.

Another fascinating dimension of the origin of the LNT concept for ionizing radiation-induced mutations is radiation target theory and the hit hypothesis. [8] Target theory was quantitative and dosimetric, with mathematical calculations related to quantum mechanics. This theory was established prior to the discovery by medical physicists and by Muller of inducible mutations. Muller's mutation findings were a major scientific advance that easily fit into the target theory concept and advanced its scientific status. The expanded set of observations gave a basis for collaboration between theoretical physicists and radiation geneticists. They developed interrelated physical science-based genetics perspectives including a biophysical model of the gene, a radiation-induced gene mutation target theory and the single-hit hypothesis of radiation-induced mutation, which, when integrated, provided the theoretical mechanism and mathematical basis for the LNT model. The detailed interactions and collaborations of the leading radiation geneticists and theoretical physicists established a conceptual framework for gene structure, target theory for the induction of mutations by radiation, the single-hit mechanism hypothesis to account for the shape of the LNT dose response and the application of this dose-response model for cancer risk assessment. Genetic target theory saw mutation as a purely physical action by an all or none law, independent of all other ionizations and energy absorptions. It had the important assumptions that damage was proportional to the energy absorbed, the dose-response model was LNT, and that effects were cumulative and deleterious. Based on a very small amount of data, this was transformed by Muller into a biological "proportionality rule" that dominated mutation literature during the 1930s.

The LNT concept became accepted by radiation geneticists and recommended by national and international advisory committees for risk assessment of ionizing radiation-induced mutational damage/cancer. It was later generalized to chemical carcinogen risk assessment and used by public health and regulatory agencies worldwide, from the mid-1950s to the present time. Muller used his Nobel Prize acclaim to communicate his concerns about genetic damage. He became an activist strongly promoting the fear from any exposure to x-rays or nuclear radiation stating that the risk was linearly proportional to dose without any threshold. [11, 24]

Recently, Japanese scientists have performed more accurate studies on the same type of fruit fly, at low dose rate and over a wide dose range. Research by Koana et al. revealed that the mutation frequencies of irradiated fruit flies exceed that of the controls only when the dose is raised above 1 Gy. [63] Another study by Koana et al. demonstrated that their mutation frequency is much lower than the mutation frequency of the controls when a low dose-rate is employed. [64] Since repair capacity is limited, use of a high dose-rate results in a greater mutation frequency for the same dose. Ogura et al. subjected flies to a dose rate of 22.4 mGy per hour (2.2 rad/h) and measured their mutation frequency at 0.5 mGy to be only 0.09%, a factor of 3.6 lower than the 0.32% mutation frequency of the control culture. [65] When the data are combined to reduce the error bars, as shown in Table 2.3 and Figure 2.7, it is clear that the mutation frequency exceeds that of the controls only when the dose exceeds 1 Gy, the threshold for excess harm. These measured data contradict the LNT model, shown over the range from 10 to 0 Gy.

Table 2.3. Binomial statistics applied to fruit fly mutation data of Ogura et al. [65]

Dose Gy	Number of lethals y	Number of chromosomes n	Mutation frequency $p = y/n$	$q = 1 - p$	Variance σ^2 n.p.q	Standard deviation σ	$2\sigma/n$ %	$p + 2\sigma/n$ %	$p - 2\sigma/n$ %
0.0005	9	10,500	0.0009	0.9991	9.441	3.07	0.06	0.15	0.03
0.1	2	1507	0.0013	0.9987	1.957	1.399	0.186	0.32	-0.06
1	6	2662	0.0023	0.9977	6.109	2.472	0.186	0.42	0.04
5	8	2055	0.0039	0.9961	7.983	2.825	0.27	0.66	0.12
10	21	2730	0.0077	0.9923	20.86	4.567	0.33	1.10	0.44
0.3	8	4169	0.0019	0.9981	7.906	2.81	0.13	0.32	0.06
7	29	4785	0.0061	0.9939	29.01	5.386	0.225	0.84	0.38

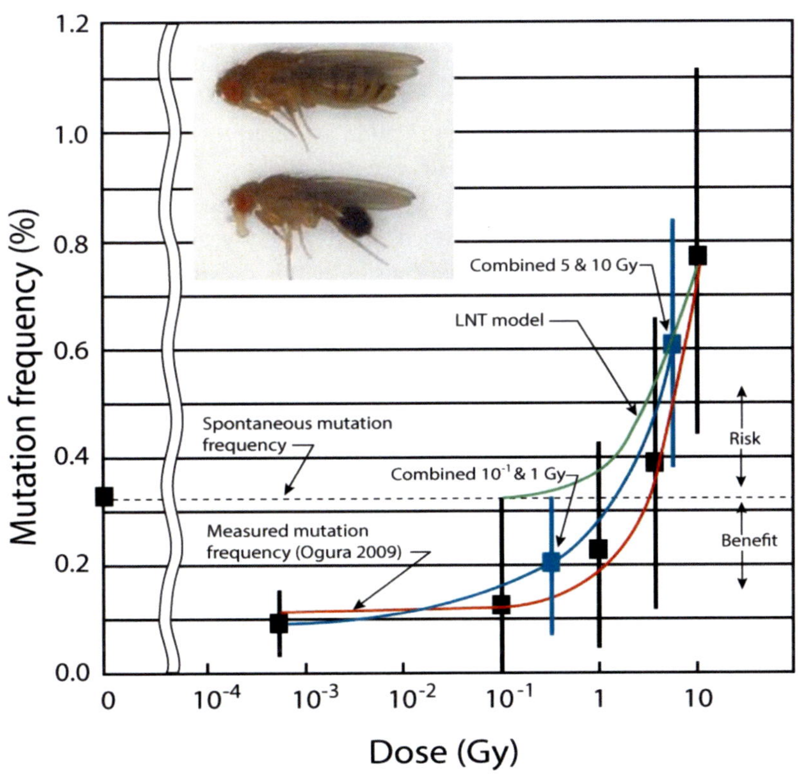

Figure 2.7. Fruit fly mutation frequency versus radiation dose [20].

A binomial distribution is assumed for the occurrence of the mutations. The error bars are two standard deviations from the mean frequency. The data points at 0.3 Gy (0.19%) and at 7 Gy (0.61%) are obtained by "pooling" the Ogura et al. [65] data at 10^{-1} and 1 Gy, and at 5 and 10 Gy, respectively. Note that the mean mutation frequency is below the spontaneous mutation frequency (0.32%) when the dose is below 1 Gy.

Recent Studies on Effects of Radiation on DNA Mutations and Cancer

A very remarkable commentary by Billen was published 24 years ago. [17] One of the crucial problems in radiation protection is the reality of the negligible dose or *de minimus* concept. The issue of a "practical zero" and its resolution is central to our understanding of the controversy concerning the existence of a "safe" dose. He explained that for very low levels of environmental mutagens and carcinogens, spontaneous or endogenous DNA damage may have an important impact on the biological consequences of the induced cellular response. The rapidly increasing evidence of this intrinsic modification of DNA decay had not gone unnoticed by radiobiologists.

Billen revealed that ~ 10,000 DNA modifications occur per hour in each (mammal) cell due to natural, intrinsic causes, while 100 or fewer DNA alterations per cell occur per cGy of low-LET radiation. Because the DNA is unstable, mechanisms evolved to repair or bypass the natural background of chemical and physical lesions caused by thermodynamic decay processes as well as the reactive molecules formed by metabolic processes leading to free radicals such as OH, peroxides and reactive oxygen species (ROS). Single-strand breaks (SSBs) occur at a rate of at least 8 x 10^3 events/cell/h, and double-strand breaks (DSBs) at 6 x 10^3 events/cell/h. This SSB rate gives 1.9 x 10^5 events/cell/day, 7 x 10^7 events/cell/year and 5 x 10^9 events/cell in an average human lifespan of 75 years. Comparing this rate with the fewer than 10 DNA alterations per cell induced by 0.1 cGy/year of low-LET background radiation indicates that the DNA damage rate caused by this radiation is more than 7 million times lower than the natural rate of DNA damage. Geneticists associate DSBs with mutations and cancer risk. This knowledge that low radiation is a negligible cause of DNA damage (cancer) has been ignored for decades by radiation protection organizations who continue to raise concerns about low level radiation and to tighten dose limits.

Feinendegen et al. point out that epidemiology fails to substantiate the claim of increased cancer incidence in humans following low level exposure to ionizing radiation. [19] Rather a decrease in cancer risk has been demonstrated repeatedly. [66-71] Nevertheless, observed data are fitted using the LNT hypothesis. [4] Despite contradicting epidemiological and experimental findings this hypothesis is also applied to predict cancer risks from low-dose irradiation. [41] What was a good intention years ago to protect workers from overexposure to radiation has been turned to producing a widespread radiation phobia now.

Feinendegen et al. maintain that plausibility of the LNT-hypothesis is based on two assumptions: (1) immediate damage to DNA from radiation increases in proportion to absorbed dose; (2) the damage is amplified and propagates to cause the cancer incidence in an

[4] Pollycove has suggested that efforts to support the LNT hypothesis have led authors of several studies to suppress and misrepresent their own contradictory data [34] p 191.

exposed population to rise in proportion to dose. [19] Assumption 2 is invalid for both epidemiological and experimental reasons. The statistical constraints of epidemiology data require very large numbers of irradiated individuals to assess the carcinogenic risks of low doses, and such large numbers are not available. Data modeling with the LNT hypothesis arrives at relative risks of cancer that are actually not observed. Furthermore, LNT does not consider the complex nonlinear dynamics of oncogenesis in the body. Recent data in experiments with various biological systems show specific responses of physiological damage control systems at various levels of biological organization and also show a low-dose induced reduction of the incidences of neoplastic transformation in culture cells and overt malignancies in animals, not observed at high-dose exposures. New findings challenge the validity of the LNT hypothesis; it cannot be maintained.

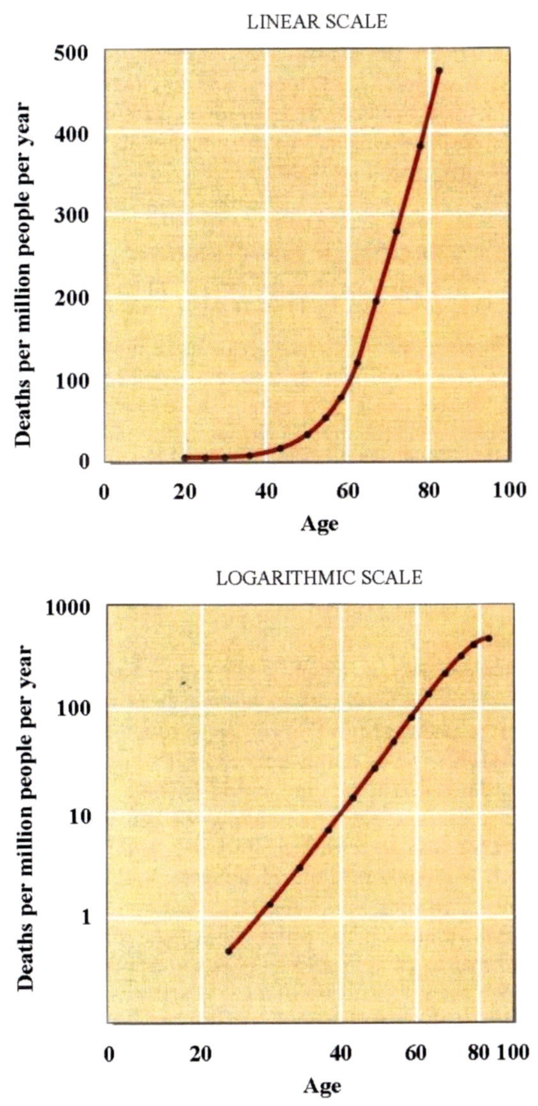

Figure 2.8. U.S. colon cancer death rate in 1986 versus age. [76] p 157 Aging of the damage-control protection systems, which include the immune system, has a very powerful effect on cancer mortality.

Research on cells, tissues, and animals indicate that there are at least three types of defending barriers to damage induction and propagation: one is static and two are metabolic-dynamic defenses. One of the latter two responds soon after perturbation, while the other involves up-regulation of adaptive defenses that are delayed by hours and last for various times up to more than 1 year after low-dose exposure. Adaptive protections can operate against both radiogenic and non-radiogenic DNA damage and its consequences. The model described by Feinendegen et al. accepts the observed experimental and epidemiological data at low doses, with their wide ranges of uncertainties. [19] It embraces both low-dose induction of radiogenic damage and radiogenic adaptive protections. It re-emphasizes not only the inconsistency of the LNT hypothesis but also the beneficial effects following low-dose irradiation that are described by Sanders, Luckey, and Calabrese and Baldwin. [71-74]

Recent experiments have determined that non-irradiated cells, depending on type and age, contain on average from about 0.1 to numerous DSBs at steady state. This value corresponds well to the calculated probability of 0.1 for a DSB to occur per average human cell/day from endogenous, non-radiogenic sources. In contrast, at background radiation level, the probability of a radiogenic DSB to occur per day was calculated to be on average only about 1 in 10,000 cells. In other words, the ratio of non-radiogenic to radiogenic DSBs produced per day in the human cell is about 1,000 [75], so radiation is not a significant carcinogen as alleged. As shown in Figure 2.8, the incidence of cancer mortality is highly dependent on age, and this is most likely due to the progressive decrease in the activity of the immune system with increasing age. [76] Since a low dose of radiation up-regulates the adaptive protection systems, it is not surprising that that immunity against cancer is stimulated by a low dose, as shown in Figure 2.9. [77]

The capacity of normal cells to repair damage to DNA and other cellular components is genetically determined and may vary individually.

Today, hundreds of genes have been described to be involved in DNA repair at high and low doses. Some genes are active only in low-dose stress responses; others again are modulated only after high doses. This reproducible data alone already contradicts the justification of the LNT-hypothesis for assessing health detriment as function of low dose. As mentioned earlier, the powerful adaptive protection systems also remove cell, tissue and organ damage, and restore health. The up-regulated immune system identifies cancer cells, regardless of their origin, and destroys them. This decreases the risk of cancer mortality.

UNSCEAR carried out a review of papers on positive biological effects of low doses of radiation presented at four conferences. One on radiation hormesis was in Oakland California in 1985; the second on low-dose radiation and the immune system was in Frankfurt in 1987; the third on low-dose irradiation and biological defense mechanisms was in Kyoto in 1992, and the fourth on low-level exposures to radiation and related agents was in Changchun, China in 1993. This review of 192 papers appears in Annex B of the UNSCEAR 1994 report, pp 185-272. Annex B is titled, Adaptive responses to radiation in cells and organisms. [78] The term "adaptive response" refers to the phenomenon in which a low "conditioning dose" is administered to an organism prior to a subsequent and much greater "challenging dose" and results in less damage to the organism from the challenging dose than would have occurred had it not received the conditioning dose. The more general effect of a low acute dose of radiation or a low dose-rate of chronic radiation is the up-regulation of many protection systems that remedy not only the damage caused by the radiation but also the endogenous damage occurring in the organism and the damage that was caused by other agents. In addition to the 192 papers that were reviewed in UNSCEAR 1994, there are many of papers on such beneficial effects that have been published prior to 1985 in medical and scientific journals, and there are many more from 1994 to the present time.

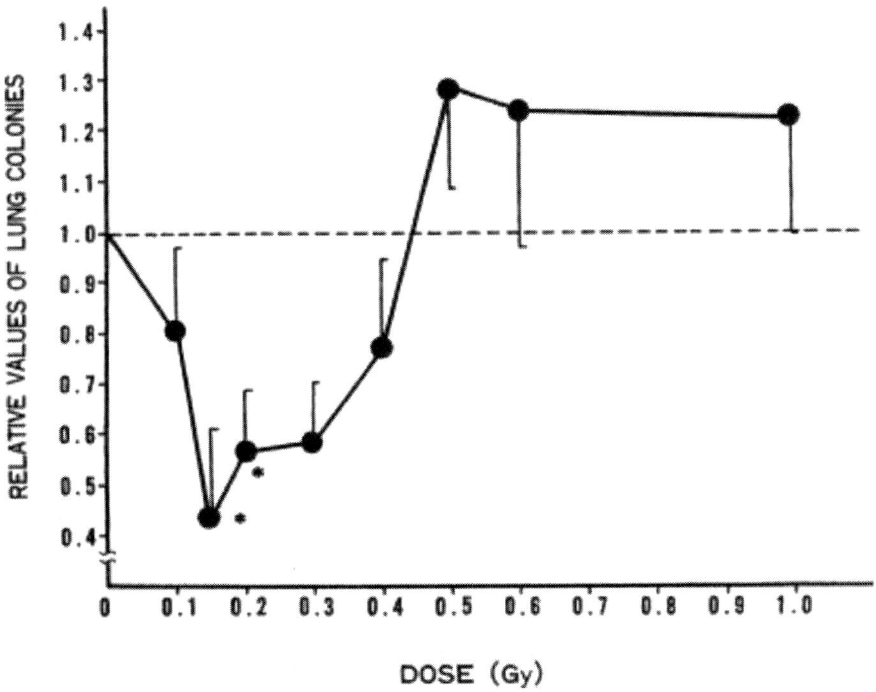

Figure 2.9. Effect of total body irradiation (TBI) dose on spontaneous lung metastasis: TBI given 12 days after tumor-cell transplantation into groin; * $p \leq 0.05$. Sakamoto [77] p 304.

MEDICAL TREATMENTS WITH LOW DOSES OF RADIATION

Medical treatments with low doses of radiation were carried out with great success for many decades until the advent of antibiotic pharmaceuticals. Hattori has drawn attention to the extensive research that has been carried out in Japan on health effects of low-level radiation. [79] However, antinuclear activities promoting fear of radiation has led to severe cutbacks in all medical applications of low doses of radiation. Today, more and more restrictions are being imposed on the use of low radiation doses in diagnostic imaging, X-rays and CT-scans. Risk-benefit assessments are required for each case, to achieve ALARA, even though there is no evidence of any excess cancer risk. Risk is assessed using the LNT hypothesis. [80] There is no other way to calculate risk. As mentioned earlier, precautions being introduced to minimize hypothetical risks are causing real risks of harm to patients due to poor quality images from inadequate exposures, physicians not ordering diagnostic studies when indicated, and patients withholding consent for such imaging studies. [42]

A review by Myron Pollycove outlines the radiobiological basis for low-dose irradiation to prevent and treat cancer. [81] From the 1970s until the present time, several hundred patients have given a series of whole or half-body radiation treatments in studies that were carried out in the U.S., Europe, Egypt and Japan, to prevent and cure cancer. Each dose fraction, either 10 or 15 cGy, was delivered within one minute. The dose was repeated three times or twice each week for 5 weeks, a total of 5 x 30 = 150 cGy. This treatment was well tolerated, with no symptomatic side effects, and most of the patients received a significant benefit. The cases were mainly non-Hodgkin's lymphoma; others were colon, ovarian and hurtle cell carcinoma. The Japanese study was funded by the federal government. [77] When the budget was exhausted this therapy stopped because the local health care plans refused to support this alternative method of treatment. It was very controversial because it jeopardized the disability pensions of the survivors of the Hiroshima and Nagasaki bombings. There are similar political barriers in other countries because this method employs low radiation doses as a cure for cancer, contradicting the raison d'être of the radiation protection organizations. [23]

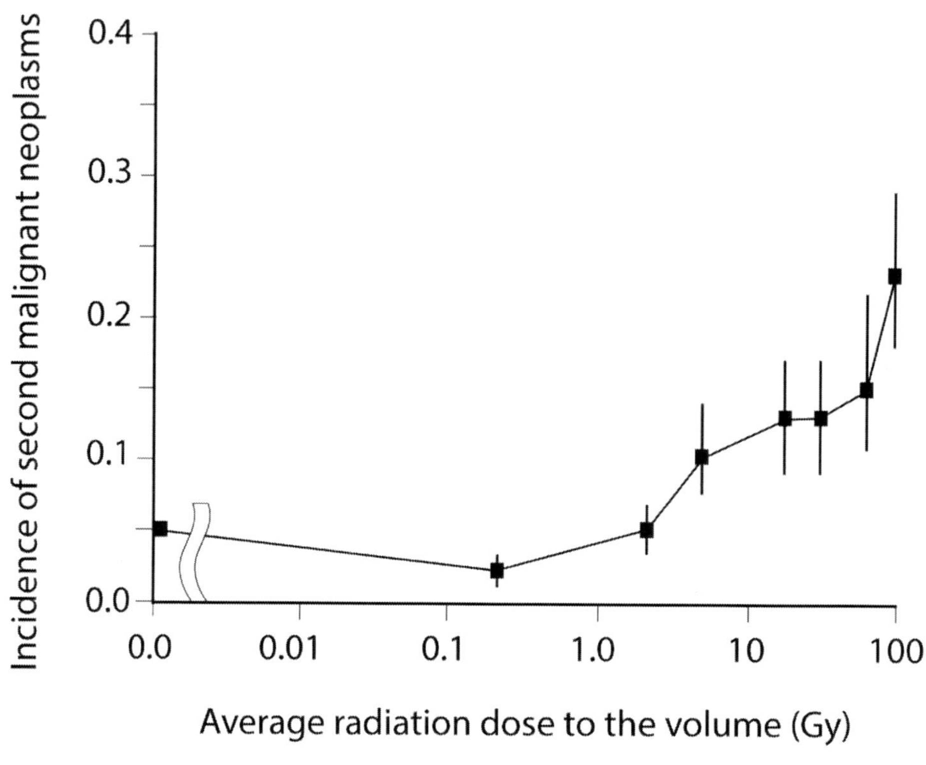

Figure 2.10. Incidence of second malignant neoplasms per unit mass as a function of dose for 5,000 survivors of childhood cancer. [82] Figure 1.

A recent paper by Tubiana et al. presented the preliminary results of a study to assess the carcinogenic efficiency against dose for a second cancer, after radiation therapy (RT). [82] This new method provided data on the risk of cancer per unit mass of healthy tissue that is exposed in RT. In each patient, isodose curves were drawn for the dose delivered in 30 sessions over six weeks. This study was on 5,000 French-UK survivors of childhood cancer, treated before 1985 and followed on average for 29 years. A total of 369 cases of second malignant neoplasms (SMN) were diagnosed and validated. Radiation treatment was reconstructed for the cases. The method included a conventional case-controlled study by individually matching each of the SMNs to three controls of the same gender, date of diagnosis and age at diagnosis, with similar follow-up to the follow-up time of the SMN case. The cumulative dose was estimated at the site of the SMN of each case and at the corresponding site of the controls. For all the cases and the controls, statistical analysis was performed for sarcoma and carcinoma, together and separately.

The method yielded no increase in risk for doses lower than 1 Gy, for all SMNs together. Also for both carcinomas and sarcomas separately, there was no excess risk from doses less than 1 Gy.

Figure 2.10 presents the incidence of SMNs per unit mass as a function of dose, calculated with the integral dose method. The data strongly suggest a hormetic effect for both carcinoma and sarcoma.

CONCLUSION

The author perceives that the radiation protection organizations and the nuclear regulators have made errors of omission by ignoring information on the following very important discoveries:

1. DNA molecules are not stable—single and double-strand breaks occur spontaneously at very high rates due to natural, internal causes.
2. Organisms have very powerful protection systems, immediate and adaptive, that deal with potential and actual cell damage and with all other injuries and health risks.
3. Many of these protection systems (more than 200 genes) are up-regulated by radiation. A low acute dose or a low chronic dose rate produces beneficial health effects that outweigh the harmful effects. The benefits include a reduced risk of cancer.
4. A variety of very important medical treatments were carried out, starting in 1896, using low doses of radiation to cure serious illnesses without causing cancers. Humanity risks losing many life-saving health benefits if it continues to ignore this evidence.
5. The 1981 study of the British radiologists demonstrated that radiation protection based on the tolerance dose concept, introduced in 1921, is effective and more than adequate.

Recommendations

The author makes the following recommendations [85]:

1. Nuclear science, medical and other professional societies should organize events to learn and discuss the facts that have been discovered about biological effects of radiation. They should question risk concepts that are based on primitive target theory.

2. Professionals should urge the nuclear regulators and the radiation protection organizations to examine all the data, both beneficial and harmful, and to comply with the requirements of The Scientific Method in setting safe limits for acute and chronic radiation exposures.
3. Regulators should develop and implement a public communication program to dispel the myth about radiation and cancer. They should stop regulating harmless radiation sources, such as radon in homes. The radon levels in many homes in Ramsar, Iran are much higher than in most homes elsewhere. No significant detrimental effects, such as increased incidence of cancer mortality, have been observed in Ramsar. [84]
4. For radiation protection, the dose-response relationship should be based on scientific facts. [83] Calculating risk of excess cancers using the invalid LNT methodology should end.
5. In medical diagnostics, physicians should not be intimidated from prescribing x-rays and CT-scans by unjustifiable concerns about cancer risks that are calculated using the LNT hypothesis. A low dose of radiation up-regulates adaptive protection systems that induce beneficial health effects, including a lower incidence of cancer mortality.
6. The radiation level for precautionary measures in the event of a radiological release should be set as high as relatively safe, based on the known thresholds for biological harm. The scientific evidence reviewed in Reference [15] suggests that the threshold for harm, due to a continuous exposure, is a dose rate of about 700 mGy per year. The UNSCEAR 1958 data for leukemia incidence among the Hiroshima atomic bomb survivors suggest that the threshold for harm for an acute exposure is about 500 mGy.
7. Nuclear safety analysts should learn the three most important lessons from the Chernobyl and Fukushima accidents that they have not yet learned:
    - Severe accidents release radioactive materials that result mainly in low radiation levels.
    - Long-term evacuation of residents would not be appropriate because the low radiation levels would not be harmful. The dose-rate limit should be based on the threshold for harm.
    - Precautionary measures should be avoided because the fear induced by these actions cause severe psychological stress that results in many premature deaths.

REFERENCES

[1] Caffrey WG and Wilson NE (1897) Medicinal Properties of Rontgen Rays. Digest: Electricity. Publisher: The Electrical World. 29(6): 67. January 9. Available at: http://whnlive.com/library/RadiationHormesis/ResearchFolder/ElectricalWorld_1897.pdf

[2] Rowland RE (1994) Radium in Humans, a Review of U.S. Studies. Argonne National Laboratory. DOE ANL/ER-3, UC-408. http://www.osti.gov/scitech/biblio/10114798

[3] Clarke R and Valentin J (2005) A History of the International Commission on Radiological Protection. *Health Physics* 88(5): 407-422.

[4] Smith PG and Doll R (1981) Mortality from Cancer and All Causes Among British Radiologists. *British Journal of Radiology* 54(639): 187-194.

[5] Murphy JB and Morton JJ (1915) The effect of roentgen rays on the rate of growth of spontaneous tumors in mice. *Journal of Experimental Medicine* 22(6): 800-803. Available at: http://www.ncbi.nlm.nih.gov/pmc/articles/PMC2125377/pdf/800.pdf

[6] Muller HJ (1927) Artificial transmutation of the gene. *Science* 66(1699): 84-87.

[7] Muller HJ (1927) The problem of genetic modification. Proceedings of the 5th International Congress of Genetics. Berlin. pp234-260.

[8] Calabrese EJ (2013) Origin of the linearity no threshold (LNT) dose-response concept. Arch Toxicol. Available at: http://link.springer.com/article/10.1007%2Fs00204-013-1104-7

[9] Calabrese EJ (2011) Commentary: Key Studies Used to Support Cancer Risk Assessment Questioned. *Environmental and Molecular Mutagenesis* 52(8): 595-606.

[10] Calabrese EJ (2011) Muller's Nobel lecture on dose–response for ionizing radiation: ideology or science? *Archives of Toxicology* 85(12): 1495-1498.

[11] Calabrese EJ (2012). Review: Muller's Nobel Prize Lecture: *When Ideology Prevailed Over Science. Toxicological Science* 126(1): 1-4.

[12] Caspari E and Stern C (1948) The influence of chronic irradiation with gamma rays at low doses on the mutation rate in Drosophila Melanogaster. *Genetics* 33: 75-95. Available at: http://www.genetics.org/content/33/1/75.full.pdf+html?sid=cb861a39-fb63-48c4-bcbe-2433bb5c8d6a

[13] Calabrese EJ (2013) How the US National Academy of Sciences misled the world community on cancer risk assessment: new findings challenge historical foundations of the linear dose response. *Arch Toxicol* DOI 10.1007/s00204-013-1105-6. Available at: http://link.springer.com/article/10.1007/s00204-013-1105-6.

[14] Calabrese EJ (2013) Evolution of the Linear No Threshold Model of Radiation Injury. Presentation to the American Association of Physicists in Medicine www.aapm.org . Available at: http://amos3.aapm.org/abstracts/pdf/77-22624-310436-101886.pdf

[15] Cuttler JM (2014) Remedy for radiation fear—Discard politicized science. Dose-Response 12(2): 170-184. http://www.ncbi.nlm.nih.gov/pmc/articles/PMC4036393/

[16] BEAR I (1956) Genetic effects of atomic radiation. U.S. National Academy of Science. Biological Effects of Atomic Radiation. *Science* 124: 1157–1164.

[17] Billen D (1990) Commentary: Spontaneous DNA Damage and Its Significance for the "Negligible Dose" Controversy in Radiation Protection. *Radiation Research* 124: 242-245. Available at: http://radiationeffects.org/2014/08/19/negligible-dose-controversy-in-radiation-protection-billen-1990/

[18] Feinendegen LE, Brooks AL and Morgan WF (2011) Biological Consequences and Health Risks of Low-Level Exposure to Ionizing Radiation: Commentary on the Workshop. *Health Physics* 100(3): 247-259.

[19] Feinendegen LE, Pollycove M and Neumann RD (2013) Hormesis by Low Dose Radiation Effects: Low-Dose Cancer Risk Modeling Must Recognize Up-Regulation of Protection. *In:* Baum RP (ed.). *Therapeutic Nuclear Medicine.* Springer. *ISBN* 973-3-540-36718-5. Available: http://radiationeffects.org/2014/08/21/hormesis-by-low-dose-radiation-effects-low-dose-cancer-risk-modeling-must-recognize-up-regulation-of-protection/

[20] Cuttler JM (2013) Commentary on Fukushima and beneficial effects of low radiation. *Dose-Response* 11(4): 432-443. Available at: http://www.ncbi.nlm.nih.gov/pmc/articles/PMC3834738/

[21] Calabrese EJ and Dhawan G (2013) The historical use of radiotherapy in the treatment of sinus infections. *Dose-Response* 11(4):469–479. Available at: http://www.ncbi.nlm.nih.gov/pmc/articles/PMC3834741/

[22] Calabrese EJ and Dhawan G (2014) Historical use of x-rays: treatment of inner ear infections and prevention of deafness. *Human and Experimental Toxicology and Pharmacology* 33(5): 542-553. Available at: http://het.sagepub.com/content/33/5/542.full

[23] Calabrese EJ (2012) NEPA, EPA and risk assessment: Has EPA lost its way? *Regulatory Toxicology and Pharmacology* 64: 267-268.

[24] Cuttler JM and Pollycove M (2009) Nuclear Energy and Health: And the Benefits of Low-Dose Radiation Hormesis. *Dose-Response* 7(1): 52-89. Available at: http://www.ncbi.nlm.nih.gov/pmc/articles/PMC2664640/

[25] Calabrese EJ (2009) The road to linearity: why linearity at low doses became the basis for carcinogen risk assessment. *Archives of Toxicology* 83: 203-225. Available at: http://radiationeffects.org/2014/08/19/the-road-to-linearity-why-linearity-at-low-doses-became-the-basis-for-carcinogen-risk-assessment/

[26] Webster EW (1992) Dose and Risk in Diagnostic Radiology: How Big? How Little? National Council on Radiation Protection and Measurements. The Lauriston S. T*aylor Lecture Series in Radiation Protection and Measurement.* Lecture No. 16.

[27] Cuttler JM (2012) Commentary on the appropriate radiation level for evacuations. *Dose-Response* 10: 473-479. Available at: http://www.ncbi.nlm.nih.gov/pmc/articles/PMC3526322/

[28] ICRP (2014) International Commission on Radiological Protection. Publications. *Annals of the ICRP and more*. Available at: http://www.icrp.org/publications.asp .

[29] EPA (2004) An Examination of EPA Risk Assessment Principles & Practices. U.S. Environmental Protection Agency. Office of the Science Advisor. Staff Paper EPA/100/B-04/001 March 2004. Available at: http://www.epa.gov/osa/pdfs/ratf-final.pdf

[30] Allison W (2009) Radiation and Reason, The Impact of Science on a Culture of Fear. Wade Allison Publishing. Available at: http://www.radiationandreason.com/

[31] Nomura S, Gilmour S, Tsubokura M, Yoneoka D, Sugimoto A, et al. (2013) Mortality risk amongst nursing home residents evacuated after the Fukushima nuclear accident: a retrospective cohort study. *PLoS One,* 8, e60192. Available: http://www.ncbi.nlm.nih.gov/pubmed/23555921

[32] Saji G (2013) A post accident safety analysis report of the Fukushima Accident – future direction of evacuation: lessons learned. *In:* Proceedings of the 21st International Conference on Nuclear Engineering. ICONE21. Jul 29 - Aug 2. Chengdu. China. ASME. http://proceedings.asmedigitalcollection.asme.org/proceeding.aspx?articleid=1829180

[33] AJW (2014) Editorial: Plight of Fukushima evacuees deserves serious policy responses. *The Asahi Shimbun*, March 8, 2014. http://ajw.asahi.com/article/views/editorial/AJ201403080030

[34] ANS (2012) President's Special Session: Low-Level Radiation and Its Implications for Fukushima Recovery. Rockwell T and Cuttler JM (ed.). 2012 American Nuclear Society Annual Meeting. Chicago. June 24-28. Available at: http://www.new.ans.org/about/officers/special_session.php

[35] UNSCEAR (2000) The Radiological Consequences of the Chernobyl Accident. Report to the (UN) General Assembly. Section 1.C.18.

[36] NEA (2002) Chernobyl: Assessment of Radiological and Health Impacts, 2002 Update of Chernobyl: Ten Years On. Nuclear Energy Agency. OECD. Available at: https://www.oecd-nea.org/rp/chernobyl/

[37] Lee TR (1996) Environmental stress reactions following the Chernobyl accident. IAEA. One decade after Chernobyl, Summing up the consequences of the accident. pp 283-310. *Proceedings of an International Conference*. Vienna. 8-12 April 1996. Available at: http://www-pub.iaea.org/MTCD/publications/PDF/Pub1001_web.pdf

[38] Lee TR (1998) Environmental Stress One Decade after Chernobyl, An Exercise in Applied Ecopsychology. the Ecopsychology Institute. http://ecopsychology.athabascau.ca/On-line, Number 6. September 1998.
Available at: http://ecopsychology.athabascau.ca/Final/research.htm

[39] Phillips ES (1906) Bibliography of X-ray Literature and Research (1896-1897). The Electrician Printing and Publishing. Available at: https://archive.org/stream/bibliographyofxr00philrich#page/n7/mode/2up

[40] Henriksen T and the Biophysics and Medical Physics Group at UiO (2013) Radiation and Health. Taylor & Francis. ISBN 0-415-27162-2 (2003 updated to 2013). U of Oslo. Available at: http://www.mn.uio.no/fysikk/tjenester/kunnskap/straling/radiation-and-health-2013.pdf

[41] Brenner D J and Hall EJ (2007) Computed tomography--an increasing source of radiation exposure. *New Engand Journal of Medicine,* 357**,** 2277-84. Available at: http://www.ncbi.nlm.nih.gov/pubmed/18046031

[42] McCollough CH (2011) Defending the use of medical imaging. *Health Physics* 100: 318-321.

[43] Ozasa K, Shimizu Y, Suyama A, Kasagi F, Soda M, et al. (2012) Studies of the mortality of atomic bomb survivors, report 14, 1950-2003: an overview of cancer and noncancer diseases. *Radiation Research* 177(3): 229-243.

[44] Cuttler JM (1999) Resolving the Controversy over Beneficial Effects of Ionizing Radiation. *Proc World Council of Nuclear Workers Conference*. Effects of Low and Very Low Doses of Ionizing Radiation on Health. Versailles. France. June 16-18. Elsevier Sci Pub. pp 463-471. AECL Report No. 12046.

[45] Socol Y and Dobrzyński L (2014) Atomic Bomb Survivors Life-Span Study: Insufficient Statistical Power to Select Radiation Carcinogenesis Model. *Dose-Response* (in press).

[46] Seiler FA and Alvarez JL (1994) The Scientific Method in Risk Assessment. Technology: *Journal of the Franklin Institute* 331A: 53-58. Available at: http://www.researchgate.net/publication/259527706_The_Scientific_Method_in_Risk_Assessment

[47] Sponsler R and Cameron JR (2005) Nuclear shipyard worker study (1980-1988): a large cohort exposed to low-dose-rate gamma radiation. *Int. J. Low Radiation* 1(4): 463-478.

[48] UNSCEAR (1958) Report of the United Nations Scientific Committee on the Effects of Atomic Radiation. United Nations. General Assembly. Official Records. Thirteenth Session. Supplement No. 17 (A/3838). New York. Available at: http://www.unscear.org/unscear/en/publications/1958.html

[49] Mitchel REJ (2007) Cancer and Low Dose Responses in Vivo: Implications for Radiation Protection. *Dose-Response* 5(4): 284-291. Available at: http://www.ncbi.nlm.nih.gov/pmc/articles/PMC2477713/

[50] Fliedner TM, Graessle DH, Meineke V and Feinendegen LE (2012) Hemopoietic response to low dose-rates of ionizing radiation shows stem cell tolerance and adaptation. *Dose-Response* 10: 644-663. http://www.ncbi.nlm.nih.gov/pmc/articles/PMC3526333/

[51] Cuttler JM (2014) Leukemia incidence of 96,000 Hiroshima atomic bomb survivors is compelling evidence that the LNT model is wrong. *Archives of Toxicology* 88(3): 847-848. Available at: http://rd.springer.com/article/10.1007/s00204-014-1207-9/fulltext.html

[52] Cuttler JM (2007) What becomes of nuclear risk assessment in light of radiation hormesis? *Dose-Response* 5: 80-90. http://www.ncbi.nlm.nih.gov/pmc/articles/PMC2477701/

[53] Kelly JF and Dowell DA (1942) Roentgen Treatment of Infections. Year Book Publishers. Chicago. http://ep.probeinternational.org/1942/12/01/roentgen-treatment-of-infections/

[54] Brucer M (1990) *A chronology of nuclear medicine*, 1600 – 1989. Heritage Publications. ISBN 0-9625674-0-X.

[55] Inkret WC, Meinhold CB and Taschner JC (1995) A Brief History of Radiation Protection Standards. *Los Alamos Science* 23:116-123. Available at: http://www.fas.org/sgp/othergov/doe/lanl/00326631.pdf

[56] Taylor LS (1980) Some non-scientific influences on radiation protection standards and practice. Proceedings of 5th Congress of the International Radiation Protection Association (IRPA). Jerusalem. pp 307-319. Available at: http://www.irpa.net/irpa5/cdrom/VOL.1/J1_64.PDF

[57] Taylor LS (1980) Some Nonscientific Influences on Radiation Protection Standards and Practice, The 1980 Sievert Lecture. *Health Physics* 39: 851-874.

[58] Walinder G (2000) Has radiation protection become a health hazard? Medical Physics Publishing. Madison WI.

[59] McGinnis R (2014) Rad Pro Calculator Software Development. MIT. Available at: http://www.radprocalculator.com/XRay.aspx

[60] Oliver CP (1930) The effect of varying the duration of X-ray treatment upon the frequency of mutation. *Science* 71:44-46. doi:10.1126/science.71.1828.44.

[61] Oliver CP (1931) An analysis of the effect of varying the duration of x-ray treatment upon the frequency of mutations. Doctor of Philosophy Thesis, University of Texas, Austin.

[62] Muller HJ (1930) Radiation and Genetics. *American Naturalist* 64:220–257. doi:10.1086/280313.

[63] Koana T, Takashima Y, Okada MO, Ikehata M, Miyakoshi J, et al. (2004) A threshold exists in the dose-response relationship for somatic mutation frequency indicated by x irradiation of *Drosophila*. *Radiation Research* 161: 391-396.

[64] Koana T, Okada MO, Ogura K, Tsujimura H and Sakai K (2007) Reduction of background mutations by low-dose x irradiation of *Drosophila* spermatocytes at a low dose rate. *Radiation Research* 167: 217-221.

[65] Ogura K, Magae J, Kawakami Y and Koana T (2009) Reduction in mutation frequency by very low-dose gamma irradiation of *Drosophila Melanogaster* germ cells. *Radiation Research* 171: 1-8.

[66] Pollycove M and Feinendegen LE (2001) Biologic response to low doses of ionizing radiation: detriment versus hormesis. Part 2: dose responses of organisms. *Journal of Nuclear Medicine* 42: 26N–32N.

[67] Tubiana M, Aurengo A, Averbeck D, Bonnin A, Le Guen B, et al. (2005) Dose-effect relationships and the estimation of the carcinogenic effects of low doses of ionizing radiation. Academy of Medicine (Paris) and Academy of Science (Paris) Joint Report No. 2. Available at: http://lowrad.wonuc.org/lowrad/lowrad-bulletin.htm

[68] Tubiana M, Feinendegen LE, Yang C, Kaminski JM (2009) The linear no-threshold relationship is inconsistent with radiation biologic and experimental data. *Radiology* 251:13–22. Available at: http://pubs.rsna.org/doi/full/10.1148/radiol.2511080671

[69] Tubiana M (1998) Problems Associated with the Effects of Low Doses of Ionizing Radiations. Académie des sciences. Report No. 38.

[70] Tubiana M (1998) The report of the French Academy of Science: Problems associated with the effects of low doses of ionizing radiations. *Journal of Radiological Protection* 18(4): 243-248. Available at: http://iopscience.iop.org/0952-4746/18/4/002/pdf/0952-4746_18_4_ 002.pdf

[71] Sanders CL (2010) Radiation Hormesis and the Linear-No-Threshold Assumption. Springer. ISBN: 978-3-642-03719-1.

[72] Luckey TD (1980) *Hormesis with Ionizing Radiation*. CRC Press.

[73] Luckey TD (1991) *Radiation Hormesis*. CRC Press.

[74] Calabrese EJ and Baldwin LA (2000) Radiation hormesis: its historical foundations as a biological hypothesis. *Human and Experimental Toxicology* 19: 41-75. Available at: http://www.radiation-hormesis.com/wp-content/uploads/2014/03/Hormesis-Historical-Investigations-Hum-Exp-Toxicol-2000-Calabrese-41-75.pdf

[75] Pollycove M and Feinendegen LE (2003) Radiation-induced versus endogenous DNA damage: possible effect of inducible protective responses in mitigating endogenous damage. *Human and Experimental Toxicology* 22: 290-306. Available at: http://www.belleonline.com/ newsletters/volume11/vol11-2.pdf

[76] Varmus H and Weinberg RA (1993) Genes and the Biology of Cancer. Scientific American Library. ISBN 0-7167-5037-6.

[77] Sakamoto K (2004) Radiobiological Basis for Cancer Therapy by Total or Half-body Irradiation. *Nonlinearity in Biology, Toxicology, and Medicine* 2: 293-316. Available at: http://www.ncbi.nlm.nih.gov/pmc/articles/PMC2657505/

[78] UNSCEAR (1994) Adaptive Responses to Radiation in Cells and Organisms. Sources and Effects of Ionizing Radiation. Report to the United Nations General Assembly, with Scientific Annexes. Annex B. http://www.unscear.org/unscear/publications/1994.html

[79] Hattori S (1998) The Research on the Health Effects of Low-Level Radiation in Japan. Proc. Pacific Basin Nuclear Conference. Banff. Canada. Available at: http://www.nuceng.ca/refer/radiation/Sadao%20Hattori_PBNC98.pdf

[80] Linet MS, Slovis TL, Miller DL, Kleinerman R, Lee C, et al. (2012) Cancer Risks Associated with External Radiation from Diagnostic Imaging Procedures. *CA: A Cancer Journal for Clinicians* 62(2): 75-100. Available at: http://onlinelibrary.wiley.com/doi/10.3322/caac.21132/pdf

[81] Pollycove M (2007) Radiobiological basis of low-dose irradiation in prevention and therapy of cancer. *Dose-Response* 5: 26-38. Available at: http://www.ncbi.nlm.nih.gov/pmc/articles/PMC2477707/

[82] Tubiana M, Diallo I, Chavaudra J, Lefkopoulos D, Bourhis J, et al. (2011) A new method of assessing the dose carcinogenic effect relationship in patients exposed to ionizing radiation, A concise presentation of preliminary data. *Health Physics* 100(3): 296-299.

[83] Calabrese EJ (2011) Commentary: Improving the scientific foundations for estimating health risks from the Fukushima incident. *Proceedings of the National Academy of Sciences of the United States of America* 108(49): 19447-19448.

[84] Ghiassi-nejad M, Mortazavi SMJ, Cameron JR, Niroomand-rad A and Karam PA (2002) Very High Background Radiation Areas of Ramsar, Iran: Preliminary Biological Studies. *Health Physics* 82(1): 87-93.

[85] Cuttler JM (2014) INEA Statement on Radiation and Health - Confirming the Evidence. International Nuclear Energy Academy. October 23, 2014. Available at: http://ineacademy.com/statements/INEAStatementonRadiationandHealth-ConfirmingEvidence.pdf

[86] Cohen BL (2008) The Linear No-Threshold Theory of Radiation Carcinogenesis Should Be Rejected. *Journal of the American Physicians and Surgeons* 13(3): 70-76. Available at: http://www.jpands.org/ vol13no3/cohen.pdf

In: Fukushima Nuclear Accident
Editor: Shizuyo Sutou
ISBN: 978-1-63463-748-0

Chapter 3

EVIDENCE AGAINST THE LINEAR NO-THRESHOLD HYPOTHESIS IN THE ATOMIC BOMB SURVIVOR CANCER DATA AND OTHER DATA AND REASONS FOR A CHANGE IN THE RADIATION SAFETY PARADIGM

Mohan Doss *
Diagnostic Imaging Department,
Fox Chase Cancer Center, Philadelphia, PA, US

ABSTRACT

The current radiation safety paradigm, recommended consistently since the 1950s by various advisory bodies, is based on the linear no-threshold (LNT) hypothesis, according to which there is no threshold dose for radiation-induced cancers. The resulting low-dose radiation (LDR) concerns have caused governments and the public to take precautionary, evasive actions when faced with the prospect of exposure to low levels of radiation. Such actions have harmed the public, e.g. in Fukushima, where there were short-term and long-term evacuation-related casualties. The fear of LDR has suppressed nuclear power, and has led to increased use of other power sources which are more expensive and hazardous, increasing costs and casualties. There is also potential for adverse impact on patient health because of delayed diagnoses from not performing diagnostic studies due to patient or physician LDR concerns, or poorer quality images from dose reduction efforts. In view of the wide-ranging harm caused by the use of the LNT hypothesis, it is imperative that its validity be re-evaluated.

The primary justification generally cited for the use of the LNT hypothesis and for LDR cancer concerns are the data from the atomic bomb survivors. Recently published update to the survivor cancer mortality data has displayed significant curvature or non-linearity in dose-response in the 0-2 Gy region that was not present in the earlier updates to the data. This non-linearity, arising from the lower than expected cancer mortality rates in the 0.3-0.7 Gy region, contradicts the LNT hypothesis. On the other hand, if a

* Correspondence to: Mohan Doss, Tel: 215 214-1707, Fax: 215 728-4755, E-mail: mohan.doss@fccc.edu.

correction is applied for a likely negative bias in the baseline cancer rates, the shape of dose-response becomes consistent with the concept of radiation hormesis, providing a possible explanation for the observed reduction of cancer rates for doses near 0.5 Gy. The updated report also claimed that a formal dose-threshold analysis indicated zero dose was the best estimate of the threshold dose. However, the analysis used a restricted functional form for dose-response that did not encompass the full range of the observed data. Performing an analysis using a more general functional form has shown that the presence of a dose-threshold cannot be excluded. Thus, the latest atomic bomb survivor data provide evidence neither for the linearity nor for the no-threshold aspect of the LNT hypothesis. Also, major flaws have been identified in other publications that have been used to support the LNT hypothesis or LDR carcinogenic concerns, negating or raising major doubts about their conclusions. Thus, there appears to be no conclusive evidence for cancers caused by LDR. On the other hand, considerable amount of evidence has accumulated over the years for a threshold dose-response or radiation hormesis. Hence, a change in the current radiation safety paradigm is warranted recognizing a threshold dose for radiation-induced cancers. This would help in reducing the LDR concerns, reduce or eliminate the harm caused by evasive actions to avoid LDR, help rejuvenate nuclear power for safer, less expensive, and cleaner energy, enable resettlement of Fukushima, and prevent evacuation-related casualties during any future radiological emergencies involving LDR exposures.

INTRODUCTION

The current radiation safety paradigm is based on the linear no-threshold (LNT) hypothesis for radiation-induced cancers. This hypothesis was adopted in the 1950s by the various advisory bodies following the observation of increased mutations in drosophila melanogaster from X-rays and the increased leukemias in atomic bomb survivors, both following exposure to high-dose radiation [1]. The consequent reduction of radiation dose limits was based on public concerns but not based on any observed harm from low-dose radiation (LDR) [2]. The LNT hypothesis has been recommended by most advisory bodies repeatedly over the years as the best approach to radiation safety and is the basis for current radiation protection regulations worldwide.

There have been major adverse consequences from using the LNT hypothesis for radiation safety. The acceptance of the concept of zero threshold implies that any increase in radiation dose would increase the risk of cancer. This has resulted in concerns which have led to precautionary actions by governments, professional organizations, professionals, and the public to reduce or avoid low doses of radiation at tremendous costs. There would be projected benefit of reduced cancers in the future if the LNT hypothesis were true.

Real harm has been caused by the precautionary actions taken to reduce or avoid low radiation doses. In Fukushima, the evacuation and the prolongation of the evacuation due to LDR concerns led to short-term and long-term casualties related to the evacuation, destroyed the local economy, and diminished the quality of life [3,4,5]. The fear of LDR has discouraged and prevented prospective studies of LDR for prevention and treatment of diseases for which LDR has shown promise in animal models [6,7], and for which presently no effective methods of prevention are available. The increased regulations arising from the use of the LNT hypothesis have led to tremendously increased nuclear power costs [8]. The fear of LDR has suppressed nuclear power and has led to increased energy costs and casualties from the use of more expensive and more hazardous power sources [9,10,11]. There is also potential for harm to patients in multiple ways because of LDR concerns [12]. Patients' refusal of CT scans [13] and physicians not prescribing CT scans because of the LDR concerns [14] can result in missed or delayed diagnoses. CT scans performed with lower doses because of these concerns can result in poorer quality images with the potential for misdiagnoses [15,16]. Considering such wide-ranging tremendous harmful consequences due to the LDR concerns from the use of the LNT hypothesis, it is imperative that the validity of the LNT hypothesis be re-examined.

The foremost and most often-quoted evidence supporting the LNT hypothesis or LDR cancer concerns are the data from the Life Span Study (LSS) cohort of the atomic bomb survivors [17,18]. For example, according to the influential BEIR VII report [19], LSS cohort serves as the single most important source of data for evaluating risks of radiation at low and moderate doses (see page 141 of the report). The report concluded that the current scientific evidence is consistent with the LNT hypothesis (see page 323 of the report). The atomic bomb survivor data have also been described as the gold standard for assessing LDR cancer risks [20] because the study includes a large non-selected population, about one third of them were exposed to doses below ~0.1 Sv, they have been followed up for more than 60 years, and both cancer incidence and mortality data are available.

1. Linearity of Dose-Response in Atomic Bomb Survivor Data

The latest report on the study of the LSS cohort by Ozasa et al. covers the period of 1950-2003 and was published in 2012 [21], with an Erratum published in 2013 [22]. The report concluded that the excess relative risk (ERR) for solid cancer mortality has a linear-dose response relationship, and zero dose is the best estimate of the threshold dose, apparently validating the continued use of the LNT hypothesis. Figure 3.1 shows the ERR for solid cancer mortality among the survivors as a function of dose from the data in Figure 4 of the publication [22], redrawn with standard error bars.

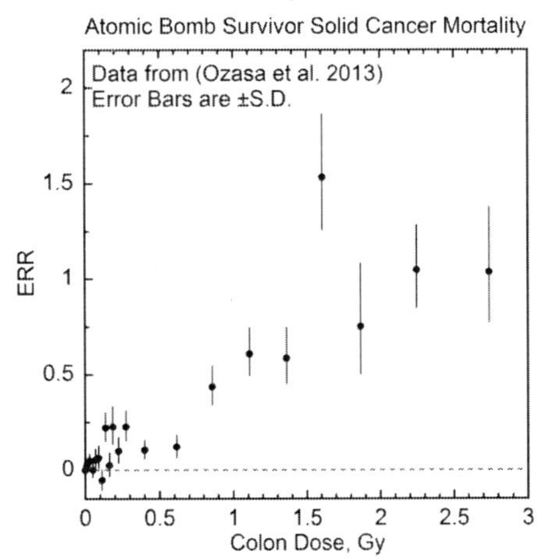

Figure 3.1. Excess relative risk (ERR) for all solid cancer mortality in relation to radiation exposure in the atomic bomb survivors. Data from [22]. Error bars are ±S.D.

The ERRs for doses just below 0.3 Gy are observed to have large fluctuations even between adjacent dose values. When these fluctuations are averaged, the ERRs are seen to generally increase with dose from 0 to 0.3 Gy, then decrease or remain steady in the 0.4-0.6 Gy range, followed by increase with dose for doses greater than about 0.6 Gy. Ozasa et al. state (on p. 234 of the publication) that the curvature of dose-response in the 0-2 Gy range has become significant in this update in contrast to previous updates, ascribing it to lower than expected cancer rates in the 0.3-0.7 Gy range (p. 238). Ozasa et al. provide no explanation for the observed significant reduction of cancers near 0.5 Gy.

Ozasa et al. state that when the whole dose range is considered, the linear model provided the best fit to the data (p. 234). It is true that adding more and more data points at higher doses above 2 Gy can reduce the significance of the dose reduction near 0.5 Gy in a statistical sense. However, the underlying assumptions of the LNT hypothesis (higher dose would lead to higher DNA damage and mutations resulting in higher cancer incidence and mortality) are not supported by the significant observed reduction of cancers in this dose region. This reduction is inconsistent with the LNT hypothesis. Thus, the dose response observed in the new update to the LSS data is certainly not linear, in spite of the assertion by Ozasa et al. that over the whole dose range, the linear model provided the best fit. Since the LSS cohort data are recognized to be the most important data for estimating LDR cancer risks, the significant non-linearity of dose-response relationship in these data should result in the rejection of the linear dose-response hypothesis, and the exploration of alternative hypotheses. However, Ozasa et al. have chosen to use the linear dose-response relationship as an integral part of the data analysis to extract the ERR values.

An alternative hypothesis for LDR health effects is that of up-regulation of protection or adaptive protection which would result in reducing cancers [23]. This effect, also known as radiation hormesis, has considerable supporting evidence in animal and human studies [24,25]. An analysis of published atomic bomb survivor data has also shown that LDR resulted in improved health with reduced mutations, leukemias, and solid cancers and increased average lifespan among the survivors [26]. The cancer mortality rate was lower in the cohort exposed to low doses when compared to the Not in City (NIC) controls. The average threshold noted for harmful effects was ~1 Sv.

The LSS cohort analysis did not use an un-irradiated cohort for obtaining baseline cancers for calculating ERRs, but rather estimated the expected numbers of background cases based on an ERR model with a linear dose response fitted to all of the data [18]. Hence, a reduction of cancers in the low dose cohorts resulting from adaptive protection would result in a negative bias in the baseline cancers used for calculating the ERRs. Additional bias may have occurred because of confounding factors that were not accounted for. The bias in the baseline cancer mortality rates in the LSS data has been estimated to be about -20% based on plausibility arguments and based on observed reduction of cancers due to LDR [27,28].

If δ is the percentage bias in the baseline cancer rate, a correction can be applied for the bias to obtain the corrected ERR using the following formula [27], as shown in Appendix A:

$$ERR(corr) = \frac{(1+ERR)\times(100+\delta)}{100} - 1$$

where ERR(corr) would be the ERR corrected for the bias.

This correction has been applied to the ERR data in Figure 3.1, assuming -20% bias in the baseline cancer rate, and the corrected ERR (from [29]) has been shown in Figure 3-2 . For doses below 0.3 Gy, the mean ERR value of the data points has been plotted to average over the observed large fluctuations, and the individual data points have been shown in grey color. The dose-response shape is seen to be consistent with that postulated by radiation hormesis hypothesis, providing a plausible explanation for the reduced level of cancers near 0.5 Gy.

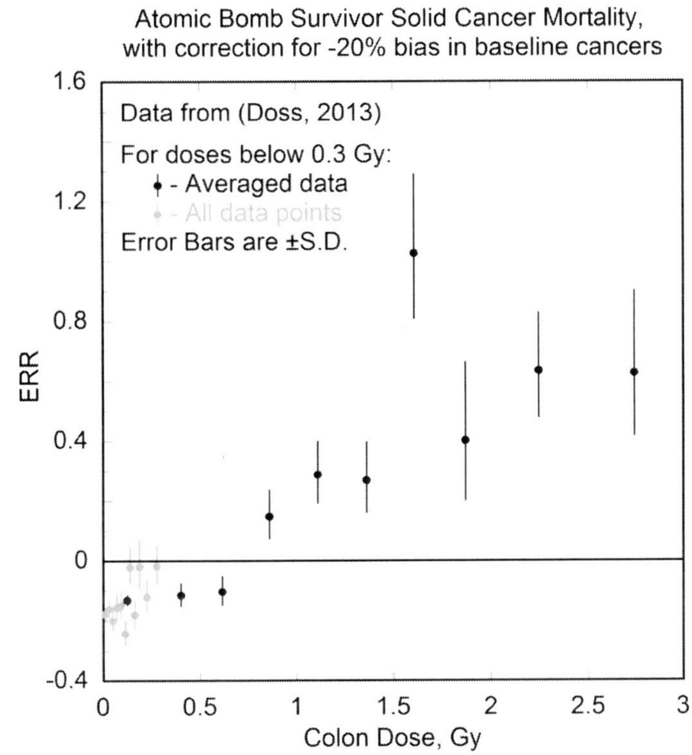

Figure 3.2. From [29], reproduced with permission from Chest. Excess relative risk (ERR) for all solid cancer mortality in atomic bomb survivors corrected for -20% bias in baseline cancer mortality rate plotted as a function of colon dose. For doses below 0.3 Gy, the black circle shows average of data below 0.3 Gy, and grey circles show the individual data points. Error bars are ±S.D.

Summarizing the discussion of linearity of dose-response data in the LSS cohort, the LNT hypothesis cannot explain the reduction of cancers near 0.5 Gy, but radiation hormesis can provide a possible explanation. The validity of the radiation hormesis hypothesis implies that LDR would reduce rather than increase cancer risk.

2. Threshold Dose in the Atomic Bomb Survivor Cancer Mortality Data

According to Ozasa et al. [21], a formal dose-threshold analysis showed that zero dose was the best estimate of the threshold dose for the carcinogenic effect of radiation. For this analysis, they used a linear dose-response function to represent ERRs for doses above an assumed threshold dose and the ERR value of zero for doses below the threshold dose (see p. 231). This functional form did not allow the ERR values to extend into negative values whereas one of the data points and 95% CI of several of the data points (see Figure 3-1) would extend into negative ERR values. Since the functional form did not encompass the full range of observed values of ERR, the dose-threshold analysis is flawed, invalidating the conclusion from the analysis that there is no threshold dose.

An analysis of the published dose-response data performed using a more general functional form has shown that a non-zero threshold dose cannot be excluded, since the lower bounds of the 95% CIs of the fit to the data were below zero for low doses [28,30] (see Figure 3-3). When the comments describing this analysis [30] were accepted for publication by Radiation Research, Ozasa et al. were given an opportunity to respond by the journal, but they chose not to respond.

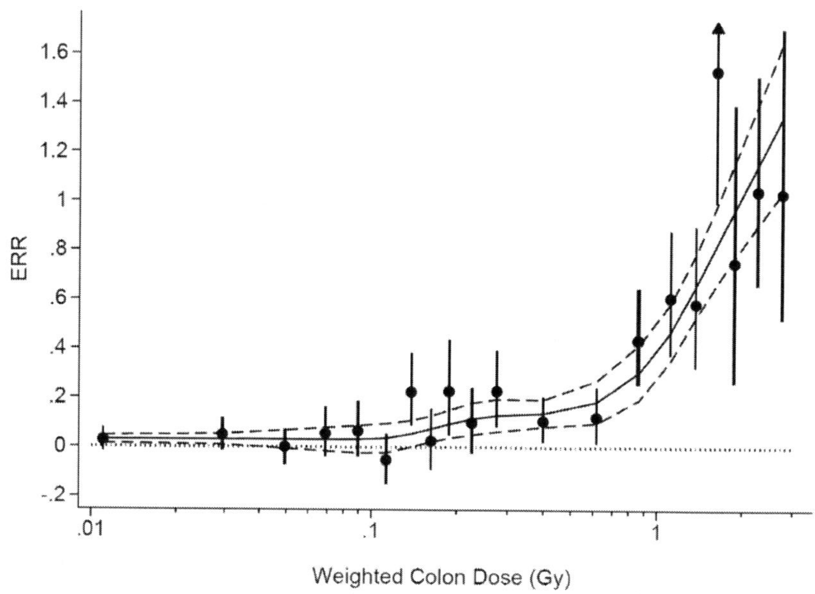

Figure 3.3. From [28], reproduced with permission from University of Massachusetts, Amherst. Excess relative risk (ERR) for all solid cancer in atomic bomb survivors in relation to radiation exposure. The black circles and error bars represent ERR and 95% CIs for the dose categories. Data are from Ozasa et al. [22]. Solid Line - fit to the ERR data using a multiple linear regression in which weighted log colon dose was entered into the model using a restricted cubic spline transformation with five knots. Regression weights were equal to the inverse of the variance of the point estimates. Dashed lines are 95% CI of the fit.

In summary, the LSS cohort data, which are the most often-quoted data to raise LDR cancer concerns, provide evidence for neither the linearity nor the no-threshold aspect of the LNT hypothesis for radiation-induced cancers. On the other hand, these data are consistent with the hypothesis of radiation hormesis.

The qualitative change in the latest update to the LSS cohort data, that they do not support the LNT hypothesis or LDR cancer concerns, is being recognized implicitly by scientists, as they refer to older data [17] to raise LDR cancer concerns and not the newer Ozasa et al. data [21] in their recent publications, e.g. [31,32,33,34,35].

3. Additional Evidence for the Presence of a Threshold Dose or Radiation Hormesis

Considerable amount of additional published evidence supports the radiation hormesis hypothesis or the presence of a threshold dose for radiation-induced cancers.

(1) The study of atomic bomb survivors in Hiroshima has shown a threshold dose of ~0.5 Sv for radiation-induced leukemias (see Figure 2-3) [36].
(2) The survival of lymphosarcoma patients who had undergone total-body irradiation (TBI) with LDR was compared to patient survival following chemotherapy in a clinical trial. Radiation dose of 0.15 Gy applied to the whole body ten times during five weeks for a total dose of 1.5 Gy had a cancer therapeutic effect performing as well as chemotherapy [37].
(3) The study of radium dial painters who ingested radium showed a threshold dose of ~10 Gy for the induction of bone sarcomas [38,39].
(4) Breast cancer mortality rates were studied in Canadian TB patients who had undergone fluoroscopic examinations. The results from this study showed reduction of breast cancer mortality in the cohort subjected to radiation dose in the range of 0.1-0.19 Gy [40,41].
(5) The populations of villages near the Mayak nuclear weapons facility were exposed to radiation due to the facility. The study of their mortality as a function of radiation dose has shown that the standardized cancer mortality rates were significantly reduced in the 0.12 and 0.50 Sv cohorts in comparison to un-irradiated controls [42].
(6) A clinical trial studied the survival of non-Hodgkin's lymphoma patients who were treated with 0.10 or 0.15 Gy of TBI or half-body irradiation (HBI) in between the standard radiation treatments to the tumor [43]. The total dose from the low-dose radiation treatments was 1.5 Gy. Survival of the patients including the TBI or HBI treatments was significantly better than the survival of the patients who had the standard radiation treatment only, indicating the cancer preventive effect of the fractionated LDR treatments. Tumors outside the HBI field had also regressed [44], showing the systemic nature of the cancer preventive effect.
(7) Cancer incidence has been studied in a cohort of apartment residents in Taiwan who had been irradiated because of contaminated building materials, with the average dose being ~50 mGy. Whereas a few cancer types were reported to be increased in the irradiated cohort in comparison to controls, other cancers had reduced, and there was significant reduction of solid cancers and all cancers in the irradiated group [45].
(8) Childhood cancer survivors who had undergone radiation therapy were followed up for a period of 30 years and the incidence of second cancers per kg of tissue was studied as a function of radiation dose to the tissue. The tissues that had 0.2 to 0.5 Gy dose from the radiation therapy had less second cancers per kg of tissue in comparison to tissues which had no radiation exposure from the therapy (see Figure 2-10) [46].
(9) Non-Hodgkin's lymphoma patients who had undergone LDR treatments were found to have better survival compared to patients who had undergone chemotherapy [44].

As the above examples show, LDR applied in many different ways has resulted in the reduction of cancers. However, there is considerable disagreement in the scientific community on this subject, with many publications claiming increase in cancers following LDR. Some of these publications will be discussed now.

4. Publications Claiming LDR Carcinogenicity or Support for the LNT Hypothesis

(1) The 15-country study of radiation workers by Cardis et al. reported increased cancer risk in the workers exposed to an average occupational radiation dose of ~20 mSv [47]. In these data, the Canadian radiation workers had a much higher radiation-induced cancer risk estimate compared to workers from other countries. Since the time of this publication, problems have been identified in the Canadian data and the data have been withdrawn from further use [48]. As a result, the Cardis et al. data would no longer show an increased risk of cancer from LDR [49].

(2) Increased childhood cancers have been reported following in-utero irradiation in case-control studies [50]. However, as described in a recent review covering a large number of studies, cohort studies (which are considered to be better than case-control studies) have failed to show the increased risk from in-utero radiation [51]. The final paragraph of the review says, referring to <0.10 Gy radiation dose to the embryo from diagnostic examinations: "Based on all the studies we have available, the risk of cancer to the embryo is very low and possibly so low that we may never be able to measure the risk."

(3) Claims of increased cancers following childhood CT scans have been made in recent publications [52,53]. There are major doubts about the validity of these studies and their conclusions for many reasons including methodological issues, the possibility of reverse causation, and pattern of observed cancers not being consistent with the current knowledge about radiation effects [54,55,56]. Hence, these studies do not provide evidence that low doses are causally associated with cancers in children [54].

(4) A follow-up report to the 2006 study of irradiated apartment residents in Taiwan claimed increased risk for leukemias and thyroid cancers [57]. However, this conclusion was based on loosening the significance level to $p<0.1$ from the $p<0.05$ used in the earlier publication [45]. An analysis [28] of the updated data has shown that the reduction in solid cancers reported in the 2006 publication continued in the follow-up report also, but was not mentioned in the article.

(5) An update on the study of Canadian TB patients who had undergone fluoroscopic examinations reported support for the LNT hypothesis based on the updated data on breast cancers [58]. However, the data for low doses and higher doses were combined into a much wider single bin compared to the earlier report, obfuscating the reduction in cancers observed in the 0.1-0.19 Gy dose cohort in the earlier study [41].

(6) In a large population-based case-control study in UK, excess risk of childhood leukemias have been reported associated with natural background radiation exposure [59]. However, the study did not include confounding factors such as breastfeeding which can affect childhood leukemias considerably. Since the 95% CI (for relative

risk per mSv for all leukemias) is 1.01 to 1.13 (see Table 3 of the publication), small changes in such confounding factors can render the increased risk not statistically significant. Thus, improved studies are needed before a definitive conclusion can be drawn.

As seen above, analyses of the publications claiming increased LDR cancer risk have negated or raised major doubts about their conclusions.

5. Current Status of the Debate on the Carcinogenicity of LDR

The opinion of the scientific community has been divided on the subject of carcinogenicity of LDR for the past several decades, with publications supporting opposite points of view, and with two recent advisory body reports - BEIR VII report and the report by French Academy of Sciences - coming to opposite conclusions [19,60].

BEIR VII report recognized the importance of the LSS data for estimating LDR cancer risk. Its conclusion was to recommend the use of the LNT hypothesis. The report also claimed support from the 15-country study of radiation workers for its conclusion. As seen above, both these data no longer support the LNT hypothesis or the carcinogenicity of LDR, and so the conclusion of the BEIR VII report can no longer be considered to be valid.

There have also been recent debates on the subject in scientific journals [61,62,63] apparently without resolving the issue. In the recently published debate on the health effects of LDR [63], the author claiming harm from LDR did not directly refer to the LSS cohort data to raise LDR cancer concerns in his opening statement, unlike in previous debates where these data played a major role [62]. Also, the evidence he quoted for LDR carcinogenicity (most of which have been discussed above) were weak and faulty and were readily refuted in the rebuttal [63].

In summary, a resolution of the longstanding debate on LDR cancer risk is imminent. The discussions on the latest update of the LSS data showed that the data no longer support the LNT hypothesis but are consistent with radiation hormesis. Additional evidence provides support for radiation hormesis hypothesis and for a threshold dose for radiation-induced cancers. The claims of increased cancers from LDR have been shown to be without merit. Thus, the debate on LDR cancer risk is essentially over. LDR does not result in increased cancers. On the other hand, LDR has led to decreased cancers in a wide variety of situations.

6. Reasons for a Change in the Radiation Safety Paradigm

The current LNT hypothesis-based radiation safety paradigm was adopted in the 1950s without any evidence for harm caused by LDR. The main evidence used by advisory bodies for justifying the current paradigm – atomic bomb survivor data – no longer support the LNT hypothesis or LDR cancer concerns. Other evidence quoted for LDR cancer concerns have not withstood scrutiny, and there is no definitive evidence for LDR carcinogenicity. On the other hand, much evidence points to the cancer preventive effect of LDR.

Tremendous harm has been caused to the public due to the current radiation safety paradigm's claim of no threshold dose. The fear of LDR has prevented study and use of LDR for prevention of aging-related diseases for which there are presently no methods of prevention, and for which animal models have shown promise. The fear has also prevented the study of LDR for cancer prevention and treatment whereas considerable evidence has shown the reduction of cancers following LDR exposure in multiple human studies. Other harm from the use of the paradigm include the devastation caused in Fukushima due to the LDR fear-based evacuation, delayed diagnoses in patients because of not performing diagnostic studies due to LDR concerns, potential misdiagnoses due to dose reduction efforts, tremendous increase in costs due to LNT hypothesis-based regulations, and suppression of nuclear power resulting in the use of less clean, more costly, and more hazardous power sources.

Changing the current radiation safety paradigm to one that recognizes the presence of a threshold dose for radiation-induced cancers would reduce the fear and concerns regarding LDR, enable study of LDR for prevention and treatment of diseases, rejuvenate nuclear power, and prevent casualties from evacuations in future radiological emergencies involving LDR.

CONCLUSION

The main evidence that has been used to justify the present radiation safety paradigm based on the LNT hypothesis are the atomic bomb survivor data, and these data, with the latest update, no longer support the hypothesis. Other published evidences supporting the LNT hypothesis or LDR carcinogenicity have not withstood careful scrutiny. On the other hand, much evidence supports the radiation hormesis hypothesis. Considerable harm has been caused to the public because of the fear of LDR arising from the current paradigm. Thus, there is no justification for continuing to use the current radiation safety paradigm. A change to a paradigm that recognizes the presence of a threshold dose is warranted. Such a change can reduce the fear of LDR, enable study of LDR for prevention and treatment of diseases, reduce or eliminate harm caused by the evasive actions due to LDR concerns, reduce the costs of radiation safety regulations, rejuvenate cleaner, safer, and less expensive nuclear power, and improve human health.

APPENDIX A: DERIVATION OF *ERR* CORRECTED FOR PERCENTAGE BIAS IN THE MEASURED BACKGROUND CANCER MORTALITY RATE

Let B and R be the measured cancer mortality rate of the baseline cohort and the radiated cohort respectively. Then, ERR is defined by the equation:

$$\frac{R}{B} = 1 + ERR \tag{1}$$

Let $B(corr)$ be the correct baseline cancer mortality rate and let δ be the percentage bias in the measured baseline cancer mortality rate. B is then given by:

$$B = B(corr) \times \left(1 + \frac{\delta}{100}\right) \tag{2}$$

Hence,

$$B(corr) = \frac{100 \times B}{(100 + \delta)} \tag{3}$$

The ERR corrected for the bias in the measured background cancer mortality rate would be defined by the equation:

$$\frac{R}{B(corr)} = 1 + ERR(corr) \tag{4}$$

Substituting Equation (3) in (4),

$$\frac{R \times (100 + \delta)}{100 \times B} = 1 + ERR(corr) \tag{5}$$

Substituting for R/B from Equation (1),

$$\frac{(1 + ERR) \times (100 + \delta)}{100} = 1 + ERR(corr) \tag{6}$$

Solving for $ERR(corr)$,

$$ERR(corr) = \frac{(1 + ERR) \times (100 + \delta)}{100} - 1 \tag{7}$$

Note: Appendix A is from [27], and is reproduced with permission from University of Massachusetts, Amherst.

References

[1] Calabrese EJ (2009) The road to linearity: why linearity at low doses became the basis for carcinogen risk assessment. *Archives of Toxicology* 83: 203-225.

[2] Sinclair WK (1981) Radiation protection: the NCRP guidelines and some considerations for the future. *The Yale Journal of Biology and Medicine* 54: 471-484.

[3] Saji G (2013) A post accident safety analysis report of the Fukushima Accident – future direction of evacuation: lessons learned. Proceedings of the 21st International Conference on Nuclear Engineering ICONE21 Jul 29 - Aug 2 Chengdu China ASME.

[4] Brumfiel G (2013) Fukushima: Fallout of fear. *Nature* 493: 290-293.

[5] http://www3.grips.ac.jp/~munro/notes%20nuclear%20valuation%20a.pdf
[6] Kojima S, Matsuki O, Nomura T, Yamaoka K, Takahashi M, et al. (1999) Elevation of antioxidant potency in the brain of mice by low-dose gamma-ray irradiation and its effect on 1-methyl-4-phenyl-1,2,3,6-tetrahydropyridine (MPTP)-induced brain damage. *Free Radical Biology and Medicine* 26: 388-395.
[7] Wei LC, Ding YX, Liu YH, Duan L, Bai Y, et al. (2012) Low-dose radiation stimulates Wnt/beta-catenin signaling, neural stem cell proliferation and neurogenesis of the mouse hippocampus in vitro and in vivo. *Current Alzheimer Research* 9: 278-289.
[8] Cohen BL (1990) The nuclear energy option : an alternative for the 90s. New York: Plenum Press.
[9] Kharecha PA, Hansen JE (2013) Prevented mortality and greenhouse gas emissions from historical and projected nuclear power. *Environmental Science & Technology* 47: 4889-4895.
[10] Bruninx K, Madzharov D, Delarue E, D'Haeseleer W (2013) Impact of the German nuclear phase-out on Europe's electricity generation—A comprehensive study. *Energy Policy* 60: 251-261.
[11] Markandya A, Wilkinson P (2007) Electricity generation and health. *Lancet* 370: 979-990.
[12] Brody AS, Guillerman RP (2014) Don't let radiation scare trump patient care: 10 ways you can harm your patients by fear of radiation-induced cancer from diagnostic imaging. *Thorax* 69: 782-784.
[13] Boutis K, Cogollo W, Fischer J, Freedman SB, Ben David G, et al. (2013) Parental knowledge of potential cancer risks from exposure to computed tomography. *Pediatrics* 132: 305-311.
[14] Pandharipande PV, Eisenberg JD, Avery LL, Gunn ML, Kang SK, et al. (2013) Journal club: How radiation exposure histories influence physician imaging decisions: a multicenter radiologist survey study. *American Journal of Roentgenology* 200: 1275-1283.
[15] Goske MJ, Strauss KJ, Coombs LP, Mandel KE, Towbin AJ, et al. (2013) Diagnostic reference ranges for pediatric abdominal CT. *Radiology* 268: 208-218.
[16] Goenka AH, Herts BR, Obuchowski NA, Primak AN, Dong F, et al. (2014) Effect of Reduced Radiation Exposure and Iterative Reconstruction on Detection of Low-Contrast Low-Attenuation Lesions in an Anthropomorphic Liver Phantom: An 18-Reader Study. Radiology: 131928.
[17] Preston DL, Ron E, Tokuoka S, Funamoto S, Nishi N, et al. (2007) Solid cancer incidence in atomic bomb survivors: 1958-1998. *Radiation Research* 168: 1-64.
[18] Preston DL, Shimizu Y, Pierce DA, Suyama A, Mabuchi K (2003) Studies of mortality of atomic bomb survivors. Report 13: Solid cancer and noncancer disease mortality: 1950-1997. *Radiation Research* 160: 381-407.
[19] NRC (2006) Health risks from exposure to low levels of ionizing radiation : BEIR VII Phase 2, National Research Council (U.S.). Committee to Assess Health Risks from Exposure to Low Level of Ionizing Radiation. Washington, D.C.: National Academies Press.
[20] Hall EJ, Brenner DJ (2008) Cancer risks from diagnostic radiology. *The British Journal of Radiology* 81: 362-378.

[21] Ozasa K, Shimizu Y, Suyama A, Kasagi F, Soda M, et al. (2012) Studies of the mortality of atomic bomb survivors, report 14, 1950-2003: an overview of cancer and noncancer diseases. *Radiation Research* 177: 229-243.

[22] Ozasa K, Shimizu Y, Suyama A, Kasagi F, Soda M, et al. (2013) Errata for Volume 177, number 3 (2012) in the article "Studies of the mortality of atomic bomb survivors, report 14, 1950-2003: an overview of cancer and noncancer diseases". *Radiation Research* 179: e0040-e0041.

[23] Feinendegen LE, Pollycove M, Neumann RD (2013) Hormesis by Low Dose Radiation Effects: Low-Dose Cancer Risk Modeling Must Recognize Up-Regulation of Protection. In: Baum RP, editor. Therapeutic Nuclear Medicine: *Springer.*

[24] Luckey TD (1980) Hormesis with ionizing radiation. Boca Raton, Fla.: CRC Press.

[25] Luckey TD (1991) Radiation hormesis. Boca Raton, Fla.: *CRC Press.*

[26] Luckey TD (2008) Atomic bomb health benefits. *Dose Response* 6: 369-382.

[27] Doss M (2012) Evidence supporting radiation hormesis in atomic bomb survivor cancer mortality data. *Dose Response* 10: 584-592.

[28] Doss M (2013) Linear No-Threshold Model vs. Radiation Hormesis. *Dose Response* 11: 480-497.

[29] Doss M Counterpoint: should radiation dose from CT scans be a factor in patient care? No. *Chest* (Accepted for publication, 2014).

[30] Doss M, Egleston BL, Litwin S (2012) Comments on "Studies of the mortality of atomic bomb survivors, report 14, 1950-2003: an overview of cancer and noncancer diseases" (Radiation Research 2012; 177:229-43). *Radiation Research* 178: 244-245.

[31] Brix G, Nekolla EA, Borowski M, Nosske D (2014) Radiation risk and protection of patients in clinical SPECT/CT. *European Journal of Nuclear Medicine and Molecular Imaging* 41 Suppl 1: S125-136.

[32] Brenner DJ (2014) What we know and what we don't know about cancer risks associated with radiation doses from radiological imaging. *The British Journal of Radiology* 87: 20130629.

[33] Brix G, Nekolla EA (2014) Response to letter by Doss: addition of diagnostic CT scan does not increase the cancer risk in patients undergoing SPECT studies. *European Journal of Nuclear Medicine and Molecular Imaging* 41 Suppl 1: 148-149.

[34] Cucinotta FA (2014) Space radiation risks for astronauts on multiple International Space Station missions. *PLoS One* 9: e96099.

[35] Steward MJ, Taylor SA, Halligan S (2014) Abdominal computed tomography, colonography and radiation exposure: what the surgeon needs to know. *Colorectal Disease* 16: 347-352.

[36] Cuttler JM (2014) Leukemia incidence of 96,000 Hiroshima atomic bomb survivors is compelling evidence that the LNT model is wrong. *Archives of Toxicology* 88: 847-848.

[37] Chaffey JT, Rosenthal DS, Moloney WC, Hellman S (1976) Total body irradiation as treatment for lymphosarcoma. *International Journal of Radiation Oncology* Biology* Physics* 1: 399-405.

[38] Rowland RE, Stehney AF, Lucas HF (1983) Dose-response relationships for radium-induced bone sarcomas. *Health Physics* 44 Suppl 1: 15-31.

[39] Sanders CL (2010) Radiation hormesis and the linear-no-threshold assumption. Heidelberg: *Springer.*

[40] Cuttler JM, Pollycove M (2003) Can Cancer Be Treated with Low Doses of Radiation? *Journal of American Physicians and Surgeons* 8: 108-111.

[41] Miller AB, Howe GR, Sherman GJ, Lindsay JP, Yaffe MJ, et al. (1989) Mortality from breast cancer after irradiation during fluoroscopic examinations in patients being treated for tuberculosis. *The New England Journal of Medicine* 321: 1285-1289.

[42] Kostyuchenko VA, Krestinina L (1994) Long-term irradiation effects in the population evacuated from the east-Urals radioactive trace area. *Science of the Total Environment* 142: 119-125.

[43] Sakamoto K (2004) Radiobiological basis for cancer therapy by total or half-body irradiation. *Nonlinearity in Biology, Toxicology and Medicine* 2: 293-316.

[44] Pollycove M (2007) Radiobiological Basis of Low-Dose Irradiation in Prevention and Therapy of Cancer. *Dose-Response* 5: 26-38.

[45] Hwang SL, Guo HR, Hsieh WA, Hwang JS, Lee SD, et al. (2006) Cancer risks in a population with prolonged low dose-rate gamma-radiation exposure in radiocontaminated buildings, 1983-2002. *International Journal Radiation Biology* 82: 849-858.

[46] Tubiana M, Diallo I, Chavaudra J, Lefkopoulos D, Bourhis J, et al. (2011) A new method of assessing the dose-carcinogenic effect relationship in patients exposed to ionizing radiation. A concise presentation of preliminary data. *Health Physics* 100: 296-299.

[47] Cardis E, Vrijheid M, Blettner M, Gilbert E, Hakama M, et al. (2005) Risk of cancer after low doses of ionising radiation: retrospective cohort study in 15 countries. *The BMJ* 331: 77.

[48] http://nuclearsafety.gc.ca/pubs_catalogue/uploads/INFO0811_e.pdf

[49] Zablotska LB, Lane RS, Thompson PA (2014) A reanalysis of cancer mortality in Canadian nuclear workers (1956-1994) based on revised exposure and cohort data. *The British Journal of Cancer* 110: 214-223.

[50] Wakeford R, Little MP (2003) Risk coefficients for childhood cancer after intrauterine irradiation: a review. International Journal Radiation Biology 79: 293-309.

[51] Brent RL (2014) Carcinogenic risks of prenatal ionizing radiation. *Seminars in Fetal and Neonatal Medicine* 19: 203-213.

[52] Mathews JD, Forsythe AV, Brady Z, Butler MW, Goergen SK, et al. (2013) Cancer risk in 680,000 people exposed to computed tomography scans in childhood or adolescence: data linkage study of 11 million Australians. *The BMJ* 346: f2360.

[53] Pearce MS, Salotti JA, Little MP, McHugh K, Lee C, et al. (2012) Radiation exposure from CT scans in childhood and subsequent risk of leukaemia and brain tumours: a retrospective cohort study. *Lancet* 380: 499-505.

[54] http://www.icrp.org/docs/John%20Boice%20Paediatric%20CT%20and%20Recent%20 Epidemiological%20Studies.pdf

[55] Cohen M (2013) Cancer risks from CT radiation: is there a dose threshold? *Journal of the American College of Radiology* 10: 817-819.

[56] Walsh L, Shore R, Auvinen A, Jung T, Wakeford R (2014) Risks from CT scans--what do recent studies tell us? *Journal of Radiological Protection* 34: E1-5.

[57] Hwang SL, Hwang JS, Yang YT, Hsieh WA, Chang TC, et al. (2008) Estimates of relative risks for cancers in a population after prolonged low-dose-rate radiation

exposure: a follow-up assessment from 1983 to 2005. *Radiation Research* 170: 143-148.

[58] Howe GR, McLaughlin J (1996) Breast cancer mortality between 1950 and 1987 after exposure to fractionated moderate-dose-rate ionizing radiation in the Canadian fluoroscopy cohort study and a comparison with breast cancer mortality in the atomic bomb survivors study. *Radiation Research* 145: 694-707.

[59] Kendall GM, Little MP, Wakeford R, Bunch KJ, Miles JC, et al. (2013) A record-based case-control study of natural background radiation and the incidence of childhood leukaemia and other cancers in Great Britain during 1980-2006. *Leukemia* 27: 3-9.

[60] Tubiana M (2005) Dose-effect relationship and estimation of the carcinogenic effects of low doses of ionizing radiation: the joint report of the Academie des Sciences (Paris) and of the Academie Nationale de Medecine. *International Journal of Radiation Oncology*Biology*Physics* 63: 317-319.

[61] Tubiana M, Feinendegen LE, Yang C, Kaminski JM (2009) The linear no-threshold relationship is inconsistent with radiation biologic and experimental data. *Radiology* 251: 13-22.

[62] Little MP, Wakeford R, Tawn EJ, Bouffler SD, Berrington de Gonzalez A (2009) Risks associated with low doses and low dose rates of ionizing radiation: why linearity may be (almost) the best we can do. *Radiology* 251: 6-12.

[63] Doss M, Little MP, Orton CG (2014) Point/Counterpoint: Low-dose radiation is beneficial, not harmful. *Medical Physics* 41: 070601.

In: Fukushima Nuclear Accident
Editor: Shizuyo Sutou
ISBN: 978-1-63463-748-0

Chapter 4

GENETIC EFFECTS OF ATOMIC BOMB RADIATION ON HUMANS

Yoshiaki Kodama*
Cytogenetic Laboratory, Department of Genetics, Radiation Effects Research Foundation, Hiroshima, Japan

ABSTRACT

In order to evaluate the trans-generational effects of atomic-bomb (A-bomb) radiation on the offspring of A-bomb survivors, a number of studies have been conducted over the past several decades. These include birth defects (malformations, stillbirths, and newborn deaths) among the early studies, and these were followed by studies of sex ratios, chromosome aberrations, blood-protein mutations, and minisatellite DNA mutations. Long-term epidemiological follow-up studies for mortality and cancer incidence are still underway in a large population of the survivors' offspring. However, none of these studies has found evidence for genetic effects resulting from parental exposures to radiation. Currently, new clinical studies are being planned and executed to observe the effects of radiation on the occurrence of multifactorial diseases. In the first prevalence studies, no significant differences were observed among the subjects which could be correlated with parental radiation doses. Presently, three studies (mortality and cancer incidence, clinical health, and DNA mutations) are underway or are being planned with the aim of clarifying the genetic effects of A-bomb radiation on humans.

INTRODUCTION

Ionizing radiation is known to be a universal mutagen and abundant data on experimental animals has indicated that radiation exposures induce mutations in germ cells [1-4]. Because of this, the genetic effects of A-bomb radiation on the offspring of A-bomb survivors has been one of the major concerns of the A-bomb survivors and their families in Hiroshima and Nagasaki, apart from the direct somatic effects of radiation exposures to the survivors

* Correspondence to: Yoshiaki Kodama, Tel: 81-82-264-5905; Fax: 81-82-263-7279; e-mail: ykodama@rerf.or.jp.

themselves. The role of possible genetic effects of radiation is an important issue, not only for A-bomb survivors and their children, but also for all humans because everyone is subjected to medical radiation exposures during the course of modern life.

The Atomic Bomb Casualty Commission (ABCC) and its successor, the Radiation Effects Research Foundation (RERF), have conducted a number of genetic studies since 1946 on large populations. These studies covered a broad range of topics such as birth defects, sex ratios, chromosome aberrations, mortality, cancer incidence, etc. (Table 4.1). To date, however, there has been no clear evidence indicating that there are significant effects from A-bomb radiation in the survivors' offspring, in spite of numerous efforts to detect such effects [e.g. 5]. This review will present detailed information from the results reported in the genetic studies conducted by the ABCC-RERF from 1946 to the present.

Table 4.1. Genetic studies of children of the A-bomb survivors)

Studies	Period	Population size
Birth defects (still births, malformations, etc.)	1948-1954	77,000
Sex ratio	1948-1962	140,000
Chromosome aberrations	1967-1985	16,000
Blood protein mutations	1977-1985	24,000
Mortality and cancer incidence*	1946-	77,000
DNA studies	1985-	1,000**
Clinical health examinations	2002-	12,000

* Cancer incidence since 1958.
** No of families.

1. Birth Defects

In 1948, the ABCC initiated a large-scale genetic study of the offspring of A-bomb survivors, and investigated birth defects in relation to radiation doses delivered to their parents [5]. These studies were continued for six years (1948-1954). At that time, there was an acute shortage of food in Japan and pregnant women had government benefits that allowed them to obtain extra food (mainly rice) when they registered their pregnancies with the city office. From that information, more than 90% of all pregnancies were confirmed in Hiroshima and Nagasaki, and a population for further studies was established. Information of exposure status (distance from the hypocenter and shielding conditions) was obtained through interviews with the pregnant women and their spouses. Since no systematic dosimetry system was available at that time, exposure conditions of the survivors (mothers and fathers) were classified by the distance from the hypocenter and degree of shielding present at the time of the bombings (Table 4-2; see footnote in the table for details), but those collecting the data on birth defects were not aware of exposure status.

In that study, about 77,000 newborns from Hiroshima and Nagasaki were tested by physicians within two weeks of birth. The frequency of malformations, stillbirths, and neonatal deaths are shown in Tables 4-3 to 4-5. In the study of malformations, among 65,431 newborns who satisfied additional conditions such as there being no biological relationship between the two parents, 594 (0.91%) had malformations, but there was no relationship to parental radiation exposures (Table 4.3). Observed major congenital malformations at birth were anencephaly, cleft palates, cleft lips with or without cleft palates, club feet, polydactyly and syndactyly. These abnormalities constituted 445 of the 595 (75%) malformations in Table 4-3, and are commonly seen in the general population in Japan. The overall frequency of malformations (0.91%) in this study concurred well with the results (456/49,654 = 0.92%) of another large study of Japanese births at the Tokyo Red Cross Maternity Hospital during the years 1922 through 1940, where radiation exposures were not involved. In addition, results from other studies (stillbirths and neonatal deaths, Tables 4.4 and 4.5) did not show any relationship between malformations and parental radiation exposures.

Since some birth defects are generally difficult to identify soon after birth (e.g. congenital heart disease), re-examinations were performed at 8-10 months after birth (Table 4-6). A total of 18,876 children were examined and 378 (2.0%) were found to have major birth defects. Again, there was no correlation with parental radiation doses. The data were re-evaluated when individual estimated doses became available using the DS86 Dosimetry System [6] (Table 4.7). However, these newer results followed the same trend and indicated that no radiation effects were observed [7].

Table 4.2. Distribution of registered births correlated with parental exposures* (Hiroshima) [5]

Maternal exposure conditions	Paternal exposure conditions					Total
	1	2	3	4	5	
1 (Not in city)	18,723	1,611	648	147	275	21,204
2 (Low doses)	5,721	1,993	416	119	145	8,394
3 (Middle doses)	2,320	451	545	81	80	3,477
4 (High doses)	417	90	49	28	19	603
5 (Symptom +)	791	127	67	26	54	1,065
Total	27,972	4,272	1,725	401	573	34,943

*Exposure status was classified by the distance from the hypocenter and the degree of shielding at the time of the bombing: A low dose is defined as >3km or 2-2.9 km with any shielding, 1.5-1.9 km with moderate or heavy shielding, 1-1.4 km with heavy shielding, and <1 km with heavy shielding. A mid-range dose is defined as 2-2.9 km with no shielding, 1.5-1.9 km with light shielding, 1-1.4 km with light or moderate shielding, and <1 km with moderate shielding. A high dose is defined as <1.9 km with no shielding. Symptomatic is defined as <3 km with acute radiation symptoms (epilation, petechiae or gingivitis). Group 5 (Symptom+) was pooled with group 4 (High doses) in later analysis.

Table 4.3. Malformations diagnosed within two weeks of birth and determined from cases/children examined, 1948–1953 (Hiroshima + Nagasaki) [5]

Maternal exposure conditions	Paternal exposure conditions			
	Not in cities	Low doses	Middle doses	High doses
Not in cities	294/31,904 (0.9%)	28/3,670 (0.8%)	12/839 (1.4%)	6/534 (1.1%)
Low doses	121/14,684 (0.8%)	62/5,994 (1.0%)	4/658 (0.6%)	4/442 (0.9%)
Middle doses	23/2,932 (0.8%)	7/703 (1.0%)	6/615 (1.0%)	1/192 (0.5%)
High doses	19/1,676 (1.1%)	3/318 (0.9%)	3/145 (2.1%)	1/145 (0.7%)

Table 4.4. Stillbirths in A-bomb survivors determined from cases/children examined, 1948–1953 (Hiroshima + Nagasaki) [5]

Maternal exposure conditions	Paternal exposure conditions			
	Not in cities	Low doses	Middle doses	High doses
Not in cities	408/31,559 (1.3%)	57/3,629 (1.6%)	15/826 (1.8%)	9/528 (1.7%)
Low doses	216/14,457 (1.5%)	107/5,924 (1.8%)	14/654 (2.1%)	11/418 (2.6%)
Middle doses	63/2,905 (2.2%)	8/696 (1.1%)	10/607 (1.6%)	2/198 (1.0%)
High doses	26/1,656 (1.6%)	4/315 (1.3%)	2/142 (1.4%)	2/144 (1.4%)

Table 4.5. Neonatal deaths in cases/children examined, 1948-1953 (Hiroshima + Nagasaki) [5]

Maternal exposure conditions	Paternal exposure conditions			
	Not in cities	Low doses	Middle doses	High doses
Not in cities	431/31,151 (1.4%)	49/3,572 (1.4%)	19/811 (2.3%)	10/519 (1.9%)
Low doses	217/14,331(1.5%)	80/5,817 (1.4%)	5/640 (0.7%)	2/407 (0.5%)
Middle doses	40/2,842 (1.4%)	3/688 (0.4%)	3/597 (0.5%)	2/188 (1.1%)
High doses	24/1,630 (1.5%)	4/311 (1.3%)	4/140 (2.9%)	1/142 (0.7%)

Table 4.6. Malformations diagnosed at 8 to10 months after birth in cases/children (Hiroshima + Nagasaki) [5]

Maternal exposure conditions	Paternal exposure conditions		
	Not in cities	Low doses	Middle to high doses
Not in cities	213/9,440 (2.3%)	21/1,021 (2.1%)	6/419 (1.9%)
Low doses	70/4,192(1.7%)	34/1,491 (2.3%)	4/320 (0.7%)
Middle to high doses	18/1,387 (1.3%)	6/295 (2.0%)	6/311 (0.7%)

Table 4.7. Birth defects (stillbirths, malformations, and neonatal deaths within two weeks of birth) among A-bomb survivors, correlated with parental radiation doses and cases/children examined, 1948–1953 (Hiroshima + Nagasaki) [7]

Maternal dose (Gy)	Paternal dose (Gy)			
	< 0.01	0.01 - 0.49	0.50 - 0.99	≥1.0
< 0.01	2,257/45,234 (5.0%)	81/1,614 (5.0%)	12/238 (5.0%)	17/268 (6.3%)
0.01-0.49	260/5,445 (4.8%)	54/1,171 (4.6%)	4/68 (6.0%)	2/65 (3.0%)
0.50-0.99	44/651 (6.8%)	1/43 (2.3%)	4/47 (9.0%)	1/17 (6.0%)
≥ 1.0	19/388 (4.9%)	2/30 (6.7%)	1/9 (11.0%)	1/15 (7.0%)

2. Sex Ratio

From observations derived from genetic studies in fruit flies, lethal mutations in the X chromosome induced by radiation were thought to alter the sex ratios among the offspring. The rationale for this hypothesis is as follows: The single X chromosome in a male is always derived from his mother. Thus, if mothers were exposed to radiation, the proportion of males among live births may decrease due to a recessive lethal mutation in the X chromosome.

Conversely, a father's exposure to radiation may result in a reduction of female births since father's X chromosome which might carry a dominant mutation is transferred only to daughters. Based on this idea, sex ratio studies on the offspring of A-bomb survivors were initiated in 1948 [5]. Early observations from the studies (1948–1953) indicated that the sex ratio decreased from 108.6 to 102.9 with increasing radiation dose in the case of maternal exposures, and increased from 108.6 to 113.6 in the case of paternal exposures. The results seemed to support the hypothesis but were not statistically significant. Sex-ratio data were further collected through 1962 even after the termination of other studies on birth defects. However, a large data collection (140,542 births in total) showed a different trend (108.0 to 108.4 in the case of maternal exposures and 108.0 to 107.0 in the case of paternal exposures) and did not support the initial observation.

3. Chromosome Aberrations

Extensive cytogenetic investigations were carried out in the offspring of A-bomb survivors to examine whether or not stable-type chromosome aberrations (reciprocal translocations and inversions) were induced in parental germ cells. The study was conducted using chromosomes in blood lymphocytes from 8,322 children for whom either or both parents were exposed to an estimated dose of 0.01 Gy or greater at distances of 2,000 m or less from the hypocenter (exposed group), and in blood lymphocytes of 7,976 children from the unexposed control group for whom either both parents were not exposed to radiation (including those who were not in Hiroshima and Nagasaki) or exposed to less than 0.01 Gy (at ≥2,500 m from the hypocenter) [8]. Results are summarized in Table 4-8. As shown, there were 43 children (0.52%) bearing chromosome aberrations in the exposed group and 51 (0.64%) in the control group. Thus, the results did not show any effect of radiation exposure to the parents. Cytogenetic data based on international collaborative studies on newborn infants provided valuable information on background frequencies in persons who carry inherent chromosome aberrations (0.47%-0.83%) [9-13]. The observed frequencies in the ABCC-RERF studies showed good concordance when compared with those data.

Table 4.8. Summary of cytogenetic studies in the offspring of A-bomb survivors* [8]

		Control group	Exposed group
Number of offspring	Male	3,682	3,914
	Female	4,294	4,408
	Total	7,975	8,322
Sex chromosome abnormality		24	19
Autosomal structural aberration		27	23
Autosomal trisomy (Down syndrome)		0	1
Total		51 (0.64%)	43 (0.52%)

*Mean age of subjects was 24 years old at examination.

Table 4-9 indicates the number of autosomal balanced structural aberrations. This study is the most relevant to radiation exposures. The results indicated that 18 children in the exposed group and 25 in the control group bore stable-type chromosome aberrations. Further chromosome tests of parents and siblings showed that most aberrations were pre-existing and inherited from either one of the parents (10 out of 11 in the exposed group and 15 out of 16 in the control group). Only one subject in each group had a de novo aberration (0.01%), which did not show an increased frequency following a parental exposure to radiation. These results agree well with the estimated spontaneous frequency (0.019%) of new structural aberrations in neonates [14]. The origin of aberrations in 16 cases (9 in control and 7 in exposed) could not be determined because the parents had either died or refused to participate in this study. There was no indication of selective loss of parents who were exposed to high doses (i.e. the dose distributions were similar in tested and untested parents).

Table 4.9. Stable chromosome aberrations in the offspring of A-bomb survivors [8]

	Control group	Exposed group
Total number of offspring with abnormalities	25 (0.31%)	18 (0.22%)
Breakdown:		
Inherited from either parent	15 (0.19%)	10 (0.12%)
Newly induced (both parents were normal)	1 (0.01%)	1 (0.01%)
Undetermined (both parents were not tested)	9 (0.11%)	7 (0.08%)
Total number of offspring examined	7,976	8,322

4. Blood Protein Mutations

The purpose of this study was to detect potential genetic effects from A-bomb radiation by screening proteins in blood samples from the F_1 children using one-dimensional gel electrophoresis and enzyme activity measurements. The former method could detect rare electrophoretic variants caused by amino-acid substitutions, and the latter could find a 50% decrease in enzymatic activities induced by deletion-type mutations.

The study took about ten years to complete (1975-1985). During those years, about 24,000 children were studied for the occurrence of mutations which altered electrophoretic mobility in 30 blood proteins (Table 4.10), and about 10,000 of these children were tested for enzyme deficiencies (Table 4.11). Although 1,233 electrophoretic variants and 47 enzyme-deficient variants were detected, many of them were found to be inherited from one of their parents with family studies. Only five electrophoretic variants (3 in controls and 2 in exposed populations) and one enzyme-deficient variant were new mutations. Thus, no statistically significant effects were found at the protein level in these studies [15-17].

Table 4.10. Results of screening for electrophoretic protein variants [15-17]

	Controls <0.01 Gy	Exposed >0.01 Gy
Children examined	12,297	11,364
Loci tested	589,506	544,779
New mutations	3	2
Mutation rate/locus/generation	0.5×10^{-5}	0.4×10^{-5}

Mean dose = 0.49 Gy (DS86), neutron RBE = 20.

Table 4.11. Results of screening for enzyme-deficient protein variants [17]

	Controls <0.01 Gy	Exposed >0.01 Gy
Children examined	5,026	4,989
Loci tested	61,741	60,529
New mutations	0	1
Mutation rate/locus/generation	0	2×10^{-5}

5. DNA Studies

5-1. Minisatellite and Microsatellite Studies

Minisatellite and microsatellite genes are composed of many short sequence units (6-100 nucleotides in the former and 1-5 nucleotides in the latter) present in tandem repeats, and both types are widely distributed in the human genome. They are highly polymorphic with respect to repeat numbers, and some minisatellite loci are known to have high spontaneous mutation rates in human germ cells [18]. Because of the very unstable nature of these sequences, it was thought that genetic effects from A-bomb radiation might be detected in relatively small numbers of children. Among minisatellite loci, 8 highly polymorphic loci were tested among 61 families (both parents and children) in the exposed group and 58 in the control group. The results indicated that mutation rates for the exposed and control groups were 2.6% and 2.8% respectively, showing no evidence for significant effects of parental exposure to radiation (Table 4.12) [19]. Forty microsatellite loci which are known to show high background mutation rates were also studied in 66 families in the exposed group and 63 families in the control group. Observed mutation rates were 0.39% in the former and 0.35% in the latter (Table 4.13) [20]. Thus the study again showed no trans-generational effects of radiation.

Results from minisatellite mutation studies following radiation exposures in humans vary among published reports. One English group described significant increases in mutation rates in families exposed to radiation in Chernobyl [21-23], Semiparatinsk [24] or Techa River [25]. Although the exposed doses in our studies are much larger than those reported in the Chernobyl subjects, our results are negative, which is in sharp contrast to those positive findings. However, studies on other persons after the Chernobyl accident also failed to indicate discernible effects from radiation exposure [26-28]. More recently, studies on survivors of childhood cancer also indicated negative results for minisatellite germline mutations [29]. The causes of these discrepancies are not currently understood.

Table 4.12. Mutations at 8 minisatellite loci in children of A-bomb survivors [19]

	Control <0.01 Gy	Exposed >0.01 Gy*
Children examined	58	61
Minisatellite loci tested	1,403	496
New mutations	39	13
Mutation rate/locus/generation	2.8%	2.6%

*Mean dose = 1.47 Gy (DS86), neutron RBE = 10.

Table 4.13. Mutations at 40 microsatellite loci in children of A-bomb survivors [20]

	Control <0.01 Gy	Exposed >0.01 Gy*
Children examined	63	66
Microsatellite loci tested	7,465	2,789
New mutations	36	11
Mutation rate/locus/generation	0.35%	0.39%

*Mean dose = 1.56 Gy (DS02), neutron RBE = 10.

5-2. Two-dimensional DNA Gel Electrophoresis Studies

To detect deletion type mutations, a two-dimensional DNA gel electrophoresis system was developed at RERF. This method provides some 2,000 DNA fragments (spots) from a genomic DNA digest on a single gel without using any probes, and allows the detection of deletion mutations in those fragments which are indicated as a 50% decrease in individual spot intensity. A total of 66 children (56,176 alleles) in the exposed group (mean dose = 1.7 Gy) and 62 children (59,942 alleles) in the control group were studied, and only one deletion type mutation was found in the control group. No mutations were detected in the exposed group [30].

5-3. DNA Microarray Studies

A pilot study was conducted to detect changes in copy number variations using a comparative genome hybridization (CGH)-microarray system (using 2,238 bacterial artificial chromosome (Bac) clones) to test DNA from B-cell lines from 80 children (40 from the control and 40 from the exposed group). As a result of this analysis, a total of 251 variants showing different copy numbers were identified. Subsequent examination of the parents, however, showed that all of the variants were also detected in the somatic cells of either of the parents. Thus, no de novo mutations were detected in this pilot study [31].

6. Mortality and Cancer Incidence

Since 1946, long-term follow-up studies have been conducted to ascertain the effects of A-bomb radiation on mortality and cancer incidence in about 77,000 offspring of the survivors. Periodical analyses of the mortality data at different ages in the populations (from birth to 20 years or older) were carried out, but none of the studies detected evidence for significantly increased risks in relation to parental radiation dose [32-34]. Cancer incidence was also assessed through population-based tumor registry data. Early-stage follow-up studies for childhood leukemia and early-onset solid cancers (younger than age 20 or ages 20-40) did not show any evidence for an effect from parental radiation exposure [35]. More recent cancer incidence data for offspring whose ages were in their 50s again indicated no significant increased risk of cancer associated with parental radiation exposures [36]. Since the ages of the participants in these studies were still relatively young (in their mid-50s or younger), further follow-up studies are still required to reach any firm conclusion regarding the effects of A-bomb radiation exposure on mortality and cancer incidence in this population.

7. Clinical Health Studies

The purpose of clinical studies of the offspring of A-bomb survivors is to determine if parental exposures to A-bomb radiation could affect the development of adult onset multifactorial diseases and subclinical conditions in their offspring. In order to address this issue, the first clinical study on "adult" offspring was conducted from 2002 to 2006 for 11,951 subjects (this was a cross-sectional study, and hence the prevalence of a specific condition is the only outcome).

Associations between the prevalence of adulthood multifactorial diseases (hypertension, diabetes mellitus, hypercholesteroemia, angina pectoris, myocardial infarction, and stroke) and parental radiation dose were investigated. The results did not show evidence for an increased prevalence of adult-onset multifactorial diseases as a whole [37], or separately according to the type of disease [38] in relation to parental radiation dose.

To date, no trans-generational effects of A-bomb radiation among the offspring of A-bomb survivors at the clinical health level have been observed. However, because the incidence of multifactorial diseases will increase with increasing ages of the participants in the study, further long-term follow-up studies will hopefully clarify this issue.

Conclusion

The epidemiological studies described here (mortality, cancer incidence, and clinical health), are currently still in progress and the cohort members are still relatively young (the mean age is about 60 years). Thus further follow-up studies are still necessary to reach any definitive conclusions regarding the genetic effects of A-bomb radiation on each epidemiological endpoint.

Recent progress in molecular biology and technological advances will provide new opportunities to detect genetic changes caused by radiation in finer detail. Rapid DNA sequencing technologies are being developed, and are one of the possible future technological options likely to provide new information at the molecular level. This could permit the design of new assays to look for trans-generational effects of radiation in humans.

Acknowledgment

The author thanks Dr. L. Kapp for his careful reading of the manuscript and valuable comments. The Radiation Effects Research Foundation (RERF), Hiroshima and Nagasaki, Japan is a public interest foundation funded by the Japanese Ministry of Health, Labor and Welfare (MHLW) and the US Department of Energy (DOE). The views of the authors do not necessarily reflect those of the two governments.

References

[1] Russell WL (1977) Mutation frequencies in female mice and the estimation of radiation hazards in women. *Proceedings of the National Academy of Sciences of the United State of America* 74: 3523–3527.

[2] Russell WL, Kelly EM (1982) Specific-locus mutation frequencies in mouse stem-cell spermatogonia at very low radiation dose rates. *Proceedings of the National Academy of Sciences of the United State of America* 79: 539-541.

[3] Nomura T (1982) Parental exposure to x rays and chemicals induces heritable tumours and anomalies in mice. *Nature* 296: 575-577.

[4] United Nations Scientific Committee on the Effects of Atomic Radiation, Hereditary Effects of Radiation (2001) *Report to the General Assembly with Scientific Annex.* United Nations, New York.

[5] Neel JV, Schull WJ (1991) The children of atomic bomb survivors. *National Academic Press*, Washington DC.

[6] Roesch WC (ed) (1987) US-Japan Joint Reassessment of Atomic Bomb Radiation Dosimetry. Radiation Effects Research Foundation (http://www.rerf.or.jp/shared/ds86/ds86a.html).

[7] Ohtake M, Schull WJ, Neel JV (1990) Congenital malformations, stillbirth, and early mortality among the children of atomic bomb survivors: Reanalysis. *Radiation Research* 122: 1-11.

[8] Awa AA, Honda T, Neriishi S, Sofuni T, Shimba H, et al (1987) Cytogenetic study of the offspring of atomic bomb survivors, Hiroshima and Nagasaki. In Obe G, Basler A (eds) Cytogenetics: basic and applied aspects, *Springer,* Berlin, pp 166-183.

[9] Jacobs PA, Melville M, Ratcliffe S, Keay AJ, Syme J (1974) A cytogenetic survey of 11,680 new born infants. *Annals of Human Genet.* 37: 359-376.

[10] Nielsen J, Sillensen I (1975) Incidence of chromosome aberrations among 11,148 newborn children, *Humangenetik* 30: 1-12.

[11] Hamerton JL, Canning N, Ray M, Smith S (1975) A cytogenetic survey of 14,069 newborn infants. *Clinical Genetics* 8: 223-243.

[12] Walzer S, Gerald PS (1977) A chromosome survey of 13,751 male newborns. In EB Hook and IH Porter (Eds.), Population Cytogenetics: Studies in Humans, *Academic Press*, New York, pp. 45-61.

[13] Buckton KE, O'Riordan ML, Ratcliffe S, Slight J, Mitchell M, et al (1980) A G-band study on chromosomes in liveborn infants. Annals of Human Genetics 43: 227-239.

[14] Jacobs PA (1981) Mutation rates of structural chromosome rearrangements in man. *American Journal of Human Genetics* 33: 44-54.

[15] Neel JV, Satoh C, Hamilton HB, Otake M, Goriki K, et al. (1980) Search for mutations affecting protein structure in children of atomic bomb survivors: Preliminary report. *Proceedings of the National Academy of Sciences of the United State of America* 77: 4221–4225.

[16] Neel JV, Satoh C, Goriki K, Fujita M, Takahashi N, et al. (1986) The rate with which spontaneous mutation alters the electrophoretic mobility of polypeptides. *Proceedings of the National Academy of Sciences of the United State of America* 83: 389–393.

[17] Neel JV, Satoh C, Goriki K, Asakawa J, Fujita M, et al. (1988) Search for mutations altering protein charge and/or function in children of atomic-bomb survivors: Final report. *American Journal of Human Genetics* 42:663–676.

[18] Jeffreys AJ, Royle NJ, Willson V, Wong Z (1988) Spontaneous mutation rates to new length alleles at tandem-repetitive hypervariable loci in human DNA. *Nature* 332: 278-281.

[19] Kodaira M, Izumi S, Takahashi N, Nakamura N (2004) No evidence of radiation effect on mutation rates at hypervariable minisatellite loci in the germ cells of atomic bomb survivors. *Radiation Research* 162: 350-356.

[20] Kodaira M, Ryo H, Kamada N, Furukawa K, Takahashi N, et al. (2010) No evidence of increased mutation rates at microsatellite loci in offspring of A-bomb survivors. *Radiation Research* 173: 205–213.

[21] Dubrova YE, Nesterov VN, Krouchinsky NG, Ostapenko VA, Neumann R et al. (1996) Human minisatellite mutation rate after the Chernobyl accident. *Nature* 380: 683-686.

[22] Dubrova YE, Nesterov VN, Krouchinsky NG, Ostapenko VA, Vergnaud G, et al. (1997) Further evidence for elevated human minisatellite mutation rate in Belarus eight years after the Chernobyl accident. *Mutation Research* 381: 267-78.

[23] Dubrova YE, Grant G, Chumak AA, Stezhka VA, Karakasian AN (2002) Elevated minisatellite mutation rate in the post-chernobyl families from Ukraine. *American Journal of Human Genetics* 71: 801-809.

[24] Dubrova YE, Bersimbaev RI, Djansugurova LB, Tankimanova MK, Mamyrbaeva ZZ, et al. (2002) Nuclear weapons tests and human germline mutation rate. *Science* 295: 1037.

[25] Dubrova YE, Ploshchanskaya OG, Kozionova OS, Akleyev AV (2006) Minisatellite germline mutation rate in the Techa River population. *Mutation Research* 602:74-82.

[26] Livshits LA, Malyarchuk SG, Kravchenko SA, Matsuka GH, Lukyanova EM, et al. (2001) Children of chernobyl cleanup workers do not show elevated rates of mutations in minisatellite alleles. *Radiation Research* 155: 74-80.

[27] Kiuru A, Auvinen A, Luokkamaki M, Makkonen K, Veidebaum T, et al. (2003) Hereditary minisatellite mutations among the offspring of Estonian Chernobyl cleanup workers. *Radiation Research* 159: 651-655.

[28] Slebos RJ, Little RE, Umbach DM, Antipkin Y, Zadaorozhnaja TD, et al. (2004) Mini- and microsatellite mutations in children from Chernobyl accident cleanup workers. *Mutation Research* 559: 143-151.

[29] Tawn EJ, Rees GS, Leith C, Winther JF, Curwen GB, et al. (2011) Germline minisatellite mutations in survivors of childhood and young adult cancer treated with radiation. *International Journal of Radiation Biology* 87: 330-40.

[30] Asakawa J, Kuick R, Kodaira M, Nakamura N, Katayama H, et al. (2004) A genome scanning approach to assess the genetic effects of radiation in mice and humans. *Radiation Research* 161: 380–90.

[31] Takahashi N, Tsuyama N, Sasaki K, Kodaira M, Satoh Y, et al. (2008) Segmental copy-number variation observed in Japanese by array-CGH. *Annals of Human Genetics* 72: 193–204.

[32] Kato H, Schull WJ, Neel JV (1966) A cohort-type study of survival in the children of parents exposed to atomic bombings. *American Journal of Human Genetics* 18: 339–373.

[33] Neel JV, Kato H, Schull WJ (1974) Mortality in the children of atomic bomb survivors and controls. *Genetics* 76: 311–336.

[34] Izumi S, Suyama A, Koyama K (2003) Radiation-related mortality among offspring of atomic bomb survivors: a half-century of follow-up. *International Journal of Cancer* 107: 292-297.

[35] Yoshimoto Y, Neel JV, Schull WJ, Kato H, Soda M, et al. (1990) Malignant tumors during the first 2 decades of life in the offspring of atomic bomb survivors. American *Journal of Human Genetics* 46: 1041–52.

[36] Izumi S, Koyama K, Soda M, Suyama A (2003) Cancer incidence in children and young adults did not increase relative to parental exposure to atomic bombs. *British Journal of Cancer* 89: 1709-1713.

[37] Fujiwara S, Suyama A, Cologne JB, Akahoshi M, Yamada M et al. (2008) Prevalence of adult-onset multifactorial disease among offspring of atomic bomb survivors. *Radiation Research* 170: 451–457.

[38] Tatsukawa Y, Cologne JB, Hsu WL, Yamada M, Ohishi W et al. (2013) Radiation risk of individual multifactorial diseases in offspring of the atomic-bomb survivors: A clinical health study. *Jounal of Radiolgical Protection* 33: 281–93.

In: Fukushima Nuclear Accident
Editor: Shizuyo Sutou
ISBN: 978-1-63463-748-0

Chapter 5

HORMESIS: INSIGHT INTO ADAPTIVE DEFENSE MECHANISMS AGAINST IONIZING RADIATION ESTABLISHED DURING EVOLUTION OF LIFE ON THE EARTH

Masanobu Sugimoto*
M & M Office, Shiromeguri, Kamakura,
Kanagawa, Japan

ABSTRACT

The administrative action for radiation contamination, including elimination of contaminants caused by the nuclear accident of Fukushima Daiichi Nuclear Power Plant following the Great Eastern Japan Earthquake in March 11, 2011, is based on the linear no-threshold (LNT) model: according to the governmental guideline, contaminants that result in more than 1 mSv radiation dose per year are to be eliminated. A body of evidence, however, has accumulated implicating that the LNT model is incorrect and that low-dose radiation is not only harmless, but sometimes is even beneficial to the human body (hormetic effect). This chapter reviews these effects of low-dose radiation focusing on the molecular defence mechanisms of the body against radiation that have been developed since the birth of life on the earth.

INTRODUCTION

After the nuclear accident of the Fukushima Daiichi Nuclear Power Plant (FDNPP) accompanying the Great Eastern Japan Earthquake (GEJE) in March 11, 2011, elimination of radiation contaminants from areas around the plant is one of the most important problems to be solved.

* C orrespondence to: Masanobu Sugimoto, Tell and Fax, +81-467-45-5902; E-mail, masasugi0467@m.jcnnet.jp.

The linear no-threshold (LNT) model assumes that long-term biological damage caused by ionizing radiation is directly proportional to the dose without a safety threshold. The LNT model is the basis for the currently established guidelines that contaminants that result in more than 1 mSv radiation dose per year should be eliminated. According to the guidelines, considerable public funds are spent for this purpose. Many mothers with children leave the area due to fear of low-dose radiation and are forced to be cut off from their husbands. Furthermore, mortality risk amongst elderly people who are nursing home residents increased by about 2.7 times after evacuation [1].

However, as discussed in detail by other authors, questions against the validity of the LNT model have been raised [2]. A body of recent scientific data indicates that low-dose radiation is not harmful, but rather favorable to prevent natural occurrence of cancer: this favorable effect of low-radiation is known as "hormesis". In this chapter I will review hormetic phenomena and discuss the biological background behind them.

1. Definition of Hormesis

According to Calabrese and Baldwin [3,4] "hormesis" is a concept that describes the nature of dose-response relationships in biological systems as showing a stimulatory response at low doses and an inhibitory response at higher doses. A search of the older literature uncovered Virchow (Virchchows Archiv; 6: 133-34, 1854) as the first person to describe the phenomenon of hormesis [5]. Virchow observed an increase in the beating activity of cilia of tracheal epithelia of postmortem mucosa by sodium and potassium hydroxide at low concentrations, and a concentration-dependent decrease to arrest at higher concentrations.

Sometimes hormesis is defined as a favorable biological response to low exposures to toxins and other stressors. However, Calabrese and Baldwin [4] criticized this definition by citing as an instance the case of a cancer chemotherapeutic that may be effective at high doses due to inhibitory effects on cell proliferation, but harmful to patients at lower doses when it may stimulate tumor cell proliferation [6]. Therefore, in this review the concept by Calabrese and Baldwin [4] is adopted.

Hormetic effects are typically observed with both chemical substances [7] and ionizing radiation [3], and studies on hormesis have a long history beginning as early as 1965. Hormesis includes general biological phenomena widely observed in the living world not only in animals, including single-cell organisms, but also in plants [3,7-9]. In the following sections, I will focus on radiation hormesis.

2. Radiation Hormesis

Since 1980 two books about radiation hormesis have been published [10,11], and several reviews have also been published on this subject [3,4,7,8]. Here, I will not repeat in detail the contents of these books and reviews, but introduce some typical instances.

2.1. Single Cell Organisms

The sensitivity of *E. coli* to X-ray irradiation correlates with the stages of replication cycle of DNA. Sensitivity to X-ray irradiation in the wild type can be attributed to the presence of nuclear targets, such as branched chromosomes plus DNA repair mechanisms [12]. Interestingly, irradiation of *E. coli* cells with ultraviolet rays or X-rays, followed by incubation under conditions in which protein synthesis can occur, results in a population of cells resistant to X-rays; however, this resistance develops only if cells are *recA+ and lexA+*, and which associates the phenomenon with induced SOS response [13]. The SOS response is a state of high-activity DNA repair, and is activated by bacteria that have been exposed to heavy doses of DNA-damaging agents. Thus, *E. coli* shows a radiation hormetic effect by augmenting synthesis of repair proteins following DNA damage.

Luckey [9] studied if ionizing radiation in a ciliated protozoan is essential for a maximum growth rate. When extraneous ionizing radiation was limited to the rate of 0.15 mrad/d (1.5 μSv/d), the reproduction rate of *Tetrahymena pyriformis* was significantly less (P<0.01) than it was at near ambient levels, 0.5 mrad/d (5 μSv/d) (Figure 1). Significantly higher growth rates (P<0.01) were observed when chronic radiation was increased. The data suggest that ionizing radiation is essential for an optimum reproduction rate by this organism.

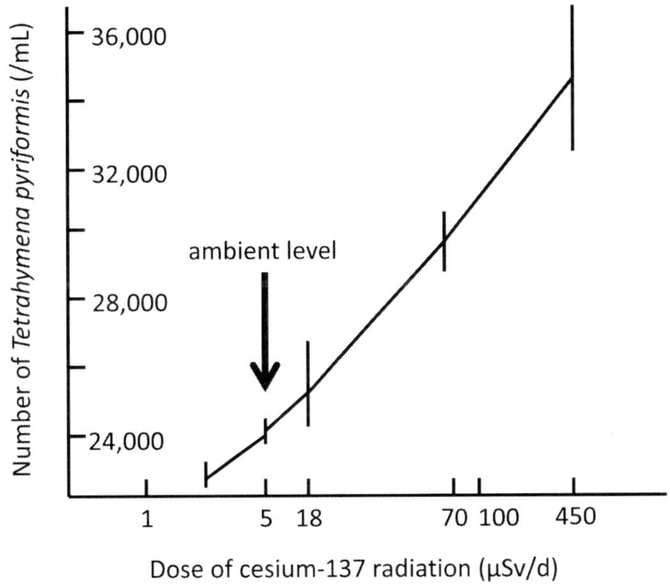

Figure 1. Dose response of *Tetrahymena pyriformis* to radiation dose from Cs-137. The figure is based on the data from the reference by TD Luckey [9]. The vertical bars indicate standard errors.

Planel et al. [14] reported that additional radiation can stimulate proliferation of single-cell organisms protozoan *Paramecium tetraurelia* and cyanobacterium *Synechococcus lividus*, which were shielded against background radiation or exposed to very low doses of gamma radiation. Radiation hormesis depends on internal factors (age of starting cells) and external factors (lighting conditions). The stimulatory effect occurred only in a limited range of doses and disappeared for doses higher than 50 mGy/y (ca. 50 mSv/y).

2.2. Mouse

Yonezawa et al. [15] reported that preirradiation with 0.05 Gy/y of X rays two months before a second exposure to a mid-lethal dose significantly increased the survival rate in both female and male ICR strain mice. This radioresistance was induced only by whole-body preirradiation (not by partial irradiation of the head or the trunk). Preirradiation with higher dose rates of 0.30 Gy/y and 0.50 Gy/y induced radioresistance by whole-body preirradiation or partial preirradiation of the trunk. They noted that these results conflict with the LNT model.

Shimada and Ogyu [16] summarized the data obtained worldwide of mice exposed by radiation continuously through their lifetime. In their analysis, data of six study groups were pooled. Above dose rate of 10 mGy/d, the life span shortened with increase of dose rate, but below this dose rate the life span tended to increase. The results are essentially consistent with the results of Yonezawa et al. [15].

Caratero et al. [17] studied the effect of continuous γ irradiation at very low doses on the lifespan of mice. Six hundred female C57BL/6 mice, 1 month old, were exposed to chronic γ radiation at very low dose rates of 7 or 14 cGy/y. The life span determined by the survival time of 50% of each population, increased in irradiated mice: 549 days in controls, 673 days in both irradiated groups. The differences were significant between control and irradiation mice. From these results they confirmed the possibility of no harm but rather a beneficial effect (hormesis) of low-dose ionizing radiation, and concluded that the paradigm, which states that high-dose effects can predict low-dose effects, cannot be systematically applied in radiation biology in general and gerontology in particular.

Ina et al. [18] reported that suppression of thymic lymphoma induction by life-long low-dose irradiation is accompanied by immune activation in C57BL/6 mice. Induction of thymic lymphomas by whole-body X irradiation with four doses of 1.8 Gy (total dose: 7.2 Gy) in C57BL/6 mice was suppressed from a high frequency (90%) to 63% by preirradiation with 0.075 Gy X rays given 6 h before each 1.8-Gy irradiation. This level was further suppressed to 43% by continuous whole-body irradiation with Cs-137 γ rays at a low dose rate of 1.2 mGy/h for 450 days, starting 35 days before the challenging irradiation. Continuous irradiation at 1.2 mGy/h (for a total dose of 7.2 Gy over 258 days) yielded no thymic lymphomas, indicating that this low-dose radiation does not induce these tumors. Further continuous irradiation up to 450 days (total dose: 12.6 Gy) produced no tumors. Continuously irradiated mice showed no loss of hair and greater body weight than nonirradiated controls. Immune activities of the mice, as measured by the numbers of CD4+ T cells, CD40+ B cells and antibody-producing cells in the spleen after immunization with sheep red blood cells, significantly increased by continuous 1.2 mGy/h irradiation alone. From these results they concluded that an adaptive response in tumor induction exists accompanying radiation-induced immune activation in tumor suppression.

2.3. Humans

Cohen [19] examined data on lung cancer mortality rates versus average radon concentration in homes of 1,601 U.S. counties to test the LNT model. With or without corrections for variations in smoking prevalence, a surprisingly strong tendency for lung cancer rates to decrease with increasing radon exposure was found, in sharp contrast to the increase expected from the LNT theory. This report stated that in spite of extensive efforts, no potential explanation for the discrepancy could be found other than the failure of LNT for lung carcinogenesis from inhaled radon decay products.

The 1952-88 cancer mortality records for inhabitants of the Misasa spa area, Japan, which has a high radon background, and a neighboring control area without a radon spa were analyzed by Mifune et al. [20] in 1992. The average outdoor radon concentration was 26 mBq/liter in Misasa and 11 mBq/liter in the control area. Standardized mortality ratios for cancers of all sites were significantly lower among inhabitants of both Misasa (male 0.538; female 0.463) and the control area (male 0.850; female 0.770), than in the whole Japanese population. Poisson regression analysis showed that the relative risks among the inhabitants of Misasa were significantly lower than in the control area for deaths from cancers at all sites (0.67) and stomach cancer (0.59). The relative risk of lung cancer death was also lower (0.55 times) in Misasa than in the control area, although the difference was not statistically significant. From these results they suggested that the LNT hypothesis for radiation risk may not be valid for exposure to low doses of radon. However, a follow-up study (1998) by the same group [21] reported that no difference was observed in the incidence of all-site cancers in the increased radon level area. Neither study recognized an increase in cancer risk in the increased radon level area, inconsistent with the LNT model.

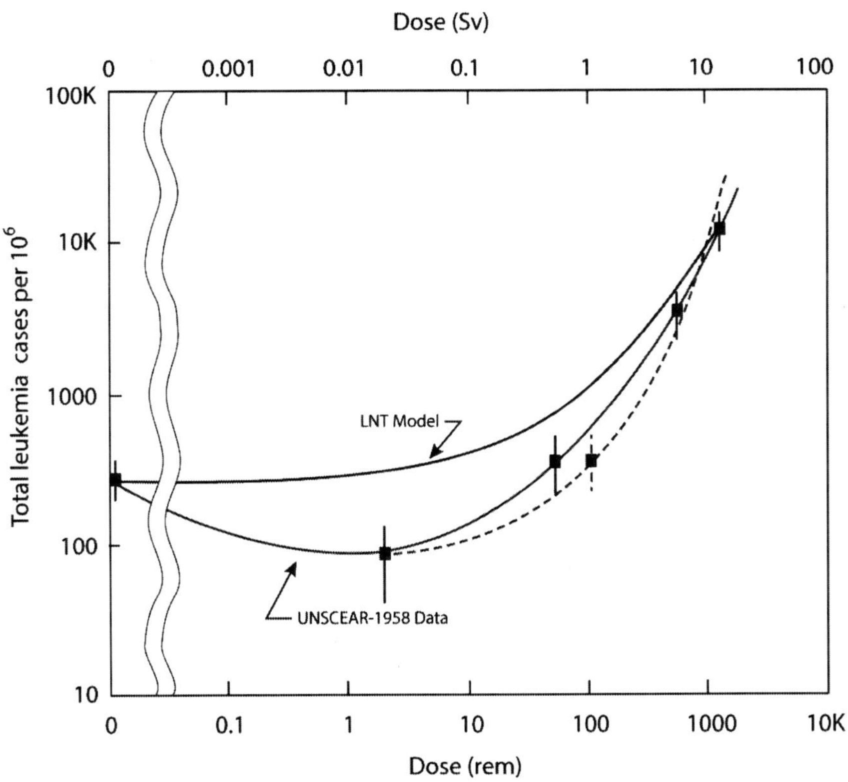

Figure 2. Leukemia incidence in Hiroshima survivors, 1950-1957 cited from the reference by Cutller [22]. The dashed line through the 100 rem dose reflects the footnote for the 50- rem dose in the United Nations Scientific Committee on the Effects of Atomic Radiation (UNSCEAR) 1958 "that almost all cases of leukemia in this zone occurred in patients who had severe radiation combinations, indicating that their doses were greater than 50 rem".

A recent study [22] on leukemia incidence in 96,000 Hiroshima atomic bomb survivors, 1950~1957, showed compelling evidence that the LNT model is wrong (Figure 2): no increase in leukemia from low dose radiation in atomic bomb survivors was observed and the threshold dose was near 50 rem (500 mSv).

All these studies on mice and humans support the idea that the LNT model is wrong; some of these studies support the idea that low-dose radiation has a homeotic effect to lower the risk of carcinogenesis.

3. Natural Radiation and Evolution of Life

3.1. Earth's Background Radiation Level Over Geologic Time

Since the birth of life on the earth, living organisms have been exposed to external background radiation from radionuclides, such as uranium (U-235), thorium (Th-232), radium (Ra-226) and potassium (K-40) and to internal radiation, such as from K-40 [23]. The current average background γ radiation level on earth is 0.28 mSv/y. Humans are furthermore exposed to approximately 0.40 mSv/y dose from K-40 in their bodies, and to an additional dose of 2.90 mSv/y derived from radon (2.0 mSv/y), 0.28 mSv/y from building materials, 0.27mSv/y from cosmic radiation and various anthropogenic sources. Radiation doses from geologic emitters has changed because of the chemical evolution of the continental crust, changes in relative abundances of U-235 and U-238, and radioactive decay of uranium, thorium, and K-40 over the past 4 billion years. Accordingly, background radiation exposure from external and internal radiation has dropped from about 7.0 to 1.35 mGy/y during the time life has existed on the earth. Karam and Keslie [24] argue that the conservative nature of mutation repair mechanisms in modern organisms suggests that these mechanisms may have evolved in the distant past and that organisms may retain some capability to efficiently repair damage from higher radiation levels than exist at present. Relevant to this, SOS response to x-irradiation [13] in *E. coli* , a prokaryote, agrees well with this idea.

3.2. How Could Living Organisms Avoid Damage by Natural Radiations?

As a theoretical conclusion, life should have developed various mechanisms to escape from radiation damage, just like it has developed various defense mechanisms, including the immune system, against invasive pathogens like bacteria and viruses.

Ionizing radiation damages DNA in two ways: directly (about 20% of total DNA damage) and indirectly by generating reactive oxygen species (ROS) (about 80% of total DNA damage). Examples of DNA damage are double-strand breaks by direct attack by radiation, oxidation of bases by ROS, replication errors. Sutou argues that most of ROS produced by respiration and radiation may be destroryed by radical scavenger systems before ROS damage DNA (Chapter 16). Living organisms have developed various mechanisms to repair DNA damage [25]: 1) base excision repair replacing damaged bases in the DNA code, 2) nucleotide excision repair replacing a string of bases if one or more base is damaged, 3) non-homologous end joining repair of double-strand breaks in the DNA double helix, 4) homologous repair of double-strand breaks in interstrand cross-links of DNA and 5) mismatch repair correcting mismatches in sequence of bases in DNA. The two kinds of excision repair and homologous end joining are especially important for DNA repair caused by ionizing radiation. The presence of the DNA repair system in the body is strong supporting evidence that the LNT model is not correct because of the following. If the body is exposed to low-dose radiation, DNA damage is small and should be easily corrected by the repair system. However, if the body is damaged by high-dose radiation, DNA damage is too serious to be corrected by the repair system. Therefore, no linear correlation should exist between degree of DNA damage and radiation dose at low-dose levels. If the total exposed doses of radiation are the same between a high-dose exposure for a short time and a low-dose exposure for a longer time, the damage by a short-time exposure is more serious than that by a longer-time exposure, because the body does not have enough time to be repaired after the former exposure. This dose-rate effect is widely accepted.

Hiyama et al. [26] reported that the nuclear accident of Fukushima Daiichi Nuclear Power Plant caused physiological and genetic damage to pale grass blue *Zizeeria maha*, a common lycaenid butterfly in Japan. Low-dose radiation does not increase carcinogenesis or deformation in humans, which was well established by long-term epidemiological studies on atomic bomb-survivors in Hiroshima and Nagasaki, but the phenomenon of the butterfly should be carefully considered from a scientific standpoint. Whether the effect of radiation on the butterfly was due to genetic mutation or other factors, such as oxidative stress caused by radiation, remains to be clarified, as discussed by Sutou (Chapter 14) and Obinata (Chapter 15) in this monograph.

3.3. Role of RECQ Helicases in Repair of Radiation-Induced DNA Damage

In this section, the role of RECQ helicases, including WRN helicase, in the repair of radiation-induced DNA damage is discussed. Incidentally, I have researched for decades Werner syndrome (WS), the patients of which show a wide range of premature aging phenotypes, including gray hair, hoarseness, diabetes and malignancy [27]. *WRN* is the causative gene of WS and encodes WRN protein that acts as a DNA helicase with exonuclease activity. The mutation in the *WRN* gene induces a wide range of premature aging phenotypes accompanied by an abnormal pattern of tumors. I proposed a hypothesis in a review [27] as to how the dysfunction of WRN protein induces multiple premature aging phenotypes because of the following: telomere dysfunction induced by *WRN* mutation causes multiple premature aging phenotypes of WS, including a shortened cellular life span and inflammation induced by ROS and diabetes mellitus. WRN protein is assumed to maintain correct function of telomeres having repeated sequences that are prone to generate disturbed structures during replication.

WRN is a member of the RECQ helicase gene family, which includes *RECQ1*, Bloom syndrome gene (*BLM*), *WRN*, Rothmund-Thomson syndrome gene (*RTS/RECQ4*), and *RECQ5*. Mutations of *BLM*, *WRN* and *RTS* result in Bloom syndrome, WS and Rothmund-Thomson syndrome, respectively. These three genetic disorders are associated with genomic instability, and therefore RECQ helicases are considered to be guardians of the genome. Diseases caused by mutations of *RECQ1* and *RECQ5* have yet to be identified.

Recently, the function of WRN exonuclease activity that is separable from helicase activity was suggested to be essential for repair of radiation-induced DNA damage [28]. This finding was made by a study using human cell lines overexpressing the N-terminal truncated WRN protein, which has helicase activity, but not exonuclease activity, *in vitro*. These cell lines expressing exogenously the N-terminal-deleted WRN protein were slightly but, nonetheless, significantly more radiosensitive in cytogenetic damage and killing cells than control cell lines expressing endogenous WRN protein.

Liu et al. [29] reported the role of WRN protein in repair of radiation-induced DNA damage. Irradiation by γ-rays leads to translocation of endogenous WRN protein from nucleoli to nucleoplasmic DNA double-strand breaks. Relocation of WRN after irradiation is perturbed by promyelocytic leukemia protein (PML) knockdown and is increased by isoform PML IV overexpression. These results support the idea that PML IV physically interacts with and regulates translocation of WRN for DNA damage repair.

Das et al. [30] reported that mammalian DNA glycosylase, NEIL1, specific for repair of oxidatively damaged bases in the genome by the base excision repair pathway, is activated by ROS and prevents toxicity due to radiation. WRN protein associates with NEIL1 in the early damage-sensing step of base excision repair. WRN stimulates NEIL1 in excision of oxidative lesions from bubble DNA substrates. Binary interaction between NEIL1 and WRN includes C-terminal residues 288-349 of NEIL1 and the RECQ C-terminal region of WRN. Exposure to oxidative stress, such as induced by radiation, increases the NEIL-WRN association accompanying their strong nuclear co-localization.

Camptothecin, the topoisomerase I-trapping agent, causes single-strand DNA damage, and WRN helicase repairs the DNA damage. WRN protein moves from nucleoli to form distinct nuclear foci, and participates in DNA repair there, similar to WRN that participates in repair of radiation-induced DNA damage in cooperation with PML IV [31].

A study on tissue distribution of WRN and other RECQ proteins showed that RECQ proteins exist abundantly, especially in the testis [32,33]. WRN is also highly expressed in the thymus, pancreas and ovary in humans. In the mouse, two isoforms of RECQ1 proteins (RECQ1α and β) were identified: RECQ1α is expressed in all tissues tested, but RECQ1β is expressed only in the testis [34]. The distribution pattern of RECQ proteins preferentially in the testis may implicate that they perform an important role in maintaining genome stability in the testis. Thus, the testis genome is assumed to be strongly guarded by RECQ family helicase proteins from external mutagenic factors, including radiation.

Living organisms on the earth have developed various repair systems against DNA damage induced by ROS, radiation and chemical agents. RECQ-type helicases exist in almost all organisms, including bacteria, yeasts and mammals, and is considered to participate deeply in adaptive response against radiation-induced DNA damage. Of interest, RECQ helicase family helicases participate deeply in keeping genome stability, especially, in the testis.

3.4. Possible Mechanisms of Hormesis

How does low-dose preirradiation induce responses in defense mechanisms against high-dose irradiation?

The exact mechanisms of radiation hormesis remains to be clarified. Most radiation-induced DNA damage (80%) is considered to be caused by production of ROS, and low-dose radiation may activate defense mechanisms against ROS. Another author of this monograph will discuss in detail the role of transcription factor NRF2 that possibly has an important role in radiation hormesis (Chapter 7).

Only one example is briefly introduced here. In cellular adaptive survival responses against nitrosative stress, the role of heme oxygenase-1 (HO-1) upregulation, especially through activation of transcription factor NRF2 is known [35]. HO-1 is a rate-limiting enzyme in the process of degrading potentially toxic free heme into biliverdin, free iron and carbon monoxide. Nitrosative stress causes cell death during inflammation and has been implicated in the pathogenesis of many human ailments. However, by way of, activating NRF2 by HO-1 upregulation, relatively mild nitrosative stress may fortify cellular defense capacities, rendering cells tolerant or adaptive to ongoing and subsequent cytotoxic challenges, a phenomenon known as 'preconditioning' or 'hormesis'.

An additional molecular mechanism of hormesis may be augmented accumulation of repair enzymes, as in the case of *E. coli* [13]. Because the function of WRN exonuclease activity is suggested to be essential for repair of radiation-induced DNA damage [27], whether accumulation of WRN protein is augmented by irradiation is an interesting question in the mechanism of hormesis of mammals.

Conclusion

A body of data show the invalidity of the linear no-threshold (LNT) model that has been used for radiation safety purposes worldwide since the 1950s. In most cases, low-dose radiation confers favorable effects on living organisms from protozoa to mammals (hormesis). Life came into existence on the earth about 3.8 billion years ago, and since then it evolved under exposure to external and internal natural ionizing radiation, presumably at higher levels than at present. Therefore, it appears reasonable that life has developed defense mechanisms against natural radiation, otherwise life would not have survived. Indeed, our body has various mechanisms to repair radiation-induced DNA damage, including RECQ helicases.

References

[1] Nomura N GS, Tsubokura M., Yoneoka D., Sugimoto A., Oikawa T., Kami M, Shibuya K. (2013) Mortality risk amongst nursing home residents evacuated after the Fukushima nuclear accident: A retrospective cohort study. *PLoS One* DOI: 10. 1371/journal.pone. 0060192.

[2] Allison W, Brodsky A, Cohen M, Cuttler JM, Dobryznski L, et al. (2014) An open letter to professional bodies regarding low dose radiation cancer risk. http://radiation effectsorg/.

[3] Calabrese EJ, Baldwin LA (2000) Radiation hormesis: its historical foundations as a biological hypothesis. *Human Experimental Toxicology* 19: 41-75.

[4] Calabrese EJ, Baldwin LA (2002) Defining hormesis. *Human Exprimental Toxicology* 21: 91-97.

[5] Henschler D (2006) The origin of hormesis: historical background and driving forces. *Human Exprimental Toxicology* 25: 347-351.

[6] Foekens JA, Sieuwerts AM, Stuurman-Smeets EM, Dorssers LC, Berns EM, et al. (1992) Pleiotropic actions of suramin on the proliferation of human breast-cancer cells in vitro. *International Journal of Cancer* 51: 439-444.

[7] Calabrese EJ, Baldwin LA (1999) Chemical hormesis: its historical foundations as a biological hypothesis. *Toxicologic Pathology* 27: 195-216.

[8] Vaiserman AM (2010) Radiation hormesis: historical perspective and implications for low-dose cancer risk assessment. *Dose Response* 8: 172-191.
[9] Luckey TD (1986) Ionizing radiation promotes protozoan reproduction. *Radiation Research* 108: 215-221.
[10] Luckey TD (1980) Hormesis with ionizing radiation. Boca Taton, FL: CRC Press, Inc.
[11] Luckey TD (1991) Radiation Hormesis. Boca Raton, FL.: CRC Press, Inc.
[12] Clark DJ (1968) Effects of ionizing radiation on synchronous cultures of Escherichia coli B-r. *Journal of Bacteriology* 96: 1150-1158.
[13] Pollard EC, Fluke DJ, Kazanis D (1981) Induced radioresistance: an aspect of induced repair. *Molecular and General Genetics* 184: 421-429.
[14] Planel H, Soleilhavoup JP, Tixador R, Richoilley G, Conter A, et al. (1987) Influence on cell proliferation of background radiation or exposure to very low, chronic gamma radiation. *Health Physics* 52: 571-578.
[15] Yonezawa M, Misonoh J, Hosokawa Y (1996) Two types of X-ray-induced radioresistance in mice: presence of 4 dose ranges with distinct biological effects. *Mutation Research* 358: 237-243.
[16] Shimada Y, Ogyu T (1992) The effect of continuous low-dose radiation on lifespan---literature search <II> (in Japanese). *Houshasenkagaku* 35: 378-383.
[17] Caratero A, Courtade M, Bonnet L, Planel H, Caratero C (1998) Effect of a continuous gamma irradiation at a very low dose on the life span of mice. *Gerontology* 44: 272-276.
[18] Ina Y, Tanooka H, Yamada T, Sakai K (2005) Suppression of thymic lymphoma induction by life-long low-dose-rate irradiation accompanied by immune activation in C57BL/6 mice. *Radiation Research* 163: 153-158.
[19] Cohen BL (1995) Test of the linear-no threshold theory of radiation carcinogenesis for inhaled radon decay products. *Health Physics* 68: 157-174.
[20] Mifune M, Sobue T, Arimoto H, Komoto Y, Kondo S, et al. (1992) Cancer mortality survey in a spa area (Misasa, Japan) with a high radon background. *Japanese Journal of Cancer Research* 83: 1-5.
[21] Ye W, Sobue T, Lee VS, Tanooka H, Mifune M, et al. (1998) Mortality and cancer incidence in Misasa, Japan, a spa area with elevated radon levels. *Japanese Journal of Cancer Research* 89: 789-796.
[22] Cuttler JM (2014) Leukemia incidence of 96,000 Hiroshima atomic bomb survivors is compelling evidence that the LNT model is wrong: Edward Calabrese's papers "Origin of the linear no threshold (LNT) dose-response concept" (Arch Toxicol (2013) 87:1621-1633) and "How the US National Academy of Sciences misled the world community on cancer risk assessment: new findings challenge historical foundations of the linear dose response" (Arch Toxicol (2013) 87:2063-2081). *Archives of Toxicology* 88: 847-848.
[23] Karam P. A. and Leslie SA http://wwwirpanet/irpa9/cdrom/VOL2/V2_69PDF.
[24] Karam P. A. and Leslie SA (1999) Calculations of background beta-gamma radiation dose through geologic time. *Health Physics* 77: 662-667.
[25] Hoeijmakers J, Jackson S, Jiricny J, Kanaar R, Krokan H (2009) DNA damage response and repair mechanisms._http://wwwdna-repairnl/DNA_Repairpdf.

[26] Hiyama A, Nohara C, Kinjo S, Taira W, Gima S, et al. (2012) The biological impacts of the Fukushima nuclear accident on the pale grass blue butterfly. *Scintific Repeport* 2: 570.

[27] Sugimoto M (2014) A cascade leading to premature aging phenotypes including abnormal tumor profiles in Werner syndrome (review). *International Journal of Molecular Medicine* 33: 247-253.

[28] Kashino G, Kodama S, Suzuki K, Matsumoto T, Watanabe M (2005) Exogenous expression of exonuclease domain-deleted WRN interferes with the repair of radiation-induced DNA damages. *Journal of Radiation Resesarch* 46: 407-414.

[29] Liu J, Song Y, Qian J, Liu B, Dong Y, et al. (2011) Promyelocytic leukemia protein interacts with werner syndrome helicase and regulates double-strand break repair in gamma-irradiation-induced DNA damage responses. *Biochemistry* (Mosc) 76: 550-554.

[30] Das A, Boldogh I, Lee JW, Harrigan JA, Hegde ML, et al. (2007) The human Werner syndrome protein stimulates repair of oxidative DNA base damage by the DNA glycosylase NEIL1. *Journal of Biological Chemistry* 282: 26591-26602.

[31] Futami K, Ishikawa Y, Goto M, Furuichi Y, Sugimoto M (2008) Role of Werner syndrome gene product helicase in carcinogenesis and in resistance to genotoxins by cancer cells. *Cancer Science* 99: 843-848.

[32] Imamura O, Ichikawa K, Yamabe Y, Goto M, Sugawara M, et al. (1997) Cloning of a mouse homologue of the human Werner syndrome gene and assignment to 8A4 by fluorescence in situ hybridization. *Genomics* 41: 298-300.

[33] Kitao S, Ohsugi I, Ichikawa K, Goto M, Furuichi Y, et al. (1998) Cloning of two new human helicase genes of the RecQ family: biological significance of multiple species in higher eukaryotes. *Genomics* 54: 443-452.

[34] Wang WS, Seki M, Yamaoka T, Seki T, Tada S, et al. (1998) Cloning of two isoforms of mouse DNA helicase Q1/RecQL cDNA; alpha form is expressed ubiquitously and beta form specifically in the testis. *Biochimica Biophysica Acta* 1443: 198-202.

[35] Surh YJ, Kundu JK, Li MH, Na HK, Cha YN (2009) Role of Nrf2-mediated heme oxygenase-1 upregulation in adaptive survival response to nitrosative stress. *Archives of Pharmacal Resesearch* 32: 1163-1176.

In: Fukushima Nuclear Accident
Editor: Shizuyo Sutou
ISBN: 978-1-63463-748-0

Chapter 6

Dose Rate Problems in Extrapolation of Hiroshima-Nagasaki Atomic Bomb Data to Estimation of Cancer Risk of Elevated Environmental Radiation in Fukushima

Hiroshi Tanooka[1,2,*]
[1]Radiation Effects Association, Tokyo, Japan
[2]National Cancer Center Research Institute, Tokyo, Japan

Abstract

Problems in evaluating the cancer risk in the Fukushima environment caused by radioactive contamination following the nuclear accidents are discussed, with special concern given to the dose-rate effects in relation to their application to the cancer data from the atomic bomb (A-bomb). Determination of the A-bomb radiation dose rate itself requires evaluation of the exposure time of the A-bomb radiation which consists of multi-complex components. By taking 1 μsec at minimum and 5 sec as median for the exposure time for survivors on the ground, the dose rate of A-bomb radiation was estimated to be in the range between 6×10^5 and 6×10^8 Gy/min, or 0.12 and 120 Gy/min, respectively, in contrast to $10^{-9} - 10^{-8}$ Gy/min in environmental radiation. From analysis of non tumor doses as a function of the dose rate of radiation, the dose rate effectiveness factor to extrapolate the A-bomb data to the cancer risk of environmental radiation was estimated as 16.5.

Keywords: Radiation, dose rate, cancer risk, non tumor dose, DDREF

---

* Correspondence to: Hiroshi Tanooka, E- mail: tanooka-h@wind.ocn.ne.jp.

INTRODUCTION

The environmental contamination by radioactive nuclear fission products that were released by the explosion of the nuclear reactor buildings that occurred in Fukushima, Japan, after the tsunami in March, 2011 [1], created a number of social problems, including the evacuation of residents from the contaminated areas in accordance with strict radiation protection regulations. These regulations are based upon the risk estimated from the cancer data on Hiroshima-Nagasaki A-bomb survivors. However, there is a large difference in the magnitude of radiation cancer risk between A-bomb and environmental radiation cases for the same dose. The main factor for this difference is the dose rate of radiation.

The effect of the dose-rate on cancer induction was shown in experimental animals in early studies by Upton and coworkers [2] and this effect was extensively surveyed with special concern given to the effect of low dose radiation by the United Nations Scientific Committee [3,4]. The International Commission of Radiological Protection (ICRP) recommended the value 2 for the dose and dose-rate effectiveness factor (DDREF) [5]. However, this factor was obtained within the dose-rate range of the A-bomb radiation at very high dose rates, whereas the dose rate of environmental radiation is at the level of 10^{-9} Gy/min. Obviously, the current DDREF value cannot be applied to the environmental radiation. Its application to the estimation of cancer risk in the Fukushima environmental cases creates an overestimation of the risk.

As an approach to revise the DDREF value, the dose-rate effect data on radiation-induced cancers were reviewed in previous articles [6,7]. The non tumor dose, D_{nt}, was defined as the highest dose at which no statistically significant increase of cancer incidence was observed above the control level, and these values surveyed in the literature were analyzed as a function of dose rate. The present article is an extension of the previous studies and further provides a method to extrapolate the A-bomb cancer data to the Fukushima situation.

1. EXPOSURE TIME AND DOSE RATE OF A-BOMB RADIATION BASED ON DS86

The dose of A-bomb radiation received by survivors depends on the distance from the hypocenter of the explosion (Figure 6-1). In estimating DDREF for cancer risk by ICRP [5], A-bomb cancer data were divided into two dose rate groups according to the low and high doses given in the same exposure time. The ratio of the two dose-response lines was 2, and this value was taken as DDREF. In this estimation, the difference of the dose rate between the two groups is considered to be 10, according to the difference of the dose range, 0-1 Gy and 1-10 Gy with the same exposure time. However, estimation of the exposure time of A-bomb radiation itself involves many problems.

Extensive revision of the dose of A-bomb radiation has been made by the US-Japan joint assessment DS86 [8] and DS02 [9]. A-bomb radiation includes various components: prompt γ rays, prompt neutrons, prompt secondary γ rays, delayed γ rays, delayed neutrons, and delayed secondary γ rays. The timing of radiation dose delivery was extensively analyzed in Chapter III of the DS86 study [10]. Among those, primary fission γ rays (exposure time, 1 μsec), prompt secondary γ rays (exposure time < 0.1 sec) and delayed γ rays (exposure time < 30 sec) contribute to the main part of the absorbed dose and hence to the major biological effects.

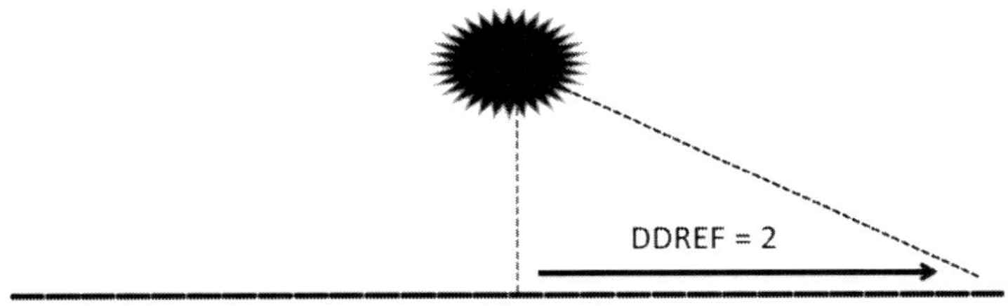

Exposure time used for calculation	Dose rate for 0.01-10 Gy dose range Gy/min
1 μsec[a]	$6 \times 10^5 - 6 \times 10^8$
5 sec[b]	0.12 – 120
Natural background	2.3×10^{-9}

Figure 6-1. Dose and dose rate of A-bomb radiation received on the ground level.The distance from the hypocenter is a major determining factor for dosimetry in cancer mortality analysis. a. Exposure time was taken from the time duration of the nuclear fission chain reaction. b. Exposure time was taken from the median value of the time interval of the dose delivery, involving delayed γ rays, estimated in the DS86 study [10]. The value 2 for DDREF was estimated by ICRP within the A-bomb dose rate range [5].

For delayed γ rays, an exposure rate is almost at the same level up to 10 sec after the chain reaction according to the measurement in the Nevada study, while it increases with time, reaches a peak at 5 sec and decreases to zero level after 20 sec at a distance of 2779 m from the hypocenter [10]. Therefore, A-bomb radiation includes different components of the dose rate within 20 sec after the explosion even at a fixed distance. As an overall estimation of the exposure time, the minimum limit is considered to be 1 μsec and a median value to be 5 sec. By applying this exposure time, the dose rate for A-bomb survivors was estimated by dividing the received dose, 0.01-10 Gy, by the exposure time, yielding $6 \times 10^5 - 6 \times 10^8$ Gy/min for the exposure time of 1 μsec and 0.12– 120 Gy/min for the exposure time of 5 sec. Accordingly, the dose rate varies over a wide range and the estimation of the dose-rate effect is not a simple task. Moreover, the dose-rate range of A-bomb radiation is far higher than the dose-rate of γ radiation in the natural environment, 1.2 mGy/yr or 2.3×10^{-9} Gy/min, or 3.8 mSv/yr including other sources of radiation in the daily life of humans.

Compared with the time scale of the biological clock, including cell cycle, DNA replication and repair, and other biochemical reactions, the exposure time of A-bomb radiation is very short and the dose rate effect appears unlikely to occur within this exposure time. The dose rate effect is thought to appear at much lower dose rates below the A-bomb dose rate.

2. Leukemia and Solid Cancers in A-Bomb Survivors

Leukemia data on A-bomb survivors are primarily important in assessing cancer risk of ionizing radiation, since leukemia is induced in the early stage of the post-exposure period with a high frequency [11], compared with solid cancers [12]. Solid cancers are a combination of various types of cancers that appear in the relatively late stage of the post-exposure period with a relatively low individual frequency compared with that of leukemia. For solid cancer mortality, 100 mGy was thought to be the detectable limit. However, the position of ICRP is that the cancer risk of radiation exists at any small dose that can be extrapolated by the linear non-threshold (LNT) model. The problem as to whether the undetectable risk should be incorporated as an existing risk into radiation protection regulations or be approved as a tolerable level of radiation is a currently important problem that needs to be argued. At least, the magnitude 2 for DDREF proposed by ICRP seems too small for extrapolation of A-bomb data to the Fukushima environmental cases.

3. Dose Rate of Environmental Radiation and Cancer Risk

Natural background radiation includes external γ radiation from the ground and cosmic rays, and internal radiation from ^{40}K β rays and α rays from radon and its decay products. It should be noted that ground γ rays, cosmic ray protons and muons, and ^{40}K β rays exert whole body low LET radiation, whereas α rays from radon and its decay products exert partial body high LET radiation to the lung. In addition, medical radiation exerts a considerably high dose of partial body low LET radiation. The contribution of these components to the overall cancer risk should be analyzed separately as discussed later.

Compared with the multi component character of A-bomb radiation, the elevated γ radiation level in the high natural background radiation areas provides more precise information on the dose-rate of radiation from a well-defined source. Epidemiological studies have been conducted in areas with elevated γ radiation backgrounds in Yanjiang, China [13] and in Kerala, India [14] in parallel with dosimetry. The level of the dose rate of γ radiation from the ground is 3 and 7 fold of the natural level respectively, that contribute to the dose rate to residents 4.05×10^{-9} and 7.55×10^{-9} Gy/min, respectively. An increase of the radiation level from the ground contaminated by radioactive cesium has been detected and its level as of September, 2014, is 5 μSv/hr, i. e., 8.3×10^{-8} Gy/min, at the Namie/Iitate district with the highest level outside the 20 km zone from the nuclear reactor according to the Fukushima Prefecture Radioactivity Measurement Map released on line. To evaluate the cancer risk of this level of radiation, the cancer data on the high natural radiation background areas are directly applicable.

The whole figure of various exposure conditions, including whole body low LET, whole body high LET, partial body low LET, and partial body high LET radiations, is schematically shown with a dose-rate range from the A-bomb radiation level down to an environmental level (Figure 6-2).

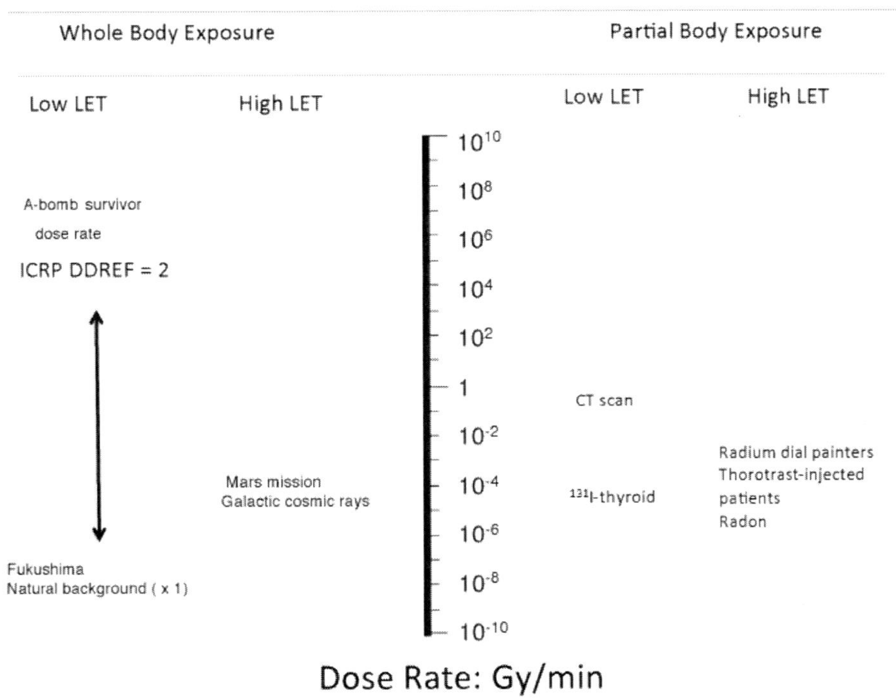

Figure 6-2. Dose rate range for radiation in human environment. Exposure conditions were classified as whole body low LET exposure, whole body high LET exposure, partial body low LET exposure, and partial body high LET exposure.

4. Application of the Non Tumor Dose Concept for Estimation of Cancer Risk

The shape of the dose-response curve varies from linear to non-linear, depending on the dose rate of radiation. Although ideally a complete set of dose response curves is desirable for different dose rates, the number of available data showing such a relation is limited, especially those at a low dose range where tumors start to increase with the increasing dose above the control level.

To get an indication on how much dose is necessary to induce cancer or is tolerable to the host, the data on non tumor dose (D_{nt}) were surveyed on experimental animals and humans in the literature (Figure 6-3b) [6,7]. The data on the whole body low LET radiation are listed in Table 6-1 with special attention to compare with A-bomb data [2, 11-14,15-29].

4.1. External Whole Body Exposure Data

Besides A-bomb cancer data, epidemiological data on cancer mortality in residents in a high natural radiation background area represent human cases with a lifetime exposure to external whole body radiation with increased dose rates. Despite the fact that the dose-response relation is not available because of the limit of life span, these data provide important information on how much an increase in background radiation humans can tolerate.

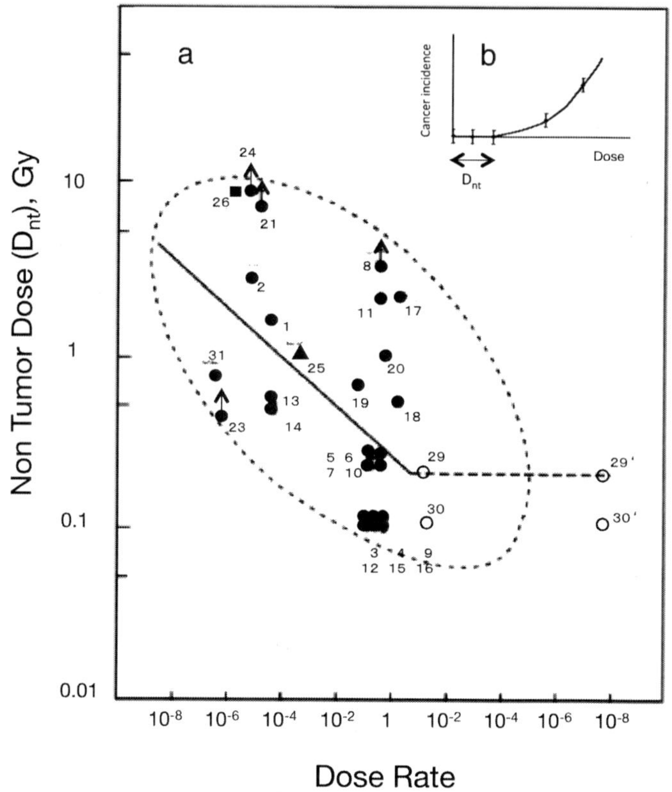

Figure 6-3. Non tumor dose, D_{nt}, plotted as a function of dose rate (Gy/min) of radiation for the whole body low LET exposure cases (a) and definition of non tumor dose (b). Attached are data numbers listed in Table 1. (●) mouse, (▲) rat, (■) dog, and (○) human; 29 and 30: A-bomb leukemia and solid cancers, exposure time 5 sec assumed; 29' and 30': exposure time 1 μsec assumed. Arrows indicate that D_{nt} is higher than these values and not used for calculating the regression line.

Laboratory experiments with animals provide substantial information with accurate dose and dose rate with well-defined source of radiation. Most of the early data were contributed by Ulrich et al, then at Oak Ridge National Laboratory. Since the dose rate range covered by each laboratory experiment is limited to a relatively narrow range, the whole range of dose rate extending to the environmental level is constructed by combination of separate experimental results.

A few animal experiments were designed to observe cancer incidence during the lifespan with radiation at very low dose rates. C57BL mice irradiated continuously at a dose rate of 2 x 10^{-5} mGy/min for the lifespan showed no tumors even when a cumulative dose reached 7.2 Gy [24], while mice irradiated with 4 repeats of acute dose with the same total dose 7.2 Gy, as the original regimen of Kaplan and Brown [30], showed a 90% incidence of thymic lymphoma. The Institute for Environmental Sciences, Japan, conducted a large scale experiment with B6C3F1 mice irradiated continuously for 22 h a day with graded dose rates up to 20 mGy/day, 400 fold of the environmental radiation level, with a total dose of 8 Gy. No detectable increase of tumor incidence was observed in the mice irradiated, although the multiplicity of tumors was increased [25]. If these results are compared with the estimated value using the LNT model from 0.5% cancer increase with 100 mSv with DDREF =2, the discrepancy is obvious. Mice irradiated with 8 Gy are expected to develop tumors with a frequency of 20%, contrary to the experimental results.

4-3. Internal Whole Body Exposure Data

Internal radiation from radioactive cesium and iodine is also of great concern after the Fukushima accident. Radioactive cesium incorporated in the body is uniformly distributed in the body, a case of whole body low LET exposure. Since the dose-response relationship for cancer from internally deposited radioactive cesium is not available, a series of precise experiments with mice orally administered with radioactive tritiated water by Yamamoto et al. [28] is most valuable for assessing the cancer risk of whole body low LET radiation from internal emitters. They demonstrated the D_{nt} = 700 mGy at an estimated dose rate of 6.4×10^{-7} Gy/min for thymic lymphoma induction. Yamamoto et al. proposed the idea of a zero equivalent point (ZEP) where the dose rate of tritium β rays exerts zero effect in cancer induction.

Radioactive iodine deposits in thyroid, a target organ. This case is fitted to the category of partial body low LET radiation, and cannot be directly compared with the whole body exposure from A-bomb radiation. Briefly, radioactive iodine injected to rats induced thyroid adenomas in a non-linear dose-response manner with D_{nt} = 3.3 Gy for dose rate of 1.7×10^{-4} Gy/min [29]. From internal high LET exposure experiments with beagle dogs by Raabe and Lloyd et al. [31, 32], the idea of a practical threshold was presented. From internal low LET exposure experiments, Brooks et al. proposed the value 20-35 for the dose-rate effectiveness factor [33]. Other human data of the dose response relation for partial body exposure were obtained for bone tumors in radium dial painters [34], liver cancers in thorotrast- administered patients [35], and secondary soft tissue sarcomas in patients after radiation therapy [36]. All of them showed a non linear and threshold-like dose response for cancer induction. These cases should be analyzed separately from whole body exposure cases. Moreover, it should be noted that the dose rate effect exists even for high LET radiation at low dose rates.

4-4. Non Tumor Dose and DDREF as a Function of Dose Rate

Focusing on whole-body exposure to low LET radiation to simulate A-bomb radiation, the non tumor dose was surveyed in human and experimental data and listed together with the corresponding dose rate in Table 6-1. A plot of D_{nt} against dose rate on the logarithmic scale gives the relation between D_{nt} and dose rate (Figure 6-3a).

Table 6-1. Whole body low LET exposure data for non tumor dose versus dose rate of radiation surveyed in literature

Exposure condition	Data Number	Subject		Radiation	Tumor	Dose rate, Gy/min	Non tumor dose, D_{nt}, Gy	Reference
External whole body exposure	1	Mouse	RFM/Un male	γ ray	myeloid leukemia	3 x 10^{-5}	1,5	Upton et al. [2]
	2	"	female	"	"	5 x 10^{-6}	2,5	[2]
	3	"	RFM/Un	"	thymic lymphoma	0,45	0,1	Ulrich et al. [15]
	4	"	"	"	Harderian tumor	0,45	0,1	"..[15]
	5	"	"	"	uterine	0,45	0,25	" [15]
	6	"	"	"	mammary	0,45	0,25	" [15]
	7	"	"	"	myeloid leukemia	0,45	0,25	" [16]
	8	"	"	"	reticulum cell sarcoma	0,45	> 3	" [15, 16]
	9	"	"	"	ovarian	0,45	0,1	" [15, 17]
	10	"	"	"	pituitary	0,45	0,25	" [15, 17]
	11	"	"	"	lung adenoma	0,45	2	" [15, 17]
	12	"	"	"	thymic lymphoma	0,45	0,1	" [18]
	13	"	"	"	thymic lymphoma	5.8 x 10^{-5}	0,5	" [18]
	14	"	"	"	ovarian	5.8 x 10^{-5}	0,5	" [18]
	15	"	BALB/c	"	lung adenoma	0,4	0,1	" [19]
	16	"	"	"	ovarian	0,4	0,1	" [19]
	17	"	"	"	thymic lymphoma	4	2	Maisin et al. [20]
	18	"	BC3F1	X ray	hepatocellular carcinoma	1,3	0,5	Di Majo et al. [21]
	19	"	"	"	solid tumor	6 x 10^{-2}	0,64	Covelli et al. [22]
	20	"	CB17	γ ray	thymic lymphoma	0,5	1	Ishii-Ohba et al. [23]
	21	"	C57BL/6	"	"	2 x 10^{-5}	> 7.2	Ina et al. [24]
	22	"	B6C3F1	"	no tumor increase	3.8 x 10^{-8}	> 0.02	Tanaka III et al. [25]
	23	"	"	"	"	7.7 x 10^{-7}	> 0.04	"
	24	"	"	"	"	1.5 x 10^{-5}	> 8	"
	25	Rat	WAG/Rij	"	mammary carcinoma	4 x 10^{-4}	1	Bartsra et al. [26]
	26	Dog	Beagle	"	myeloproliferative disease	[a]2 x 10^{-6}	8,6	Thompson [27]
	27	Human	Yangjiang, China	"	no tumor increase	[a]4.05 x 10^{-9}	> 0.15[b]	Wei et al. [13]
	28	"	Kerala, India	"	no tumor increase	7.55 x 10^{-9}	> 0.28[b]	Nair et al. [14]
	29	"	Hiroshima-Nagasaki	A–bomb γ ray + (neutron)	leukemia	(0.12-120)[c] (6x10^{5}-6x10^{8})[d]	0,2	Shimizu et al [11]
	30	"	"	"	solid cancer	"	0,1	Preston et al. [12]
Internal whole body exposure	31	Mouse	BC3F1	Tritium β ray, oral	thymic lymphoma	6.4 x 10^{-7}	0,71	Yamamoto et al. [28]
Internal partial body exposure	32	Rat	Long-Evans	^{131}I β ray, injected	thyroid adenoma	1.7 x 10^{-4}	3,3	Lee et al. [29]

[a] The increase of radiation from the ground was incorporated in total dose.

[b] Exposure time: 70 years.

[c] Exposure time assumed to be 1 μsec.

[d] Exposure time assumed to be 5 sec.

A-bomb data were well fitted on the regression line when the dose rate of A-bomb radiation was estimated from the median exposure time of 5 sec. The figure shows an inverse relation between D_{nt} and dose rate, i. e., the lower the dose rate the higher the D_{nt} level and provides a measure of the cancer risk or the tolerance level corresponding to the dose rate. Regression line fitted to the data is:

$D_{nt} = 0.258X^{-0.141}$, $R^2 = 0.320$ for whole body low LET exposure,

where D_{nt} is non tumor dose in Gy and X is dose rate between 10^{-9} and 1 Gy/min [7].

From this equation, D_{nt} values can be calculated for the corresponding dose rate. The ratio of D_{nt} values between A bomb radiation (0.258 Gy) and environmental radiation (4.26 Gy) is 16.5. Therefore, the DDREF value is 16.5 for application of the cancer risk of A-bomb radiation to environmental radiation.

For other exposure conditions, regression lines are given by the previous analysis [7],

$D_{nt} = 0.0207X^{-0.0733}$, $R^2 = 0.781$ for whole body high LET exposure
$D_{nt} = 2.69X^{-0.0857}$, $R^2 = 0.147$ for partial body low LET exposure
$D_{nt} = 0.0439X^{-0.167}$, $R^2 = 0.303$ for partial body high LET exposure.

The above four equations cover all possible exposure conditions and cancer risks of radiation.

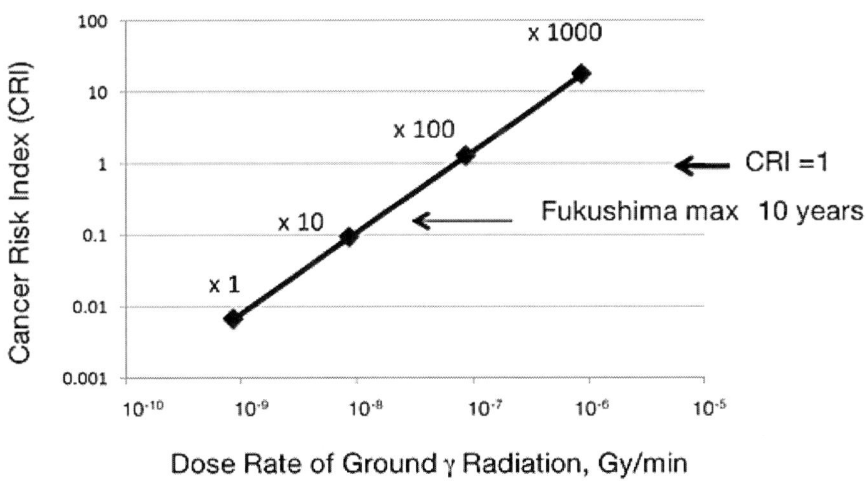

Figure 6-4. Cancer risk index, CRI, defined as the ratio of total dose of radiation to the non tumor dose, versus dose rate. The graph shows how much increase of radiation from the contaminated ground approaches the threshold of cancer risk.

4.5 Cancer Risk Index

Although D_{nt} values can be used as a measure of the cancer risk of radiation, the ratio of D_{nt} to a total dose of radiation received by the subject provides a clearer index and is defined as cancer risk index (CRI) [37]. Figure 6-4 is a plot of CRI against a dose rate of γ radiation from the ground, assuming that low LET radiation is received for 70 years for calculation of total dose. Figure 4 shows that CRI reaches 1 when γ radiation from the ground increases 100 fold. The highest level of contamination in the Namie/Iitate area outside the restricted zone in Fukushima is 5 μSv/h.The procedure to evaluate the CRI value for this condition is illustrated in Table 6-2, such as 1) classify exposure condition: whole body low LET exposure, 2) estimate dose rate in Gy/min, 3) estimate total dose for exposure time (10 years taken for calculation), 4) estimate D_{nt} from the dose rate by the equation, and 5) divide total dose by D_{nt} to obtain CRI. In the Fukushima case, CRE is 0.17. However, if the exposure time is 70 years, CRI comes close to 1. On the contrary, calculation from the LNT model with DDREF = 2 and 0.5% cancer increase with 100 mSv gives 1.2% cancer increase for 10 years, or 8.4% for 70 years in Namie/Iitate if the residents lived there.

Table 6-2. Practice to estimate cancer risk index (CRI) in Fukushima

Maximum dose rate level in Namie/Iitate, Fukushima	Non tumor dose[a] Gy	(A) Total dose, 10 yr Gy	(B) CRI (B/A)
5 μSv/hr[b]	2.57	0.44	0.17

[a] Non tumor dose, D_{nt}, was calculated by the equation $D_{nt} = 0.258\ X^{-0.141}$ for whole body low LET exposure.

[b] Maximum radiation level outside the 20 km zone in Fukushima in September, 2014.

5. Further Problems

This chapter focused on the topic on whole body low LET radiation with attention given to the applicability of A-bomb data to other radiation exposure cases. Even in this category, the radiation cancer risk varies in a wide range depending upon the dose rate of radiation. Furthermore, the dose rate problem extends to other exposure categories, i. e., whole body high LET radiation such as heavy particles in galactic cosmic rays in space travel, partial body low LET radiation such as medical radiation in cancer therapy and diagnosis, and partial body high LET radiation such as heavy charged particles in internal exposure to radon, radium, and thorotrast α rays. The cancer risk of radiation should be assessed separately in each category. Further experimental data corresponding to each exposure condition are needed to obtain the whole scheme of cancer risk from radiation.

Acknowledgments

I wish to thank Dr. Tetsuji Imanaka, Kyoto University, for valuable information on the dose rate of A-bomb radiation.

REFERENCES

[1] Tanaka S (2012) Accident at the Fukushima Dai-Ichi nuclear power stations of TEPCO-Outline & lessons learned. *Proceedings of Japan Academy of Science*, Ser. B 88: 471–484.

[2] Upton AC, Randolph ML, Conklin JW (1970) Late effects of fast neutrons and gamma-rays in mice as influenced by the dose rate of irradiation: *Induction of neoplasia. Radiation Research* 41: 467-491.

[3] United Nations Scientific Committee on the Effects of Atomic Radiation (UNCEAR) (1993) Influence of dose and dose-rate on stochastic effects of radiation. In: Sources and Effects of Ionizing Radiation. Report to the General Assembly, Annex F. New York: United Nations.

[4] United Nations Scientific Committee on the Effects of Atomic Radiation (2000) Biological effects at low radiation doses: Models, mechanisms and uncertainties. In: Sources and Effects of Ionizing Radiation: Genetic and Somatic Effects of Ionizing Radiation. Report to the General Assembly, Annex I. New York: United Nations.

[5] International Commission on Radiological Protection (1991) 1990-Recommendation of the International Commission on Radiological Protection. Publication 60. *Annals of the ICRP*. Oxford: Pergamon Press.

[6] Tanooka H (2001) Threshold dose-response in radiation carcinogenesis: an approach from chronic β-irradiaiton experiments and a review of non-tumour doses. *International Journal of Radiation Biology* 77: 541-551.

[7] Tanooka H (2011) Meta-analysis of non-tumour doses for radiation-induced cancer on the basis of dose-rate. *International Journal of Radiation Biology* 87: 645-652.

[8] Roesch WC, editor (1987) US-Japan Joint Reassessment of Atomic Bomb Radiation Dosimetry in Hiroshima and Nagasaki, DS86. *Radiation Effects Research Foundation.* Hiroshima: Radiation Effects Foundation.

[9] Young RW, Kerr GD editors (2005) Reassessment of the Atomic Bomb Radiation Dosimetry for Hiroshima and Nagasaki (DS02) vol. 1 and vol. 2. Hiroshima: *Radiation Effects Research Foundation.*

[10] Kerr GD, Pace III JV, Mendelsohn E, Loewe WE, Kaul DC et al. (1987) Transport of initial radiations in air over ground. In: Roesch WC editor, US-Japan Joint Reassessment of Atomic Bomb Radiation Dosimetry in Hiroshima and Nagasaki. Chapter III. Vol. 1. Hiroshima: *Radiation Effect Research Foundation*.

[11] Shimizu Y, Kato H, Schull WJ (1990) Studies of the mortality of A-bomb survivors. Mortality, 1950-1985: Part 2: Cancer mortality based on the recently revised doses (DS86). *Radiation Research* 121: 120-140.

[12] Pierce DA, Preston DL (2000) Radiation-related cancer risks at low doses among atomic bomb survivors. *Radiation Research* 154: 178-186.

[13] Chen D, Wei LX (1991) Chromosome aberration, cancer mortality and hormetic phenomena among inhabitants in areas of high background radiation in China. *Journal of Radiation Research* 32 suppl. 2: 46-53.

[14] Nair MK, Nambi KSV, Amma NS, Gangadharan P, Jayalekshmi P et al. (1999). Population study in the high natural background radiation area in Kerala, India. *Radiation Research* 152: S145-148.

[15] Ullrich RL, Jernigan MC, Cosgrove GE, Satterfield LC, Bowles ND et al. (1976) The influence of dose and dose-rate on the incidence of neoplastic disease in RFM mice after neutron irradiation. *Radiation Research* 68: 115–131.

[16] Ullrich RL, Storer JB (1979) Influence of γ-irradiation on the development of neoplastic disease in mice. I. Reticular tissue tumors. *Radiation Research* 80: 303–316.

[17] Ullrich RL, Storer JB (1979) Influence of γ-irradiation on the development of neoplastic disease in mice. II. Solid tumors. *Radiation Research* 80: 317–324.

[18] Ullrich RL, Storer JB (1979) Influence of γ-irradiation on the development of neoplastic disease in mice. III. Dose-rate effects. *Radiation Research* 80: 325–342.

[19] Ullrich RL (1983) Tumor induction in BALB/c mice after fission neutron or γ-irradiation. *Radiation Research* 93: 506–515.

[20] Maisin JR, Wambersie GB, Gerber GB, Guelette J, Mattelin G et al. (1983) The effects of a fractionated gamma irradiation on life shortening and disease incidence in BALB/c mice. *Radiation Research* 94: 359–373.

[21] Di Majo V, Coppola M, Rebessi S, Bassani B, Alati T et al. (1986) Radiation-induced mouse liver neoplasms and hepatocyte survival. *Journal of National Cancer Institute* 77: 933–939.

[22] Covelli V, Coppola M, Di Majo V, Rebessi S, Bassani B et al. (1988) Tumor induction and life shortening in BC3F1 female mice at low doses of fast neutrons and X rays. *Radiation Research* 113: 362–374.

[23] Ishii-Ohba H, Kobayashi S, Nishimura M, Shimada y, Tsuji H et al. (2007) Existence of a threshold-like dose for γ ray induction of thymic lymphomas and no susceptibility to radiation-induced solid tumors in SCID mice. *Mutation Research* 619: 124–133.

[24] Ina Y, Tanooka H, Yamada T, Sakai K (2005) Suppression of thymic lymphoma induction by life-long low dose-rate irradiation accompanied by immune activation in C57BL/6 mice. *Radiation Research* 163, 153-158.

[25] Tanaka IB III, Tanaka K, Ichinohe K, Matsushita S, Matsumoto T et al. (2007). Cause of death and neoplasia in mice continuously exposed to very low dose rates of gamma rays. *Radiation Research* 167: 417-437.

[26] Bartsra RW, Bentvelzen PAJ, Zoetelief J, Mulder AH, Broerse JJ et al. (2000) The effects of fractionated gamma irradiation on induction of mammary carcinoma in normal and estrogen-treated rats. *Radiation Research* 153: 557–569.

[27] Thompson RC (1989) Life-span effects of ionizing radiation in the beagle dog. A summary account of four decades of research funded by the US Department of Energy and its predecessor agencies. *Pacific Northwest Laboratory Report* PNL-6822 UC-408. pp. 1–323.

[28] Yamamoto O, Seyama T, Itoh H, Fujimoto N (1998) Oral administration of tritiated water (HTO) in mouse. III. Low dose-rate irradiation and threshold dose-rate for radiation risk. *International Journal of Radiation Biology* 73: 535–541.

[29] Lee W, Chiacchierini RP, Shleien B, Telles NC (1982) Thyroid tumours following ^{131}I or localized X irradiation to the thyroid and pituitary glands in rats. *Radiation Research* 92: 307-319.

[30] Kaplan HS, Brown MB (1952) A quantitative dose-response study of lymphoid-tumor development in irradiated C57 black mice. *Journal of National Cancer Institute* 13: 185–208.

[31] Raabe O (2011) Toward improved radiation safety standards. *Health Physics* 101: 84-93.

[32] Lloyd RD, Taylor GN, Fisher DR, Schlenker RA, Miller SC (2000) Effective thresholds for induction of skeletal malignance by radionuclides. *Health Physics* 79: 722-727.

[33] Brooks AL, Eberlein PE, Couch LA, Boecker BB (2009) The role of dose-rate on risk from ineternally-deposited radionuclides and the potential need to separate dose-rate effectivensee factor (DREF) from the dose and dose-rate effectiveness factor (DDREF). *Health Physics* 97: 458-469.

[34] Rowland RE, Stehney AF, Lucas HF (1978) Dose-response relationships for female radium dial workers. *Radiation Research* 76: 368–383.

[35] Anderson M, Storm HH (1992) Cancer incidence among Danish thorotrast-exposed patients. *Journal of National Cancer Institute* 84: 1318–1325.

[36] Tubiana, M. Diallo, I. Chavaudra, J, Lefkopoulos D, Bourhis J et al. (2011) A new method of assessing the dose-carcinogenic effect relationship in patients exposed to ionizing radiation. A concise presentation of preliminary data. *Health Physics* 100: 296-299.

[37] Tanooka H (in press) Dose Rate: A Critical Factor in Determining the Cancer Risk of Radiation at an Environmental Level. In Mishra KP editor, Radiation Exposure: Monitoring Systems, Long-Term Health Effects and Methods for Protection. New York: *Nova Scientific*.

In: Fukushima Nuclear Accident
Editor: Shizuyo Sutou

ISBN: 978-1-63463-748-0

Chapter 7

ROLE OF THE KEAP1-NRF2 PATHWAY IN PROTECTION AGAINST IONIZING RADIATION

Atsushi Maruyama and Ken Itoh*
Department of Stress Response Science,
Hirosaki University Graduate School of Medicine,
Zaifu-cho, Hirosaki, Japan

ABSTRACT

From the beginning of life, organisms on earth are assumed to have encountered harsh environments enriched with reactive substances, such as electrophiles and reactive oxygen species, that are partly generated by environmental stresses, such as solar UV and ionizing irradiation. However, they adapted to the stresses by acquiring the intricate cytoprotective mechanisms of antioxidant enzymes and proteins, such as catalase and glutathione. Nrf2 plays an important role in oxidative stress response in mammalian cells by regulating the expression of a battery of cytoprotective genes related to glutathione metabolism, antioxidant proteins and drug detoxifying enzymes, etc. Recent studies have shown that Nrf2 activators can ameliorate various oxidative stress-related diseases in animal models by restoring redox homeostasis. Nrf2 activator dimethyl fumarate is now clinically used to treat multiple sclerosis.

Ionizing radiation also activates Nrf2. Recent studies have shown that the Nrf2-mediated gene regulatory network confers resistance against ionizing radiation in normal cells and cancer cells, giving Nrf2 a double-faced character to human health. In normal tissues, Nrf2 not only regulates cytoprotective genes, but also activates the regulatory nexus, including the Notch pathway that increases hematopoiesis in mice after total body irradiation.

Thus, Nrf2 can be a promising molecular target that ameliorates susceptibility to acute ionizing radiation.

* Correspondence to: Ken Itoh, MD, PhD, Tel: +81-172-39-5158, Fax: +81-172-39-5157, Email: itohk@cc.hirosaki-u.ac.jp

Introduction

Substances are chemically divided into electrophilic, neutral and nucleophilic substances. Electrophiles are molecules that have electron-deficient centers and thereby make adducts with nucleophilic biomolecules, including proteins, DNA, and lipids. During evolution, organisms evolved an intricate defense system against reactive electrophiles mainly involving glutathione. Reactive Oxygen Species (ROS) are generated in the organism by environmental stresses, such as UV radiation, ionizing irradiation and infection, as well as by normal physiological activities, such as cellular respiration.

Organisms use ROS as cellular signaling molecules generated from mitochondria or membrane-bound NADPH oxidase. Both ROS and electrophiles are reactive with cellular thiols and an excessive amount of them provokes redox unbalance leading to oxidative damage of cellular components.

Nuclear factor erythroid 2-related factor 2 (Nrf2) activated by ROS and electrophiles restores redox homeostasis or further mediates health-promoting effects of low amounts of these stresses.

In this chapter, we introduce the Kelch-like ECH-associated protein 1 (Keap1)-Nrf2 system and we discuss the role of Nrf2 in the protection against ionizing irradiation and its possible role in the hormetic effect of low-dose irradiation.

1. Keap1-Nrf2 Stress Response Pathway

1.1. Brief Overview of the Keap1-Nrf2 Response

Nrf2 belongs to the Cap'n'collar (CNC) transcription factor family, which is characterized by a highly conserved basic region, leucine-zipper structure and a CNC domain just upstream of the basic region [1]. It regulates the expression of many antioxidant genes to protect against toxicities of environmental electrophiles and ROS. Under homeostatic conditions, Keap1, which acts an adaptor for a Cul3-based E3 ligase, degrades Nrf2 by the ubiquitin proteasome pathway (Figure 7-1) and represses Nrf2 activity. Electrophiles or ROS attenuates the Keap1-mediated Nrf2 turnover, leading to nuclear Nrf2 accumulation and heterodimerization with its obligate DNA binding partner, small Maf proteins, to bind to antioxidant responsive elements (ARE) in the gene regulatory region of target genes.

Nrf2 regulates a battery of genes that participate in regulating the synthesis and conjugation of glutathione, antioxidant proteins and enzymes, drug metabolizing enzymes or drug transporters, proteasome subunits, NADPH-generating enzymes, and enzymes that participate in nucleotide synthesis (Figure 7-2) [1].

Although *Nrf2* knockout (KO) mice are viable under normal laboratory conditions, basal as well as inducible expressions of Nrf2 target genes are decreased and they are susceptible to various chemicals and oxidative stress [1]. However, *Keap1* KO mice, in which Nrf2 is constitutively activated, die postnatally because of hyperkeratosis in the esophagus and forestomach [1]. Electrophiles activate Nrf2 because Keap1 is inactivated after modification of its reactive cysteine [1]. Pharmacological activation of Nrf2 confers on mice protection against various environmental stresses and diseases [2].

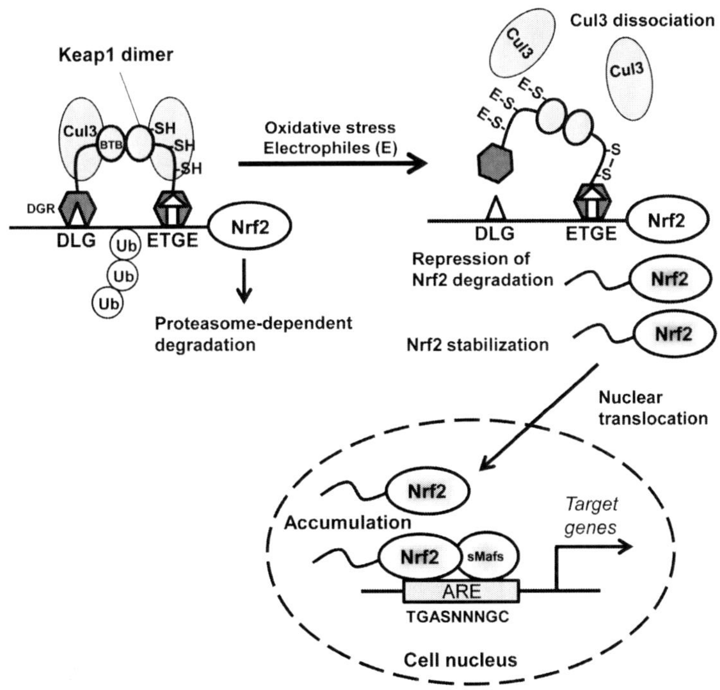

Figure 7.1. Regulation of Keap1-Nrf2 pathway.

Details are described in the text.

BTB: bric-a-brac, tramtrack, broad complex domain, DGR: double glycine repeat domain, Ub: ubiquitin, DLG: DLG motif, ETGE: ETGE motif, Cul3: cullin3, ARE: antioxidant responsive element, and sMafs: small maf proteins.

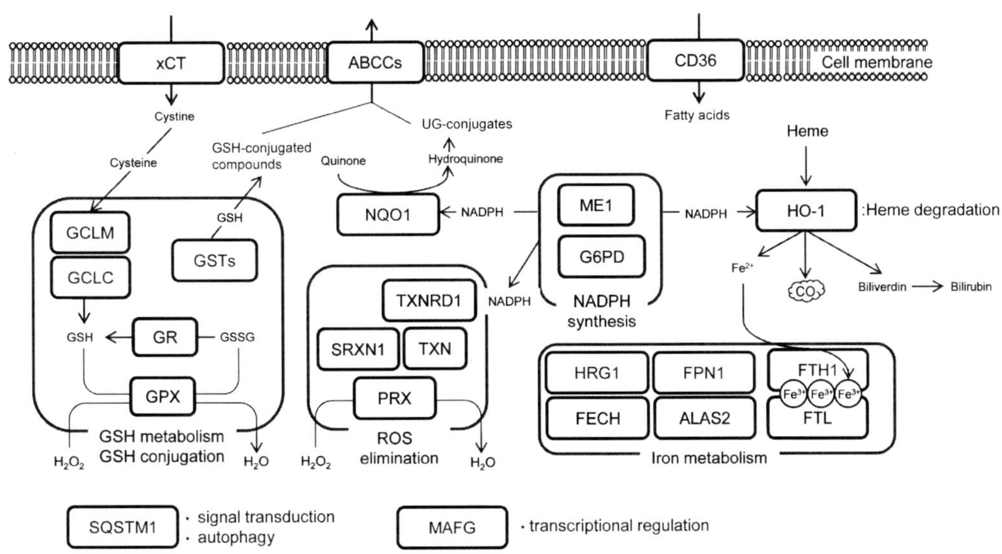

Figure 7.2. Nrf2 target genes mainly of antioxidant function.

Reported Nrf2 target proteins and those activities are shown and surrounded by rectangles.

xCT: cystin transporter, ABCCs: ATP binding cassette protein family C proteins, CD36: scavenger receptor CD36, GCLM: glutamate-cysteine ligase, modifier subunit, GCLC: glutamate-cysteine ligase, catalytic subunit, GR: glutathione reductase, GPX: glutathione peroxidase, NQO1: NAD(P)H quinone oxidoreductase 1, TXNRD1: thioredoxin reductase 1, TXN: thioredoxin, SRXN1: sulfiredoxin 1, PRX: peroxiredoxin, ME1: malic enzyme 1, G6PD: glucose-6-phosphate dehydrogenase, HO-1: heme oxygenase-1, HRG1: heme regulated gene 1, FPN1, ferroportin 1, FECH: ferrochelatase, ALAS2: delta-aminolevulinate synthase 2, FTH1: ferritin heavy chain, FTL: ferritin light chain, SQSTM1: sequestosome 1, MAFG: small MAF protein G, GSH: glutathione, NADPH: reduced nicotinamide adenine dinucleotide phosphate, UG-conjugates: uridine diphosphate glucuronide acid-conjugates.

1.2. Historical Aspects of the Keap1-Nrf2 Pathway and Relation to Hormesis

The physiological function of Nrf2 was first described as a coordinate regulator of electrophile-detoxifying phase II drug detoxifying genes [3]. Phenolic antioxidants, such as butylated hydroxyanisol (BHA), are used widely as food preservatives in the USA. BHA protects mice against various classes of carcinogens [4].

Most carcinogens are electrophiles or are converted *in vivo* to electrophiles, to bind to DNA [5]. We showed that the expression of several isoforms of glutathione-*S*-transferase is strongly induced in the liver or the intestine of wild-type mice, but not in *Nrf2* KO mice upon exposure to BHA [3]. Today, BHA is widely believed to be metabolized to electrophiles (i.e., *tert*-butyl hydroquinone (*t*BHQ)) *in vivo* and to activate Nrf2. Thus, we believe our results were the first *in vivo* demonstration that Nrf2 is essential for gene induction of electrophile-detoxifying enzymes in response to electrophiles. We showed later that Nrf2 mediates chemoprotective effects of the Nrf2 activator, named oltipraz, against benz[a]pyrene-induced or nitrosamine-induced carcinogenesis [6,7]. In the toxicology field, hormesis refers to the biphasic dose response in which a low dose exerts beneficial effects but a high dose exerts deleterious effects [8].

In the field of biology or medicine, hormesis generally refers to preconditioning events in which a subtoxic dose of a substance provokes an endogenous cytoprotective response and confers resistance against later higher amounts of the same toxicants. In the Nrf2 case, fewer toxic electrophiles, such as *t*BHQ, confer resistance against tumor-promoting effects of carcinogens.

In hormesis studies, a frequently observed phenomenon is that exposure to low levels of one type of weak toxin can protect cells against more than one class of stress. Consistent with this rule, Nrf2 up-regulation by natural and synthetic activators protects not only against carcinogens, but also against various environmental electrophiles or several diseases [2]. Forman et al. proposed to name the Nrf2-mediated hormesis-like event by neutraceuticals as "para-hormesis", because a hermetic initiator is not mild stress, but a health-promoting phytochemical [9].

1.3. Regulatory Mechanisms of Nrf2 Activation by Electrophiles

Keap1 shows structural similarity to the Drosophila Kelch protein and has two canonical protein interaction domains: the bric-a-brac, tramtrack, broad (BTB) complex and Kelch (also called double glycine repeat or DGR) (Figure 7-1) [10]. The Keap1 DGR domain directly interacts with two Nrf2 Neh2 motifs (i.e., DLG and ETGE motifs), such that the Keap1 dimer recognizes one molecule of Nrf2 (two-site recognition model or hinge & latch mechanism) [11]. The ETGE motif is a high affinity binding site (hinge site) by two orders and the DLG motif has lower affinity to Keap1 (latch site). The BTB domain is required for Keap1 homodimerization and therefore for repression of Nrf2 [12]. Keap1 acts as an adaptor molecule for Cul3-based E3 ligase and degrades Nrf2 by the ubiquitin proteasome pathway [13]. Keap1 has 25 and 27 cysteines in humans and mice, respectively [10]. Several of these cysteines, including cysteine 273 and cysteine 288, of mouse Keap1 are highly reactive due to adjacent basic amino acids. Talalay et al. showed later that the potencies of the chemicals to act as phase 2 inducers *in vivo* parallel their potencies to bind to recombinant Keap1 protein *in vitro*, indicating that Keap1 is a long-sought sensor molecule for electrophiles [14]. Binding of electrophiles to Keap1 leads to conformational change of Keap1 that disrupts Nrf2 degradation activity of Keap1. Currently, these cysteines are thought to show certain specificities against different electrophiles depending on the strength of the electrophilicity and the three dimensional structures of the chemicals [15].

1.4. Keap1-Independent Pathway for Nrf2 Activation

In addition to the Keap1-mediated mechanism, several kinases participate in activation of Nrf2 in a cell type-specific manner [16]. Among them, the AKT pathway activates Nrf2 by repressing GSK3β (Glycogen synthase kinase 3β)-mediated Nrf2 degradation by β–TrCP (β-transducin repeat containing protein) [17]. Recently, p62 has been identified as an Nrf2 activator [18-20]. Importantly, p62 has the STGE motif similar to Nrf2 ETGE that is the high affinity binding site for Keap1. Serine phosphorylation of the STGE motif is postulated to switch low affinity binding sites for Keap1 to higher affinity sites. Ichimura et al. showed that the p62 STGE motif is phosphorylated in an mTORC1 (mammalian target of rapamycin complex 1)-dependent manner during selective autophagy, such as mitophagy and xenophagy, leading to Keap1 sequestration into autophagosomes and Nrf2 activation [21]. The p62 is also indispensable for Keap1-independent activation of Nrf2 in response to arsenite [22].

1.5. Nrf2 Protects against Ischemic Reperfusion Injury — Its Possible Contribution to Ischemic Preconditioning

Mild ischemia protects neurons against later more severe ischemic injury, which is called ischemic preconditioning [23]. However, prophylactic treatment of Nrf2 inducers protects mice against ischemic reperfusion injuries [24,25]. Karen et al. showed that pretreatment with a physiologically relevant dose of H_2O_2 protects neurons in the *in vitro* model of ischemia reperfusion injuries (i. e., oxygen-glucose deprivation followed by reperfusion) [26]. Takagi et al. showed that brain ischemia-reperfusion activates Nrf2 and its target gene heme oxygenase-1 (*HO-1*) in the penumbra [27]. Bell et al. showed that the *in vivo* preconditioning protocol activates Nrf2 target genes related to glutathione synthesis, such as glutamate-cysteine ligase, catalytic subunit (*Gclc*) and cystine transporter (*xCT*) [28]. Ischemic reperfusion in the myocardium also activates Nrf2 [29]. Thus, Nrf2 appears to participate in ischemic preconditioning in the brain or the heart, which requires further verification by future experiments.

2. Nrf2 Protection against Ionizing Radiation

The Keap1-Nrf2 system contributes to oxidative stress response and to detoxification of toxic compounds. Although the cellular anti-oxidative stress system participates in the sensitivity of ionizing radiation, McDonald et al. first showed by using *Nrf2* KO mice that Nrf2 confers radioresistance [30]. However, recent studies showed that the Keap1-Nrf2 system confers on cancer cells resistance against chemotherapy and radiation. In this section, we explain the putative mechanism of how Nrf2 is activated in response to radiation exposure and then introduce both the "good face" and the "bad face" of Nrf2 in response to ionizing radiation.

2.1. Activation of Nrf2 by Ionizing Irradiation

2.1.1. Cell type-Specific Sensitivity of Nrf2 Activation by Ionizing Irradiation

McDonald et al. showed that in human breast cancer cell line MCF7, Nrf2 activation by ionizing irradiation does not occur until after 3 days, but only occurs at 5 days after irradiation [30]. This delayed response appears to be also true for fibroblast cell line NIH3T3 and dendritic cell line DC2.4. They showed that ROS formation is activated at day 5, but not at day 1, after mouse embryonic fibroblast (MEF) cells were exposed to irradiation, indicating that delayed ROS production participates in Nrf2 activation. Tsukimoto et al. showed that low-dose γ-ray irradiation (~0.1 Gy) induces Nrf2 activation in mouse macrophage cell line RAW264.7 at 4 hours after irradiation and its activation seemed to be dependent upon ERK1/2 pathway [31]. Yoshino et al. also described that 5 Gy, but not 1-2Gy, X-ray irradiation activates Nrf2 at 6 hours after irradiation and induces *HO-1* gene expression in human macrophage cell line THP1 [32]. Lee et al. showed that low-dose X-ray irradiation (~0.5 Gy) activates Nrf2 and its family factor Nrf1 as early as 2 hours after irradiation through generation of ROS in human skin fibroblasts [33]. Therefore, the dynamics and activation mechanism may differ depending on the cell type.

2.1.2. Generation of ROS and Putative Mechanism of Nrf2 Activation by Ionizing Irradiation

Cell damage caused by ionizing radiation is primarily mediated by two mechanisms: 1) the energy of ionizing radiation is accumulated in biomolecules, including nucleic acids, proteins and lipids, resulting in their disruption; 2) ionizing radiation disrupts water molecules, which is termed radiolysis, and disrupted intracellular water converts into ROS leading to damage of biomolecules [34].

Although Fourquet et al. showed that H_2O_2 promotes intramolecular disulfide formation in Keap1, this may be context-dependent [35]. Thus, still under debate is whether ROS directly reacts with Keap1 in the physiological condition, and we surmise that ROS-mediated oxidation products, including lipid peroxidation products, are candidates for Keap1 inactivation by ionizing radiation.

Ionizing radiation is a lipid peroxidation inducer that oxidizes polyunsaturated fatty acids of cellular membranes, and then finally converts the lipid peroxidation products into reactive aldehydes. The most characterized and relevant lipid peroxidation product is 4-hydroxynonenal (HNE) [36] that has anti-proliferative and pro-apoptotic effects in cultured cells and tissues. HNE has a long half-life compared with ROS, and thus it is considered to be a second messenger of oxidative stress [37]. HNE is a highly reactive electrophile and several studies have shown that HNE is a potent NRF2 inducer [38,39]. Levonen et al. showed that HNE can react with cysteine residues in the Keap1 protein, and that it may disrupt the Keap1-dependent degradation of Nrf2 [40]. Further study by McMahon et al. showed that HNE directly reacts with at least two specific cysteine residues in Keap1, leading to Nrf2 protein stabilization [41]. These data indicate that lipid peroxidation products by ROS are potential activators of Nrf2 in the context of ionizing radiation exposure.

2.2. Protective Effect of Nrf2 Activation against Ionizing Radiation in Normal Cells and *In Vivo*: Good Face of Nrf2 Activation

2.2.1. Protective Effect of Nrf2 Activating Phytochemicals against Ionizing Irradiation

Plant-derived polyphenols and electrophilic compounds show a protective effect by reducing ROS toxicity after radiation exposure and promote cell survival. Ethanol extract of maize silks, dried cut stigmas of maize female flowers, activates Nrf2 in the liver and and has a protective effect on radiation-induced oxidative stress in mice [42]. Dietary intake of curcumin contained in turmeric ameliorates radiation-induced pulmonary fibroblasts and increases mouse survival, while it does not protect Lewis lung carcinoma pulmonary metastases against killing by radiation [43]. However, radiation-induced cognitive dysfunction by radiotherapy has been reported. A recent study by Xie et al. explored the relationship between the Nrf2 pathway and heavy ion irradiation-induced cognitive deficits [44]. They observed that whole body heavy iron irradiation disrupts spatial learning, memory and structure of the brain in mice, but Nrf2 activator curcumin significantly reduced irreversible cerebral damage mediated by irradiation and attenuated oxidative stress.

Paeoniflorin is a bioactive monoterpene glucoside extracted from *Paeonia lactiflora* root. This glucoside has a protective activity against γ–irradiation induced oxidative injury by Nrf2 activation in human endothelial cells [45]. Sulforaphane is one of the most famous Nrf2 activators derived from broccoli and its sprout, and repeated stimulation by sulforaphane protects fibroblasts from ionizing radiation [46].

Epigallocatechin-3-gallate (EGCG), which is the most abundant catechin in green tea and is an Nrf2 inducer. EGCG treatment prevents HaCaT cell damage from ionizing radiation [47]. Ghosh et al. showed that pre-treatment with black tea extract significantly reduces radiation-induced damage, including loss of cell viability, ROS generation, mitochondrial dysfunction and apoptosis in normal lymphocytes compared with K562 lymphoma cells [48]. Consequently, up-regulation of *Nrf2* mRNA is observed by black tea extract administration. Pro-oxidant 1, 4-naphthoquinone alleviates radiation-induced apoptosis by activating Nrf2 and the calcium-induced ERK pathway in mice [49]. The selenium compound Se-methylselenocysteine accumulates in plants, including broccoli, garlic and onion. Se-methylselenocysteine has a preventive effect on γ–irradiation-induced oxidative stress in rat lungs [50].

2.2.2. Protective Effects of Synthetic Nrf2 Activators Against Ionizing Irradiation

Recently, the effects of synthetic Nrf2 activators against ionizing radiation have been investigated. Kim et al. showed that Nrf2 activation by synthetic triterpenoides, bardoxolone methyl (BARD) and 2-cyano-3,12-dioxooleana-1,9 (11)-dien-28-oic acid ethyl amide protect non-tumorigenic normal immortalized human colonic epithelial cells (HCEC) and colon tissue of mice against ionizing radiation-induced damage [51]. Reisman et al. described that synthetic triterpenoid RTA408 protects mice from radiation-induced dermatitis through up-regulation of Nrf2-target genes [52]. Synthetic chalcone derivative 2-trifluoromethyl-2'-methoxychalone (TMC) is also a potent Nrf2 activator [53]. Kim et al. recently showed that TMC-mediated Nrf2 activation increases hematopoietic stem progenitor cell function and mitigates ionizing radiation-induced myelosuppression and mortality in mice [54]. This result indicates that Nrf2 participates in promoting hematopoietic reconstitution after irradiation-induced myeloablative damage. Interestingly, Nrf2 expression decreases with aging. Mathew et al. described that age-related reduction of macrophage migration inhibition factor (MIF) affects irradiation-induced lung injury in mice: Nrf2 expression is attenuated in *MIF* KO mice, leading to increased susceptibility to irradiation-induced lung injury through increased inflammation [55]. However, Nrf2 activator administration restores the Nrf2 level in mice and human fibroblast IMR-90, and persistent lung fibrosis associated with aging is improved [56]. These studies suggest that impaired Nrf2 expression by aging recovers by Nrf2 activators. Taken together, these data indicate that Nrf2 activation has a potent protective role against ionizing radiation in normal cells and animals, although the protective effect of Nrf2 activation in humans should be carefully investigated in future studies.

2.3. "Bad Face" of Nrf2 Activation in Response to Ionizing Radiation

Radiation therapy is the most widely accepted and effective cancer treatment and it induces apoptosis in cancer cells through induction of direct and indirect DNA damage. Recent studies showed that Nrf2 is overexpressed in various types of cancer cells [57,58]. Overexpression of Nrf2 increases a battery of cytoprotective and antioxidant genes, resulting in resistance against chemotherapy and ionizing radiation. One reason for Nrf2 overexpression is loss-of-function Keap1 mutations (Figure 7.3). Somatic mutations of the *Keap1* gene have been identified in many tumor tissues and cancer cells [59-62]. These mutations impair the Keap1 function as an Nrf2 inhibitor, leading to Nrf2 activation, which contributes to resistance against chemotherapy and radiation therapy in cancer cells [59,63-

65]. In some cases, *Keap1* gene expression is decreased by hypermethylation of DNA at its promoter region [66-68].

Another reason for Nrf2 overexpression is gain of function of Nrf2. Somatic mutations of the *Nrf2* gene have also been reported in various human cancers, including lung, esophagus and skin [69,70]. All Nrf2 mutations accumulate are found in the DLG motif and ETGE motif, which are domains responsible for interaction with Keap1 [57]. These mutations also lead to stabilization of Nrf2 by reducing the interaction affinity between Nrf2 and Keap1. Interestingly, suppression of the Nrf2-mediated antioxidant defense system confers sensitivity to ionizing radiation and chemotherapeutic drugs [71, 72]. These findings indicate that Nrf2 plays a critical role in the development of chemoresistance and radioresistance in cancer.

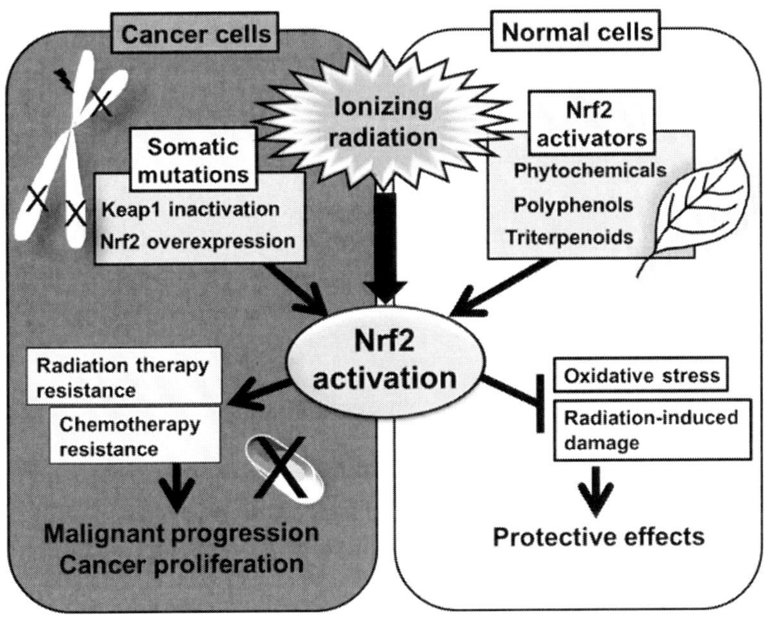

Figure 7.3. Bad face and good face of Nrf2 activation. Details are described in the text.

3. Molecular Mechanism of Nrf2 in Ionizing Radiation Response

Because Nrf2 is a transcription factor, the protective effect of Nrf2 activation against ionizing radiation is dependent upon the activity of its target genes. In this section, we describe representative Nrf2 target genes and related pathways that protect against ionizing radiation.

3.1. Heme Oxygenase-1 (HO-1)

The *HO-1* gene is a representative Nrf2 target gene that is overexpressed in various types of cancers. *HO-1* gene expression responds very sensitively to its substrate, heme, as well as to multiple environmental stressors, including oxidative stress, heavy metals, nitric oxide, UV irradiation, and inflammation [73].

HO-1 is an inducible isoform of heme oxygenase, which is a monooxygenase that catalyzes the oxidative degradation of heme into biliverdin, ferrous iron and carbon monoxide (CO). Bilirubin, which is converted from biliverdin, functions as an antioxidant. CO acts as a second messenger that controls inflammatory response and apoptosis. HO-1 is a microsomal enzyme, but a small fraction has been detected in the nucleus. HO-1 expression is inducible in many tissues, including skin, lung, liver, kidney, brain and macrophages. The wide distribution enables multiple biological roles of HO-1 that participates in heme detoxification, iron recycling, oxidative stress response and regulation of inflammation. Therefore, HO-1 is a potential therapeutic target for many diseases [74].

HO-1 contributes to radioresistance of cancer or immortalized cells. Berberat et al. found that HO-1 is overexpressed in human pancreatic cancer cells, although the expression level varied among cells [75]. They showed that HO-1 is induced by γ–irradiation and oxidative stress. Knockdown of *HO-1* by small interference RNA results in reduced pancreatic cancer proliferation and oxidative stress tolerance. Lister et al. showed that Nrf2 is overexpressed in pancreatic cancer cells and ductal adenocarcinomas and that knockdown of Nrf2 causes increase in sensitivity to γ–irradiation [76]. Liu et al. showed that HIV-TAT-HO-1 fusion protein has a protective effect on HaCaT cells against ionizing radiation [77]. These studies indicate that inhibition of HO-1 and Nrf2 might be a strategy to increase radiosensitivity of cancer cells. Lee et al. showed that 4-(2-Cyclohexylethoxy)aniline (IM3829), an inhibitor of Nrf2, is a promising candidate as a radiosensitizer for lung cancer [71].

HO-1 expression is also associated with protection against radiation. Ewing et al. demonstrated that induction of HO-1 by cobalt protoporphyrine IX inhibits irradiation-induced apoptosis [78]. On the other hand, HO-1 inhibitor tin protoporphyrine IX (SnPP) did not affect survival after irradiation. Thus, they concluded that pre-HO-1 induction protects endothelial cells from irradiation-induced damages. Intriguingly, Otterbein et al. recently showed that HO-1 directly participates in irradiation-induced DNA repair through its product CO [79]. These data suggest that HO-1 is relevant to protect cells from ionizing radiation through not only oxidative stress response but also DNA repair.

3.2. NAD(P)H Quinone Oxidoreductase 1 (NQO1)

Basal and inducible expressions of *NQO1* are regulated by Nrf2 and their expression levels correlate highly with Nrf2 activation in various tissues. NQO1 is a flavin-containing enzyme that catalyzes two-electron reduction of quinone to hydroquinone [80]. To catalyze this reaction, NQO1 requires NADH or NADPH as a cofactor. NQO1 is mainly localized in cytosol, but a small portion is in the nucleus [80]. NQO1 is overexpressed in human solid tumors, including colon, breast, pancreas and lung, due to constitutive activation of Nrf2. Garate et al. showed that NOQ1 is highly expressed in melanoma cells and up-regulates the expression of cyclins, resulting in cell cycle progression and proliferation of melanoma cells [81]. These data indicate that Nrf2 overexpression contributes to cancer proliferation through NQO1.

NQO1 is a potential target to control sensitivity to irradiation. We previously showed that X-ray irradiation induces several of Nrf2 target genes including *NQO1* and *HO-1* in human placental/umbilical cord blood hematopoietic stem cells (HSC). Interestingly, the *NQO1* gene expression was negatively correlated with the survival fraction of HSCs [82]. β-lapachone contains quinone that originated from lapacho trees in South Africa. This compound is an anticancer drug and shows strong cytotoxicity toward a variety of cancer cell lines. Choi et al. showed that ionizing radiation synergistically increases β-lapachone-induced cell death by NQO1 up-regulation, indicating that this combination treatment is potential cancer therapy [83]. *NQO1* KO mice are viable and develop normally. However, a majority of *NQO1* KO mice develop myeloproliferative disease by γ-radiation exposure. This disease shows increased neutrophil numbers, bone marrow hypercellularity, enlarged lymph nodes and spleen, disrupted follicular structure, loss of red pulp in spleen, and granulocyte and megakaryocyte invasion of the spleen. A higher incident of lymphoma and lung adenocarcinoma was also observed in *NQO1* KO mice after γ-irradiation. These data indicate that NQO1 confers protective effects against ionizing radiation in mice [84].

3.3. Notch Signaling Pathway

The Notch family constitutes a group of transmembrane proteins that are important for intercellular communications (Figure 7.4) [85]. At the time of binding to the ligands that are also transmembrane proteins of the adjacent cells, Notch undergoes serial proteolytic cleavage by ADAM family metalloproteinases and γ-secretase. The Notch intracellular domain (NICD) released from the plasma membrane enters the nucleus to form a complex with Notch binding partner recombination signal binding protein for the immunoglobulin kappa J region (Rbpjκ) and activates a set of target genes by binding to its target sequence. Thus, the Notch signaling pathway participates in cell fate specification, differentiation and maintenance of stem and progenitor cells. *Notch1* was first identified as an Nrf2 target gene because the expression of the *Notch1* gene was down-regulated in *Nrf2* KO MEFs [86]. Furthermore, Notch1 target genes such as *hes1* and *p21* were also down-regulated and the response to Notch1 ligands jagged 1 (JAG1) and Delta-like-1 (DLL1) is also reduced in *Nrf2* KO MEFs. Wakabayashi et al. [86] further showed that during partial hepatectomy Notch1 levels are reduced in *Nrf2* KO liver and that liver regeneration is significantly delayed in *Nrf2* KO mice. Wakabayashi et al. also found that *Nrf2* gene expression is also activated by the Notch signaling pathway in the liver as the gene regulatory region of *Nrf2* has binding sites for Rbpjκ [87]. They showed that NICD overexpression in mouse hepatocytes leads to hepatomegaly, bile duct hyperbranching and increased resistance to acetaminophen hepatotoxicity.

Notch signaling is also important for hematopoiesis. As described in section 2-2-2, TMC-mediated Nrf2 activation increases hematopoietic stem progenitor cell function and mitigates ionizing radiation-induced myelosuppression and mortality in mice [54]. TMC activates JAG1 expression in stromal cells and in the activated Notch signaling pathway that protects against total body irradiation. Therefore, Nrf2 increases tissue regeneration and hematopoiesis by increasing the activity of Notch signaling pathway.

3.4. DNA Repair Pathway

Double-strand breaks (DSBs) in DNA are caused by cellular stresses, such as oxygen radicals, chemical mutagens and ionizing radiation. DNA damage response (DDR) is initiated by sensing DNA damage by the sensor protein followed by mediator and activator activation [88]. DDR mediators include p53-binding protein 1 (53BP1) and Breast Cancer 1 (BRCA1). 53BP1 is recruited to the site of DNA DSBs. As described in section 2-2-2, Kim et al. showed that BARD administration before irradiation increases clonogenic survival of HCECs [51]. They showed that BARD induces DNA damage response, including rapid Rad51 and BRCA1 nuclear translocation, but it is reduced by *Nrf2* knockdown. They showed by chromatin immunoprecipitation analysis that Nrf2 binds to the regulatory region of the *53BP1* gene and that BARD increases 53BP1. Irradiation-induced oxygen radicals also induce oxidation of DNA, the most common oxidized DNA being 8-hydoxydeoxyguanosine (8-OHdG). The 8-oxoguanine DNA glycosylase (OGG1) catalyzes the rate limiting reaction to remove 8-OHdG from DNA. The expression of the *OGG1* gene is also regulated by Nrf2 though the promoter ARE in the *OGG1* gene [89].

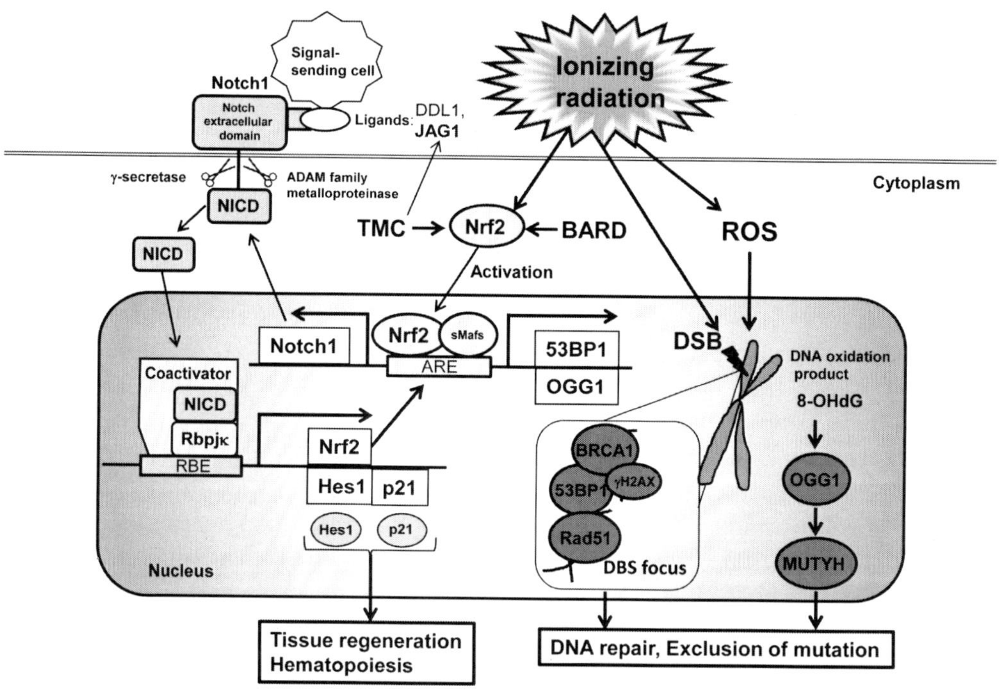

Figure 7.4. Nrf2-mediated activation Notch pathway and DNA double-strand break (DSB) response. NICD: Notch Intracellular Domain, DDL1: Delta-like-1, JAG1: Notch1 ligands jagged 1, BRCA1: Breast Cancer 1, γH2AX: phosphorylated histone H2AX, 53BP1: p53-binding protein 1, OGG1: 8-oxoguanine DNA glycosylase, MUTYH: MutY homolog, ROS: reactive oxygen species, TMC: 2-trifluoromethyl-2'-methoxychalone, BARD: bardoxolone methyl, ARE: antioxidant responsive element, RBE: Rbpjκ binding element, DSB: double strand break, Hes: hairy and enhancer of split-1.

3.5. Crosstalk with p53

Several studies showed the crosstalk between Nrf2 and p53 pathways. p53 is a transcription regulator of anticancer function and has a critical role in apoptosis, DNA repair, cell cycle regulation and genome instability in response to various stimuli, including ionizing radiation and oxidative stress; the encoded gene is often mutated in various tumors [90]. We showed that the Nrf2 and p53 coordinately attenuates N-nitrosobutyl(4-hydroxybutyl)-amine-induced urinary bladder carcinogenesis in mice [7]. Chen et al. showed that $p21^{Cip/WAF1}$, which is a p53 target gene, directly binds to the DLG motif of Nrf2 and promotes Nrf2 activation by competing with Keap1 and Nrf2 interaction in response to oxidative stress [91]. The protective function of NQO1 against ionizing radiation may be mediated by tumor suppressor p53, which is also a very labile protein and undergoes Mdm2 and ubiquitin-dependent degradation. NAD(P)H competitive NQO1 inhibitor dicumarol inhibits accumulation of γ-irradiation-induced p53 in HCT116 human colon cancer cells [92]. NQO1 selectively interacts with p73α in the 20S proteasome and stabilizes p53 after γ-irradiation [93]. Therefore, although the anti-oxidative stress property of NOQ1 is important, p53 stabilization by NQO1 may be crucial to protect cells against ionizing radiation. Thus, we surmise that the Nrf2-p53 circuit may cooperatively protect cells from ionizing radiation damage.

CONCLUSION

In this review, we argued that Nrf2 not only regulates a battery of antioxidant and detoxification genes, but also cellular damage and DNA repair response, as well as cellular proliferation of organisms to survive against harsh environments including ionizing irradiation. Such roles of Nrf2 in humans remain to be clarified, but Nrf2 activation by phytochemicals and therapeutic drugs should be a promising strategy to promote health in humans.

REFERENCES

[1] Itoh K, Mimura J, Yamamoto M (2010) Discovery of the negative regulator of Nrf2, Keap1: a historical overview. *Antioxidants and Redox Signaling* 13: 1665-1678.

[2] Kensler TW, Wakabayashi N, Biswal S (2007) Cell survival responses to environmental stresses via the Keap1-Nrf2-ARE pathway. *Annual Review of Pharmacology and Toxicology* 47: 89-116.

[3] Itoh K, Chiba T, Takahashi S, Ishii T, Igarashi K, et al. (1997) An Nrf2/small Maf heterodimer mediates the induction of phase II detoxifying enzyme genes through antioxidant response elements. *Biochemical and Biophysical Research* *Communications* 236: 313-322.

[4] Wattenberg LW (1978) Inhibition of chemical carcinogenesis. *Journal of the National Cancer Institute* 60: 11-18.

[5] Miller JA (1970) Carcinogenesis by chemicals: an overview--G. H. A. Clowes memorial lecture. *Cancer Research* 30: 559-576.
[6] Ramos-Gomez M, Kwak MK, Dolan PM, Itoh K, Yamamoto M, et al. (2001) Sensitivity to carcinogenesis is increased and chemoprotective efficacy of enzyme inducers is lost in nrf2 transcription factor-deficient mice. *Proceedings of the National Academy of Sciences of the United States of America* 98: 3410-3415.
[7] Iida K, Itoh K, Maher JM, Kumagai Y, Oyasu R, et al. (2007) Nrf2 and p53 cooperatively protect against BBN-induced urinary bladder carcinogenesis. *Carcinogenesis* 28: 2398-2403.
[8] Mattson MP (2008) Hormesis defined. *Ageing Research Reviews* 7: 1-7.
[9] Forman HJ, Davies KJ, Ursini F (2014) How do nutritional antioxidants really work: nucleophilic tone and para-hormesis versus free radical scavenging in vivo. *Free Radical Biology and Medicine* 66: 24-35.
[10] Itoh K, Wakabayashi N, Katoh Y, Ishii T, Igarashi K, et al. (1999) Keap1 represses nuclear activation of antioxidant responsive elements by Nrf2 through binding to the amino-terminal Neh2 domain. *Genes & Development* 13: 76-86.
[11] Tong KI, Katoh Y, Kusunoki H, Itoh K, Tanaka T, et al. (2006) Keap1 recruits Neh2 through binding to ETGE and DLG motifs: characterization of the two-site molecular recognition model. *Molecular and Cellular Biology* 26: 2887-2900.
[12] Zipper LM, Mulcahy RT (2002) The Keap1 BTB/POZ dimerization function is required to sequester Nrf2 in cytoplasm. *Journal of Biological Chemistry* 277: 36544-36552.
[13] Kobayashi A, Kang MI, Okawa H, Ohtsuji M, Zenke Y, et al. (2004) Oxidative stress sensor Keap1 functions as an adaptor for Cul3-based E3 ligase to regulate proteasomal degradation of Nrf2. *Molecular and Cellular Biology* 24: 7130-7139.
[14] Dinkova-Kostova AT, Holtzclaw WD, Cole RN, Itoh K, Wakabayashi N, et al. (2002) Direct evidence that sulfhydryl groups of Keap1 are the sensors regulating induction of phase 2 enzymes that protect against carcinogens and oxidants. *Proceedings of the National Academy of Sciences of the United States of America* 99: 11908-11913.
[15] Kobayashi M, Li L, Iwamoto N, Nakajima-Takagi Y, Kaneko H, et al. (2009) The antioxidant defense system Keap1-Nrf2 comprises a multiple sensing mechanism for responding to a wide range of chemical compounds. *Molecular and Cellular Biology* 29: 493-502.
[16] Bryan HK, Olayanju A, Goldring CE, Park BK (2013) The Nrf2 cell defence pathway: Keap1-dependent and -independent mechanisms of regulation. *Biochemical Pharmacology* 85: 705-717.
[17] Rada P, Rojo AI, Chowdhry S, McMahon M, Hayes JD, et al. (2011) SCF/{beta}-TrCP promotes glycogen synthase kinase 3-dependent degradation of the Nrf2 transcription factor in a Keap1-independent manner. *Molecular and Cellular Biology* 31: 1121-1133.
[18] Liu Y, Kern JT, Walker JR, Johnson JA, Schultz PG, et al. (2007) A genomic screen for activators of the antioxidant response element. *Proceedings of the National Academy of Sciences of the United States of America* 104: 5205-5210.
[19] Kosaka K, Mimura J, Itoh K, Satoh T, Shimojo Y, et al. (2010) Role of Nrf2 and p62/ZIP in the neurite outgrowth by carnosic acid in PC12h cells. *Journal of Biochemistry* (Tokyo) 147: 73-81.

[20] Komatsu M, Kurokawa H, Waguri S, Taguchi K, Kobayashi A, et al. (2010) The selective autophagy substrate p62 activates the stress responsive transcription factor Nrf2 through inactivation of Keap1. *Nature Cell Biology* 12: 213-223.
[21] Ichimura Y, Waguri S, Sou YS, Kageyama S, Hasegawa J, et al. (2013) Phosphorylation of p62 activates the Keap1-Nrf2 pathway during selective autophagy. *Molecular Cell* 51: 618-631.
[22] Lau A, Zheng Y, Tao S, Wang H, Whitman SA, et al. (2013) Arsenic inhibits autophagic flux, activating the Nrf2-Keap1 pathway in a p62-dependent manner. *Molecular and Cellular Biology* 33: 2436-2446.
[23] Stenzel-Poore MP, Stevens SL, Xiong Z, Lessov NS, Harrington CA, et al. (2003) Effect of ischaemic preconditioning on genomic response to cerebral ischaemia: similarity to neuroprotective strategies in hibernation and hypoxia-tolerant states. *The Lancet* 362: 1028-1037.
[24] Shih AY, Li P, Murphy TH (2005) A small-molecule-inducible Nrf2-mediated antioxidant response provides effective prophylaxis against cerebral ischemia in vivo. *Journal of Neuroscience* 25: 10321-10335.
[25] Satoh T, Kosaka K, Itoh K, Kobayashi A, Yamamoto M, et al. (2008) Carnosic acid, a catechol-type electrophilic compound, protects neurons both in vitro and in vivo through activation of the Keap1/Nrf2 pathway via S-alkylation of targeted cysteines on Keap1. *Journal of Neurochemistry* 104: 1116-1131.
[26] Bell KF, Al-Mubarak B, Fowler JH, Baxter PS, Gupta K, et al. (2011) Mild oxidative stress activates Nrf2 in astrocytes, which contributes to neuroprotective ischemic preconditioning. *Proceedings of the National Academy of Sciences of the United States of America* 108: E1-2; author reply E3-4.
[27] Takagi T, Kitashoji A, Iwawaki T, Tsuruma K, Shimazawa M, et al. (2014) Temporal activation of Nrf2 in the penumbra and Nrf2 activator-mediated neuroprotection in ischemia-reperfusion injury. *Free Radical Biology and Medicine* 72: 124-133.
[28] Bell KF, Fowler JH, Al-Mubarak B, Horsburgh K, Hardingham GE (2011) Activation of Nrf2-regulated glutathione pathway genes by ischemic preconditioning. *Oxidative Medicine and Cellular Longeviry* 2011: 689524.
[29] Xu B, Zhang J, Strom J, Lee S, Chen QM (2014) Myocardial ischemic reperfusion induces de novo Nrf2 protein translation. *Biochimica et Biophysica Acta* 1842: 1638-1647.
[30] McDonald JT, Kim K, Norris AJ, Vlashi E, Phillips TM, et al. (2010) Ionizing radiation activates the Nrf2 antioxidant response. *Cancer Research* 70: 8886-8895.
[31] Tsukimoto M, Tamaishi N, Homma T, Kojima S (2010) Low-dose gamma-ray irradiation induces translocation of Nrf2 into nuclear in mouse macrophage RAW264.7 cells. *Journal of Radiation Research* 51: 349-353.
[32] Yoshino H, Kiminarita T, Matsushita Y, Kashiwakura I (2012) Response of the Nrf2 protection system in human monocytic cells after ionising irradiation. *Radiation Protection Dosimetry* 152: 104-108.
[33] Lee EK, Kim JA, Park SJ, Kim JK, Heo K, et al. (2013) Low-dose radiation activates Nrf1/2 through reactive species and the Ca(2+)/ERK1/2 signaling pathway in human skin fibroblast cells. *BMB Reports* 46: 258-263.
[34] Riley PA (1994) Free radicals in biology: oxidative stress and the effects of ionizing radiation. *International Journal of Radiation Biology* 65: 27-33.

[35] Fourquet S, Guerois R, Biard D, Toledano MB (2010) Activation of NRF2 by nitrosative agents and H_2O_2 involves KEAP1 disulfide formation. *Journal of Biological Chemistry* 285: 8463-8471.

[36] Barrera G, Pizzimenti S, Dianzani MU (2008) Lipid peroxidation: control of cell proliferation, cell differentiation and cell death. *Molecular Aspects of Medicine* 29: 1-8.

[37] Forman HJ (2010) Reactive oxygen species and alpha,beta-unsaturated aldehydes as second messengers in signal transduction. *Annals of the New York Academy of Sciences* 1203: 35-44.

[38] Pettazzoni P, Ciamporcero E, Medana C, Pizzimenti S, Dal Bello F, et al. (2011) Nuclear factor erythroid 2-related factor-2 activity controls 4-hydroxynonenal metabolism and activity in prostate cancer cells. *Free Radical Biology and Medicine* 51: 1610-1618.

[39] Huang Y, Li W, Kong AN (2012) Anti-oxidative stress regulator NF-E2-related factor 2 mediates the adaptive induction of antioxidant and detoxifying enzymes by lipid peroxidation metabolite 4-hydroxynonenal. *Cell & Bioscience* 2: 40.

[40] Levonen AL, Landar A, Ramachandran A, Ceaser EK, Dickinson DA, et al. (2004) Cellular mechanisms of redox cell signalling: role of cysteine modification in controlling antioxidant defences in response to electrophilic lipid oxidation products. *Biochemical Journal* 378: 373-382.

[41] McMahon M, Lamont DJ, Beattie KA, Hayes JD (2010) Keap1 perceives stress via three sensors for the endogenous signaling molecules nitric oxide, zinc, and alkenals. *Proceedings of the National Academy of Sciences of the United States of America* 107: 18838-18843.

[42] Bai H, Hai C, Xi M, Liang X, Liu R (2010) Protective effect of maize silks (Maydis stigma) ethanol extract on radiation-induced oxidative stress in mice. *Plant Foods for Human Nutrition* 65: 271-276.

[43] Lee JC, Kinniry PA, Arguiri E, Serota M, Kanterakis S, et al. (2010) Dietary curcumin increases antioxidant defenses in lung, ameliorates radiation-induced pulmonary fibrosis, and improves survival in mice. *Radiation Research* 173: 590-601.

[44] Xie Y, Zhao QY, Li HY, Zhou X, Liu Y, et al. (2014) Curcumin ameliorates cognitive deficits heavy ion irradiation-induced learning and memory deficits through enhancing of Nrf2 antioxidant signaling pathways. *Pharmacology, Biochemistry and Behavior*126: 181-186.

[45] Yu J, Zhu X, Qi X, Che J, Cao B (2013) Paeoniflorin protects human EA.hy926 endothelial cells against gamma-radiation induced oxidative injury by activating the NF-E2-related factor 2/heme oxygenase-1 pathway. *Toxicology Letters* 218: 224-234.

[46] Mathew ST, Bergstrom P, Hammarsten O (2014) Repeated Nrf2 stimulation using sulforaphane protects fibroblasts from ionizing radiation. *Toxicology and Applied Pharmacology* 276: 188-194.

[47] Zhu W, Xu J, Ge Y, Cao H, Ge X, et al. (2014) Epigallocatechin-3-gallate (EGCG) protects skin cells from ionizing radiation via heme oxygenase-1 (HO-1) overexpression. *Journal of Radiation Research* 55: 1056-1065.

[48] Ghosh D, Dey SK, Saha C (2014) Antagonistic effects of black tea against gamma radiation-induced oxidative damage to normal lymphocytes in comparison with cancerous K562 cells. *Radiation and Environmental Biophysics*. 53: 695-704.

[49] Khan NM, Sandur SK, Checker R, Sharma D, Poduval TB, et al. (2011) Pro-oxidants ameliorate radiation-induced apoptosis through activation of the calcium-ERK1/2-Nrf2 pathway. *Free Radical Biology and Medicine* 51: 115-128.

[50] Shin HS, Yang WJ, Choi EM (2013) The preventive effect of Se-methylselenocysteine on gamma-radiation-induced oxidative stress in rat lungs. *Journal of Trace Elements in Medicine and Biology* 27: 154-159.

[51] Kim SB, Pandita RK, Eskiocak U, Ly P, Kaisani A, et al. (2012) Targeting of Nrf2 induces DNA damage signaling and protects colonic epithelial cells from ionizing radiation. *Proceedings of the National Academy of Sciences of the United States of America* 109: E2949-2955.

[52] Reisman SA, Lee CY, Meyer CJ, Proksch JW, Sonis ST, et al. (2014) Topical application of the synthetic triterpenoid RTA 408 protects mice from radiation-induced dermatitis. *Radiation Research* 181: 512-520.

[53] Kumar V, Kumar S, Hassan M, Wu H, Thimmulappa RK, et al. (2011) Novel chalcone derivatives as potent Nrf2 activators in mice and human lung epithelial cells. *Journal of Medicinal Chemistry* 54: 4147-4159.

[54] Kim JH, Thimmulappa RK, Kumar V, Cui W, Kumar S, et al. (2014) NRF2-mediated Notch pathway activation enhances hematopoietic reconstitution following myelosuppressive radiation. *Journal of Clinical Investigation* 124: 730-741.

[55] Mathew B, Jacobson JR, Siegler JH, Moitra J, Blasco M, et al. (2013) Role of migratory inhibition factor in age-related susceptibility to radiation lung injury via NF-E2-related factor-2 and antioxidant regulation. *American Journal of Respiratory Cell and Molecular Biology* 49: 269-278.

[56] Hecker L, Logsdon NJ, Kurundkar D, Kurundkar A, Bernard K, et al. (2014) Reversal of persistent fibrosis in aging by targeting Nox4-Nrf2 redox imbalance. *Science Translational Medicine* 6: 231ra247.

[57] Taguchi K, Motohashi H, Yamamoto M (2011) Molecular mechanisms of the Keap1-Nrf2 pathway in stress response and cancer evolution. *Genes to Cells* 16: 123-140.

[58] Zhou S, Ye W, Shao Q, Zhang M, Liang J (2013) Nrf2 is a potential therapeutic target in radioresistance in human cancer. *Critical Reviews in Oncology/Hematology* 88: 706-715.

[59] Padmanabhan B, Tong KI, Ohta T, Nakamura Y, Scharlock M, et al. (2006) Structural basis for defects of Keap1 activity provoked by its point mutations in lung cancer. *Molecular Cell* 21: 689-700.

[60] Konstantinopoulos PA, Spentzos D, Fountzilas E, Francoeur N, Sanisetty S, et al. (2011) Keap1 mutations and Nrf2 pathway activation in epithelial ovarian cancer. *Cancer Research* 71: 5081-5089.

[61] Nioi P, Nguyen T (2007) A mutation of Keap1 found in breast cancer impairs its ability to repress Nrf2 activity. *Biochemical and Biophysical Research Communications* 362: 816-821.

[62] Shibata T, Kokubu A, Gotoh M, Ojima H, Ohta T, et al. (2008) Genetic alteration of Keap1 confers constitutive Nrf2 activation and resistance to chemotherapy in gallbladder cancer. *Gastroenterology* 135: 1358-1368, 1368 e1-4.

[63] Singh A, Bodas M, Wakabayashi N, Bunz F, Biswal S (2010) Gain of Nrf2 function in non-small-cell lung cancer cells confers radioresistance. *Antioxidants and Redox Signaling* 13: 1627-1637.

[64] Ohta T, Iijima K, Miyamoto M, Nakahara I, Tanaka H, et al. (2008) Loss of Keap1 function activates Nrf2 and provides advantages for lung cancer cell growth. *Cancer Research* 68: 1303-1309.

[65] Cho JM, Manandhar S, Lee HR, Park HM, Kwak MK (2008) Role of the Nrf2-antioxidant system in cytotoxicity mediated by anticancer cisplatin: implication to cancer cell resistance. *Cancer Letters* 260: 96-108.

[66] Wang R, An J, Ji F, Jiao H, Sun H, et al. (2008) Hypermethylation of the Keap1 gene in human lung cancer cell lines and lung cancer tissues. *Biochemical and Biophysical Research Communications* 373: 151-154.

[67] Barbano R, Muscarella LA, Pasculli B, Valori VM, Fontana A, et al. (2013) Aberrant Keap1 methylation in breast cancer and association with clinicopathological features. *Epigenetics* 8: 105-112.

[68] Hanada N, Takahata T, Zhou Q, Ye X, Sun R, et al. (2012) Methylation of the KEAP1 gene promoter region in human colorectal cancer. *BMC Cancer* 12: 66.

[69] Shibata T, Kokubu A, Saito S, Narisawa-Saito M, Sasaki H, et al. (2011) NRF2 mutation confers malignant potential and resistance to chemoradiation therapy in advanced esophageal squamous cancer. *Neoplasia* 13: 864-873.

[70] Kim YR, Oh JE, Kim MS, Kang MR, Park SW, et al. (2010) Oncogenic NRF2 mutations in squamous cell carcinomas of oesophagus and skin. *Journal of Pathology* 220: 446-451.

[71] Lee S, Lim MJ, Kim MH, Yu CH, Yun YS, et al. (2012) An effective strategy for increasing the radiosensitivity of Human lung Cancer cells by blocking Nrf2-dependent antioxidant responses. *Free Radical Biology and Medicine* 53: 807-816.

[72] Shibata T, Ohta T, Tong KI, Kokubu A, Odogawa R, et al. (2008) Cancer related mutations in NRF2 impair its recognition by Keap1-Cul3 E3 ligase and promote malignancy. *Proceedings of the National Academy of Sciences of the United States of America* 105: 13568-13573.

[73] Sikorski EM, Hock T, Hill-Kapturczak N, Agarwal A (2004) The story so far: Molecular regulation of the heme oxygenase-1 gene in renal injury. American Journal of Physiology. *Renal Physiology* 286: F425-441.

[74] Gozzelino R, Jeney V, Soares MP (2010) Mechanisms of cell protection by heme oxygenase-1. *Annual Review of Pharmacology and Toxicology* 50: 323-354.

[75] Berberat PO, Dambrauskas Z, Gulbinas A, Giese T, Giese N, et al. (2005) Inhibition of heme oxygenase-1 increases responsiveness of pancreatic cancer cells to anticancer treatment. *Clinical Cancer Research* 11: 3790-3798.

[76] Lister A, Nedjadi T, Kitteringham NR, Campbell F, Costello E, et al. (2011) Nrf2 is overexpressed in pancreatic cancer: implications for cell proliferation and therapy. *Molecular Cancer* 10: 37.

[77] Liu P, Xue J, Gu Q, Wu J, Cao H, et al. (2014) HIV-TAT mediated protein transduction of heme oxygenase-1 protects HaCaT cells from ionizing radiation. *Saudi Medical Journal* 35: 234-241.

[78] Ewing P, Wilke A, Eissner G, Holler E, Andreesen R, et al. (2005) Expression of heme oxygenase-1 protects endothelial cells from irradiation-induced apoptosis. *Endothelium* 12: 113-119.

[79] Otterbein LE, Hedblom A, Harris C, Csizmadia E, Gallo D, et al. (2011) Heme oxygenase-1 and carbon monoxide modulate DNA repair through ataxia-telangiectasia

mutated (ATM) protein. *Proceedings of the National Academy of Sciences of the United States of America* 108: 14491-14496.

[80] Siegel D, Yan C, Ross D (2012) NAD(P)H:quinone oxidoreductase 1 (NQO1) in the sensitivity and resistance to antitumor quinones. *Biochemical Pharmacology* 83: 1033-1040.

[81] Garate M, Wani AA, Li G (2010) The NAD(P)H:Quinone Oxidoreductase 1 induces cell cycle progression and proliferation of melanoma cells. *Free Radical Biology and Medicine* 48: 1601-1609.

[82] Kato K, Takahashi K, Monzen S, Yamamoto H, Maruyama A, et al. (2010) Relationship between radiosensitivity and Nrf2 target gene expression in human hematopoietic stem cells. *Radiation Research* 174: 177-184.

[83] Choi EK, Terai K, Ji IM, Kook YH, Park KH, et al. (2007) Upregulation of NAD(P)H:quinone oxidoreductase by radiation potentiates the effect of bioreductive beta-lapachone on cancer cells. *Neoplasia* 9: 634-642.

[84] Iskander K, Barrios RJ, Jaiswal AK (2008) Disruption of NAD(P)H:quinone oxidoreductase 1 gene in mice leads to radiation-induced myeloproliferative disease. *Cancer Research* 68: 7915-7922.

[85] Li Y, Hiroi Y, Liao JK (2010) Notch signaling as an important mediator of cardiac repair and regeneration after myocardial infarction. *Trends in Cardiovascular Medicine* 20: 228-231.

[86] Wakabayashi N, Shin S, Slocum SL, Agoston ES, Wakabayashi J, et al. (2010) Regulation of notch1 signaling by nrf2: implications for tissue regeneration. *Science Signaling* 3: ra52.

[87] Wakabayashi N, Skoko JJ, Chartoumpekis DV, Kimura S, Slocum SL, et al. (2014) Notch-Nrf2 axis: regulation of Nrf2 gene expression and cytoprotection by notch signaling. *Molecular and Cellular Biology* 34: 653-663.

[88] Kakarougkas A, Jeggo PA (2014) DNA DSB repair pathway choice: an orchestrated handover mechanism. *British Journal of Radiology* 87: 20130685.

[89] Singh B, Chatterjee A, Ronghe AM, Bhat NK, Bhat HK (2013) Antioxidant-mediated up-regulation of OGG1 via NRF2 induction is associated with inhibition of oxidative DNA damage in estrogen-induced breast cancer. *BMC Cancer* 13: 253.

[90] El-Deiry WS (2003) The role of p53 in chemosensitivity and radiosensitivity. *Oncogene* 22: 7486-7495.

[91] Chen W, Sun Z, Wang XJ, Jiang T, Huang Z, et al. (2009) Direct interaction between Nrf2 and p21(Cip1/WAF1) upregulates the Nrf2-mediated antioxidant response. *Molecular Cell* 34: 663-673.

[92] Asher G, Lotem J, Cohen B, Sachs L, Shaul Y (2001) Regulation of p53 stability and p53-dependent apoptosis by NADH quinone oxidoreductase 1. *Proceedings of the National Academy of Sciences of the United States of America* 98: 1188-1193.

[93] Asher G, Tsvetkov P, Kahana C, Shaul Y (2005) A mechanism of ubiquitin-independent proteasomal degradation of the tumor suppressors p53 and p73. *Genes & Development* 19: 316-321.

In: Fukushima Nuclear Accident
Editor: Shizuyo Sutou
ISBN: 978-1-63463-748-0

Chapter 8

EFFECTS OF CHRONIC LOW DOSE-RATE GAMMA RAY IRRADIATION IN MICE

Ignacia B. Tanaka III*, Satoshi Tanaka and Shingo Nakamura
Department of Radiobiology, Institute for Environmental Sciences
Obuchi, Rokkasho, Aomori, Japan

ABSTRACT

The chapter summarizes the results from experiments performed at the Institute for Environmental Sciences (IES) directed towards studying the late biological effects of chronic low dose-rate radiation on SPF B6C3F1 mice using bodyweight, neoplasm incidence, cause of death and life span as parameters. The mice were chronically exposed to ^{137}Cs gamma rays at low dose-rates of 21, 1.1 and 0.05 mGy/day, after which they were kept until pre-determined sacrifice or until they died a natural death.

In the life span study, a total of 4,000 mice were chronically exposed to ^{137}Cs gamma rays at low dose-rates of 21, 1.1 and 0.05 mGy/day for 400 days. The overall major cause of death for both sexes, regardless of radiation dose, was neoplasms, accounting for >86.7% of all deaths. In males, the most common neoplasms were malignant lymphoma and liver neoplasms, accounting for over 30% and 17% of the total deaths, respectively. Malignant lymphomas accounted for over 52% of the total deaths in females. Decreased incidence for pituitary neoplasms in females exposed to 21 mGy/day was attributed to competing risks. The number of multiple primary neoplasms per mouse was significantly increased for both sexes irradiated at 21 mGy/day. Life spans of mice of both sexes irradiated with 21 mGy/day ($P < 0.0001$) and of females irradiated with 1.1 mGy/day ($P < 0.05$) were significantly shorter than those of the control group. No significant life shortening was found in mice exposed to 0.05 mGy/day. Despite an overall tendency towards earlier deaths (shorter life spans) in relation to increased radiation dose, fatal tumor incidences were not affected by radiation exposure.

In the adiposity study on female B6C3F1 mice exposed to 21 mGy/day, increased body weight, adipose tissue weight, serum levels of leptin, and lipid contents of the liver and sera were observed in irradiated mice after 31(individually-housed) and 43 (group-housed) weeks of irradiation exposure.

---

* Correspondence to: Ignacia B. Tanaka III (tanakaib@ies.or.jp)

INTRODUCTION

Proponents of nuclear power or nuclear energy maintain that it is a safe, sustainable and low-carbon emitting and economical source of energy. Nuclear accidents such as the Three-Mile Island accident, the Chernobyl disaster, and more recently, the Fukushima Daiichi Nuclear disaster, however, make the argument for nuclear power opponents that it is a threat to people and the environment.

Majority of the available information on the effect of radiation on human health are based on epidemiological data collected from those exposed to relatively high doses of ionizing radiation such as those of the atomic bomb survivors in Hiroshima and Nagasaki, survivors of the Chernobyl disaster, recipients of diagnostic and therapeutic radiation, and, occupational exposure. Published research data on the biological effects of radiation are results of acute or sub-acute exposures to high doses of radiation. Although radiation has long been recognized as a carcinogen, the relationship between radiation dose and cancer risk remain controversial particularly at low doses. Despite the controversy surrounding the linear non-threshold (LNT) hypothesis since it was first applied in 1956 by the National Committee on Radiation Protection and Measurements (NCRP) [1], regulatory agencies, including the International Committee on Radiation Protection (ICRP, 2005) [2], Biological Effects of Ionizing Radiation (BEIR VII, 2006) [3] and United Nations Scientific Committee on the Effects of Atomic Radiation (UNSCEAR, 2010) [4] continue to use the LNT hypothesis to describe the relationship of radiation dose to cancer incidence. The United Nations Scientific Committee on the Effects of Atomic Radiation [4, 5] refer to doses up to 0.1 mGy/minute (averaged over one hour or less; 144 mGy/24 h/day) as low dose rates and doses up to 200 mGy as low dose. The International Committee on Radiation Protection (ICRP) Publication No. 99 (2005) [2], however, defines effective doses in the order of 1 Sv, 100 mSv, 10 mSv, 1 mSv and 0.1 mSv as moderately high, moderate, low, very low and extremely low, respectively.

To date, very little research has been done on the biological effects of exposure to very low doses and low dose rates of radiation. Economics is the one major hurdle in the conduct of experiments in the study of the late biological effects of low doses of radiation mainly because it requires long-term (life span) follow-up as well as large numbers of animals in order to achieve statistically valid data due to the very low doses involved. Hence, risk estimation at low doses has traditionally been derived by extrapolating data obtained from high dose and high dose rate exposures. Human epidemiological studies on the health effects of radiation are further confounded by factors such as diet, alcohol consumption, tobacco, chronic disease conditions, infectious disease, socio-economic status as well as genetic susceptibility.

The Institute for Environmental Sciences (IES) in Rokkasho, Aomori Prefecture in Japan was established in 1990 primarily to study the effects of very low doses of radiation. To date, the Institute has completed several experiments on thousands of laboratory mice extending over several years. Currently it is the only specific pathogen free (SPF) rodent facility that can chronically irradiate large numbers of mice at doses of 0.05, 1 and 21 mGy/day. The low dose-rates used in IES are listed Table 8.1 as compared to various known doses. In addition to the low dose rate irradiators, radiation facilities for medium dose rates, such as 200 and 400 mGy/day, and acute high dose rate of 0.89 Gy/minute are also available.

Table 8.1. Comparative dose rates used in IES

Dose Rates at IES			Comparative Doses
mGy/day	μGy/h	μGy/min	
0.05	2	0.04	approximately 20x the natural background radiation (excluding food); DOE administrative control (= 20 mGy/yr)
1.1	45	0.76	approximate daily dose in the International Space Station (ISS);
21	900	15.2	Positive control; Accidental exposure
200	9,091	152	average single dose (acute) of Atomic Bomb Survivors within 2,500 meters from the epicenter; Medium dose rates; Accidental exposure
400	18,182	303	Medium dose rates; Accidental exposure

MATERIALS AND METHOD

Mice and Animal Husbandry

This chapter covers only experiments on specific-pathogen-free (SPF) B6C3F1 mice (Fub Co., Japan) used in IES. B6C3F1 mice are first generation hybrids from the cross between the inbred strains C57BL/6 female with a C3H male without the mammary tumor virus (MTV^-) [6].

The F1 hybrid was chosen because of its longer lifespan (median lifespan of ~31 months and 10% survival at ~37 months), broad spectra of spontaneous neoplasms, decreased susceptibility to common murine viral infections (especially Sendai virus) [7] and well published historical background data as it is the most commonly used mice for toxicity and carcinogenicity studies. The hybrid has a uniform genotype with the major advantage that the level of heterozygosity closely resembles that of outbred stock [6].

Each mouse was assigned an animal number and were housed in plastic cages individually or in groups of 5 [8]. Sawdust bedding, food pellets, water (chlorinated, 10 ppm) and equipment were sterilized prior to transfer into the facility. Ten percent of the animals from each sex in each group were randomly selected and weighed every 28 days during their entire lifespan. Body weights of all the mice were measured before the start and immediately after completion of irradiation exposure, and at the time of necropsy. Mice were inspected twice a day (morning and afternoon) on weekdays but only once a day during the irradiation period, weekends and holidays.

All experiments were conducted according to legal regulations in Japan and following the Guidelines for Animal Experiments of the Institute for Environmental Sciences.

Monitoring of SPF Status

Studies were conducted under SPF environmental conditions [8].

Irradiation

Irradiation of treated groups began at 8 or 9 weeks of age with ^{137}Cs gamma rays at daily doses of 21 mGy, 1.1 mGy and 0.05 mGy, as described previously [8].

Continuous radiation exposure was done for 22 hours a day from 12:00 h to 10:00 the following day. The remaining 2 hours (10:00~12:00 h) were used to check the animals and for the conduct of husbandry procedures such as cleaning, cage changes, and provision of fresh feed and drinking water. After the mice received their pre-determined doses of radiation, they were transferred to the animal rooms and kept until the scheduled sacrifice or until they die spontaneously [8].

Pathology

Necropsy was performed as soon as possible after death based on a standard protocol. Terminal body weight, organ weights and gross abnormalities observed were recorded. Organs and tissues were collected, fixed in 10% neutral buffered formalin, processed and stained with hematoxylin and eosin (H&E) for histopathological examination. Additional tissue samples were collected from neoplasms and other organs and tissues with grossly apparent lesions. Special histochemical procedures were done when deemed necessary for diagnostic purposes [9].

Neoplasms were classified based on the proposed nomenclatures of WHO/IARC [10] and the NTP [11]. Tumors originating from the hematopoietic tissue were classified based on morphology into malignant lymphomas, leukemias (erythroid and granulocytic), histiocytic sarcomas and mast cell tumors. Further sub-classification of lymphomas was not done.

A cause of death (COD) was assigned to each animal based on the evaluation of clinical and pathological findings such that each mouse was counted only once. The COD was determined based on "in life observations" (clinical signs) and pathological findings, where in the case of neoplasms includes biological behaviour such as invasion, metastases and systemic occurrence [12]. It is based on available information about the disease processes in that specific animal and not on the general behaviour of the disease process in that species. Lesions (neoplastic and non-neoplastic) are considered fatal if they compromise vital functions and are designated as "fatal" when it is regarded as being directly or indirectly responsible for the animal's death (COD) or as "incidental" when judged not to have caused death and that death had risen from an unrelated cause [12]. No diagnosis was made and the animal's cause of death is stated as "unknown" when the cause of death was uncertain or could not be determined.

All primary neoplasms are counted into the overall incidence regardless of its contribution to the cause of death, whereas multiple neoplasms of the same type in one organ and metastases to other organs were considered only once.

I. Life Span Study

A total of 4,000 8 week-old SPF B6C3F1 mice was divided into 4 groups (1 non-irradiated control and 3 irradiated) of 500 males and 500 females each. The 3 irradiated groups were exposed to ^{137}Cs gamma rays for approximately 400 days, beginning at 8 weeks of age, at dose-rates of 21 mGy/day, 1.1 mGy/day and 0.05 mGy/day. The mice were housed in groups - five animals to a cage.

Neoplastic Incidence

The complete list of neoplastic pathologies found in SPF B6C3F1 mice chronically exposed to very low dose rates of gamma ray irradiation was published in Radiation Research in 2007 [9]. Majority of the reports on similar experiments investigating the effects of radiation focus mainly on the incidence of neoplastic and non-neoplastic lesions, rarely if ever, identifying the cause of death.

Overall, 93.2% of the animals died with neoplasms, majority having 1 or 2 primary neoplasms, while some have as many as 7 primary neoplasms each. Despite the shortened lifespans, a significant increase in the number of primary neoplasms, 2.34 (males) - 2.54 (females) neoplasms per mouse, was observed in the 21 mGy/day groups of both sexes as compared to an average of 1.75 (females) - 1.97 (males) neoplasms per mouse in non-irradiated mice (Figure 8.1). No significant increase in the number of primary neoplasms was observed in mice exposed to 0.05 mGy/day. The maximum number of 7 primary neoplasms was observed in 4 (0.8%) female mice exposed to 21 mGy/day (Figure 8.1) [9]

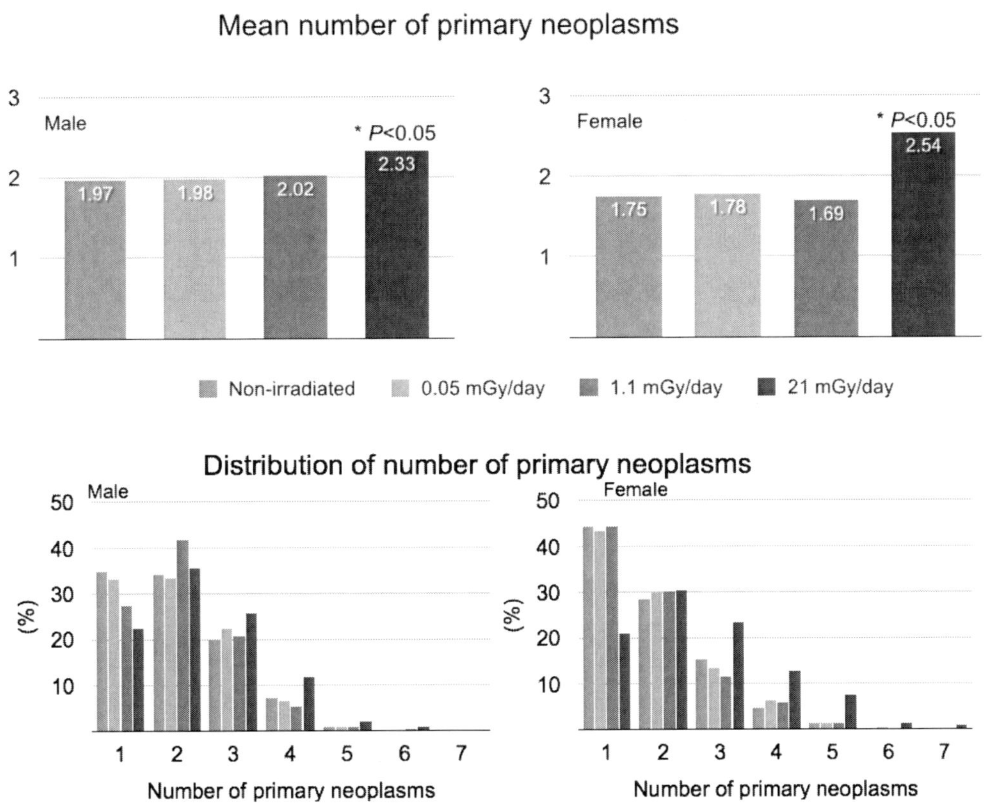

Figure 8.1. Mean number (top) of primary neoplasms and distribution (bottom) of the number of primary neoplasms in B6C3F1 mice chronically exposed to low dose-rates of gamma rays for 400 days.

Over 45.8% of the male mice had neoplasms originating from the liver, namely hepatocellular adenomas and carcinomas (Figure 8.2.1). Hematopoietic neoplasms, mostly malignant lymphomas, were found in 39.7% and over 37.3% had respiratory neoplasms. Peto's trend tests show significant radiation dose-related increase in incidences for hemangiosarcoma, hepatocellular adenomas, hepatoblastomas, myeloid leukemia, bronchiolo-alveolar carcinoma, Harderian gland neoplasms and neoplasms originating from the urinary tract [9]. While only the incidence rate for the malignant vascular tumor, hemangiosarcoma, was significantly increased in the 21 mGy/day group, total incidence rates for both hemangioma (benign) and hemangiosarcomas (malignant), were significantly increased [9].

Malignant lymphomas were found in over 50% of the female mice (Figure 8.2.1) and incidence rates were not significantly different between non-irradiated and irradiated groups. Up to 38.6% had ovarian neoplasms and 30.9% had respiratory neoplasms. Significantly increased incidence rates were observed for hemangiosarcomas, hepatocellular adenomas, and for tumors originating from the adrenal glands, ovaries, lungs, Harderian glands and the urinary system (very low incidence at 2.2%) [9].

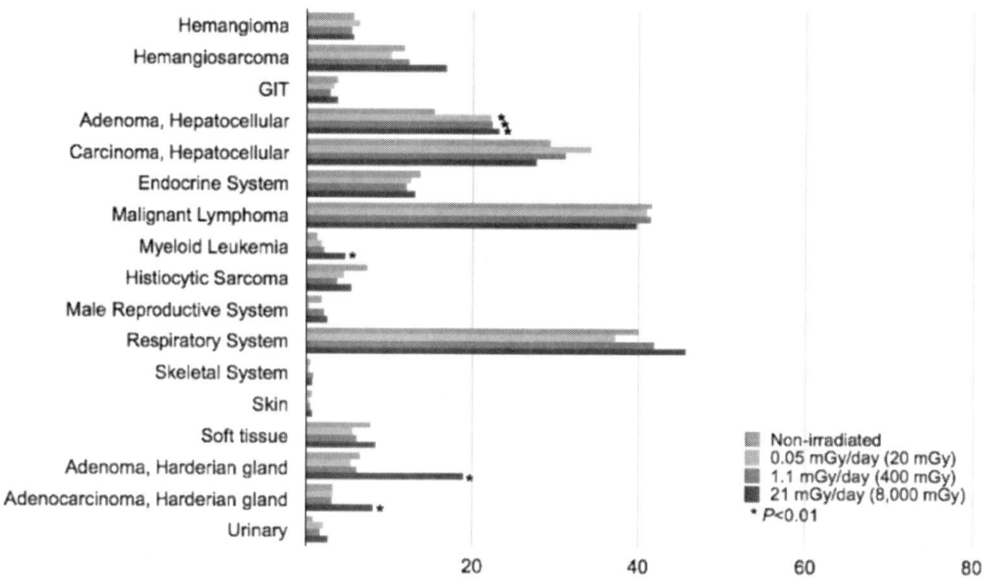

Figure 8.2.1. Neoplasm incidence (%) in male B6C3F1 mice chronically exposed to low dose-rates of gamma rays for 400 days.

Peto's trend tests show that increased incidences for hemangiosarcoma, hepatocellular adenomas, tumors originating from the adrenals, ovary, lungs, bone, soft tissue, urinary tract and Harderian glands were dose related and hence attributable to radiation exposure [9]. In females exposed to 21 mGy/day, the decreased incidences for the non-lethal primary neoplasms originating from the thyroid gland was considered radiation-induced, whereas those originating from the pituitary glands were due to competing risks [9].

Of the many biological features of neoplasms that affect the life span, only incidence have been taken into consideration for the estimation of radiation risk in man [13]. Neoplasm incidence in mice is also used as a biomarker for aging since the number of lesions develops with increasing frequency as the animal ages [14]. Overall, male mice tend to have more primary neoplasms than female mice since males, on the average, have longer life spans and increased age also increases the possibility of tumor development. Data suggest that correlations between neoplasm incidences are mainly due to the relative high lethality of neoplasms [15] hence decreasing the chances of two lethal tumors developing in the same animal.

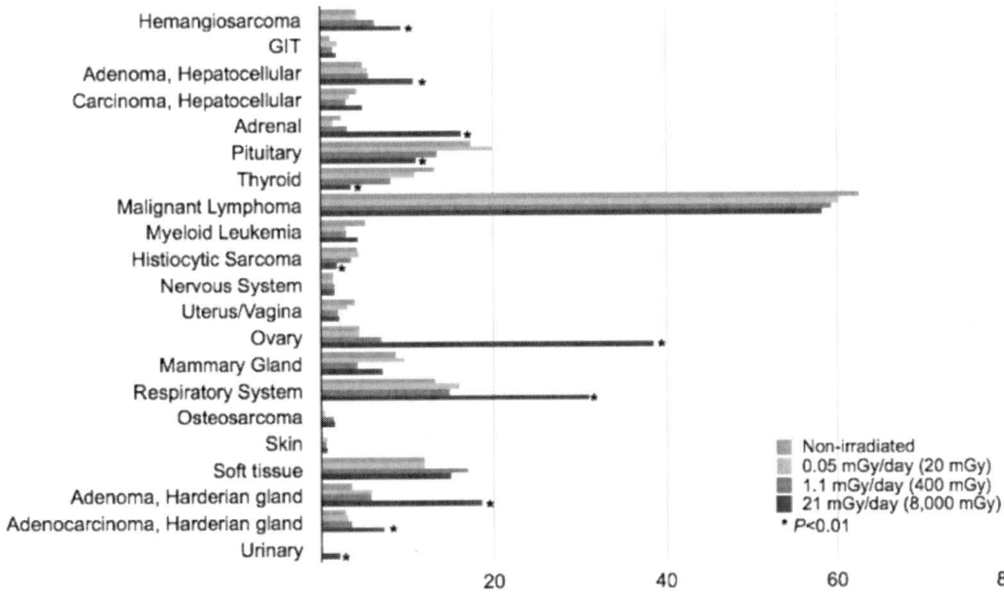

Figure 8.2.2. Neoplasm incidence (%) in female B6C3F1 mice chronically exposed to low dose-rates of gamma rays for 400 days.

Cause of Death (COD) and Age at the time of Death

Over 86.7% of all deaths in the mice were caused by neoplasms. No significant difference was observed in the incidence of the major causes of death between groups and no radiation effect was observed.

In males (Figure 8.3.1), malignant lymphomas accounted for 31.9 ~ 35.0% of the deaths, followed by neoplasms originating from the liver (18.2~23.5%) and vascular system (7.1~13.4%). Although the mortality rate for myeloid leukemia was very low at 4.6% for males exposed to 21 mGy/day, the increase was significant when compared to the non-irradiated controls (1.4%) [9]. The frequency of hemangiosarcoma as COD in male mice exposed to 21 mGy/day [9] was also significantly increased. The significant decrease in the frequency of hepatocellular carcinoma as COD in male mice exposed to 21 mGy/day was attributed to competing risks [9]. Similarly, female mice died mostly from malignant lymphomas (52~55%), followed by soft tissue neoplasms (6.5~10.3%) (Figure 8.3.2). Significant trend tests show that the number of female mice dying from malignant lymphoma decreased as the radiation dose increased. Unlike the males, there was no significant difference in the frequency of myeloid leukemia as COD in female mice. Although the frequency of hepatocellular carcinoma, bronchiolo-alveolar carcinoma, osteosarcoma and malignant fibrous histiocytoma as COD were significantly increased in female mice exposed to 21 mGy/day [9], it is difficult to satisfactorily conclude that they are related to radiation exposure because of the very low incidence. Deaths due to malignant granulosa cell tumor were significantly increased in females exposed to 21 mGy/day [9].

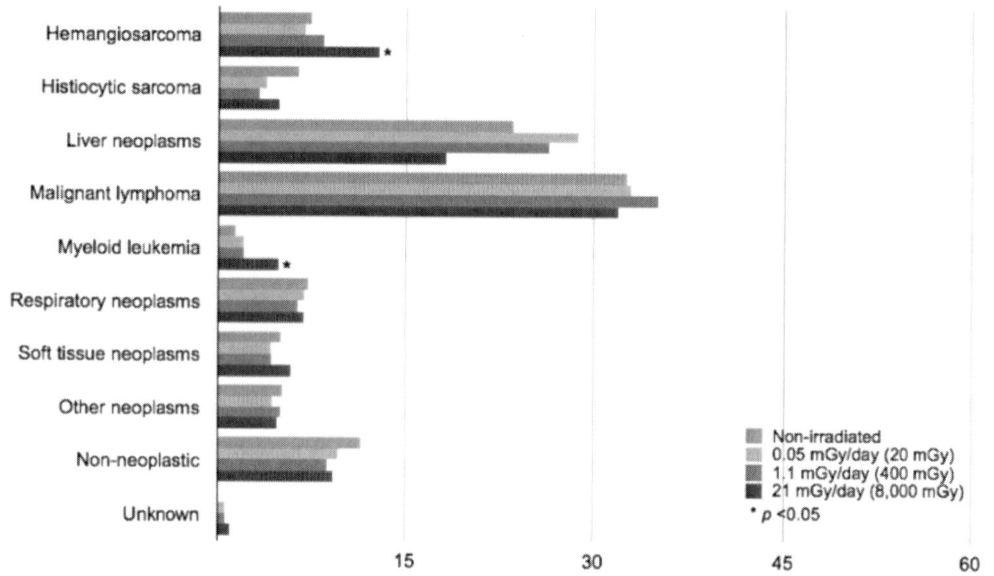

Figure 8.3.1. Causes of death (COD, %) in male B6C3F1 mice chronically exposed to low dose-rates of gamma rays for 400 days.

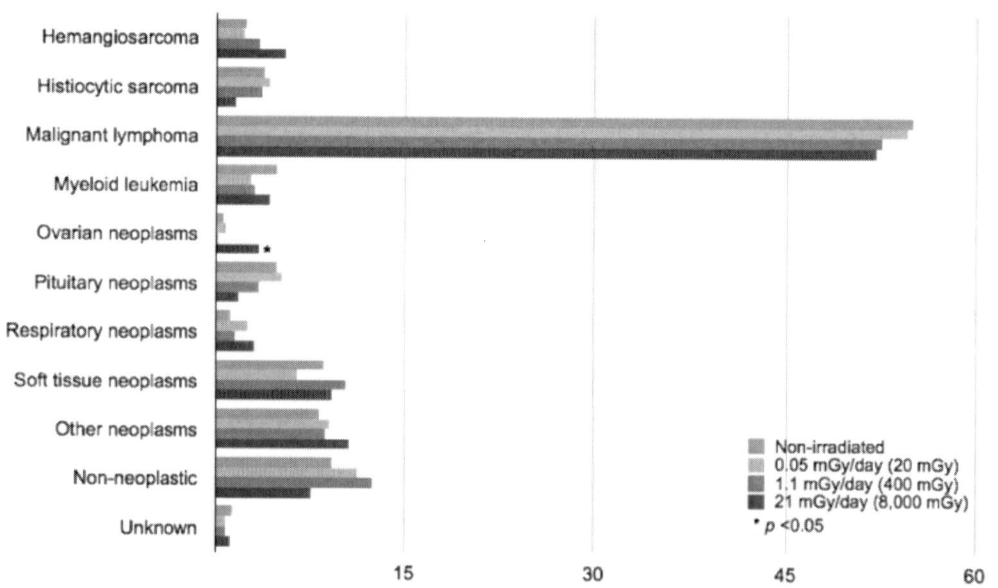

Figure 8.3.2. Causes of death (COD, %) in female B6C3F1 mice chronically exposed to low dose-rates of gamma rays for 400 days.

Non-neoplastic diseases (Figures 8.3.1, 8.3.2), such as interstitial pneumonia, hyaline glomerulopathy, hemorrhagic enteritis, renal amyloidosis and septicemia, accounted for 7.6

~12.5% of the total deaths in both sexes and was not significantly different among the non-irradiated and irradiated groups. Deaths from unknown causes accounted for 0.6 to 1.4% of the total deaths [9].

Life Span

Results showed that the mean life span of irradiated mice, both male and female, became progressively shorter as the radiation dose increased. Overall mean survival (life span) was significantly decreased in both sexes exposed to 21 mGy/day and in females exposed to 1.1 mGy/day of gamma ray irradiation [8]. Similarly, mean survival for major causes of death showed a significant decrease in both sexes of mice exposed to 21 mGy/day. The average life span of male mice exposed to 21 mGy/d was significantly shorter by 100.7 days at 812 days compared to the non-irradiated controls at 912.7 days. For female mice irradiated at 1.1 and 21 mGy/day, their life spans were significantly shorter by 20.7 and 119.6 days, respectively, when compared to the non-irradiated controls at 860.5 days [8].

Table 8.2. Kaplan-Meier estimates of mean survival of major causes of death on B6C3F1 male mice [*Radiat. Res. (2007) 167: 417-437*]

Dose rate (Total dose)	**Non-irradiated** (0 mGy)	**0.05 mGy/day** (20 mGy)	**1.1 mGy/day** (400 mGy)	**21 mGy/day** (8000 mGy)
All lethal neoplasms				
Incidence (%)	441 (88.6)	444 (89.7)	453 (90.6)	448 (89.8)
Mean lifespan (±SE)	933.5±8.2	929.6±8.1	914.6±8.3	829.4±7.8 [b]
Malignant lymphoma				
Incidence (%)	162 (32.5)	163 (32.9)	175 (35.0)	159 (31.9)
Mean lifespan (±SE)	1095.3±11.5	1094.9±12.4	1082.7±12.8	977.6±10.9 [b]
Liver neoplasms				
Incidence (%)	117 (23.5)	142 (28.7)	132 (26.4)	91 (18.2)
Mean lifespan (±SE)	1141.7±12.8	1112.6±13.3	1082.8±10.8	1064.1±13.0
Lung neoplasms				
Incidence (%)	36 (7.2)	34 (6.9)	32 (6.4)	34 (6.8)
Mean lifespan (±SE)	1207.2±8.0	1183.1±7.2	1212.7±9.6	1123.5±9.7 [b]
Hemangiosarcoma				
Incidence (%)	37 (7.4)	34 (6.9)	42 (8.4)	64 (12.8) [a†]
Mean lifespan (±SE)	1221.8±9.1	1305.7±11.4	1109.0±5.9	1092.0±11.8 [b]
Non-neoplastic diseases				
Incidence (%)	57 (11.4)	48 (9.7)	44 (8.8)	46 (9.2)
Mean lifespan (±SE)	1220.6±16.2	1199.5±10.5	1214.8±11.8	1103.3±9.5

[a] P<0.01 Fisher's exact test. [b] P<0.01Log-rank test. [†] P<0.01Peto's trend test.

There was no significant shortening of lifespan in mice of both sexes exposed to 0.05 mGy/day. The survival curves (Figure 8.4) for male and female mice exposed to 21 mGy/day showed a shift toward the left. There was also a slight shift of the curve towards the left at around 600 to 800 days of age in females irradiated with doses of 0.05 and 1.1 mGy/day [8].

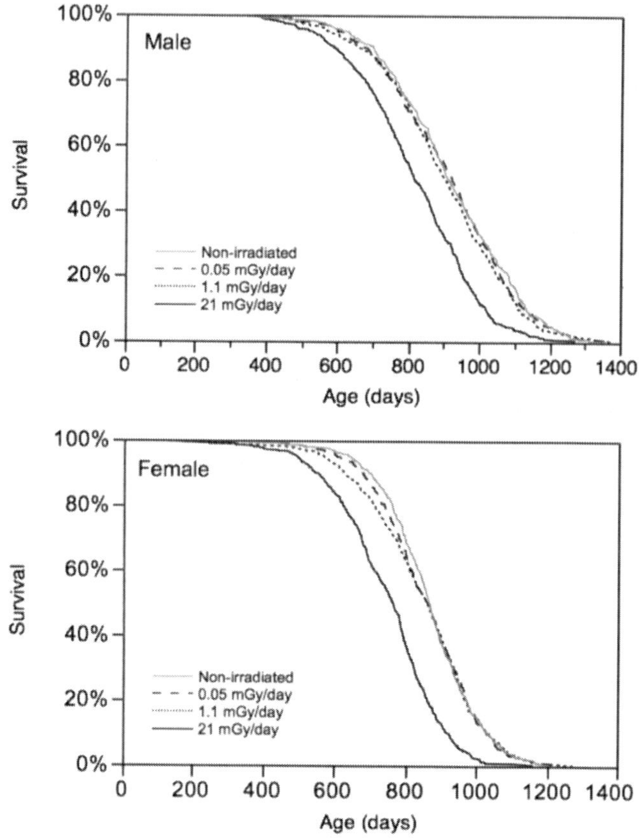

Figure 8.4. Survival curves for B6C3F1 mice exposed to low dose rates of gamma rays for 400 days. [*Radiat. Res. (2003) 160: 376-379]*.

Table 8.3. Kaplan-Meier estimates of mean survival of major causes of death on B6C3F1 female mice [*Radiat. Res. (2007) 167: 417-437*]

Dose rate (Total dose)	**Non-irradiated** (0 mGy)	**0.05 mGy/day** (20 mGy)	**1.1 mGy/day** (400 mGy)	**21 mGy/day** (8000 mGy)
All lethal neoplasms				
Incidence (%)	447 (89.4)	435 (87.9)	431 (86.7)	456 (91.2)
Mean lifespan (±SE)	877.6±6.2	870.4±6.8	865.5±7.2	758.6±6.5 [a]
Malignant lymphoma				
Incidence (%)	275 (55.0)	270 (54.5)	261 (52.5)	260 (52.0) [†]
Mean lifespan (±SE)	939.7±7.7	934.1±8.4	933.7±8.8	831.6±8.6 [a]
Soft tissue neoplasms				
Incidence (%)	43 (8.6)	32 (6.5)	51 (10.3)	46 (9.2) [†]
Mean lifespan (±SE)	1080.2±6.8	1089.1±6.5	1057.8±6.5	941.1±5.8 [a]
Non-neoplastic diseases				
Incidence (%)	46 (9.2)	56 (11.3)	62 (12.5)	38 (7.6)
Mean lifespan (±SE)	1124.4±10.7	1112.7±13.1	1127.3±17.5	958.1± 6.1

[a] P<0.01 Log-rank test. [†] P<0.01Peto's trend test.

Analysis by the Gompertz function, showed progressive increase in the displacement of death rates in both male and female groups irradiated with 21 mGy/day. The displacement was greater at an early age rather than at a later age resulting in a shift of the time axis that indicates premature or precocious aging [8]. Kaplan-Meier estimates of mean survival for major causes of death are summarized in Tables 8.2 (males) and 8.3 (females).

After correction for competing risks, significant radiation-induced life shortening in male mice exposed to 21 mGy/day were caused by malignant lymphoma, lung neoplasms, and hemangiosarcomas, whereas, female mice exposed to 21 mGy/day significant life shortening was due to malignant lymphoma and soft tissue neoplasms [9]. With regards to other neoplasms that caused death, no significant increase in the frequency was observed in irradiated mice when compared to non-irradiated mice [9].

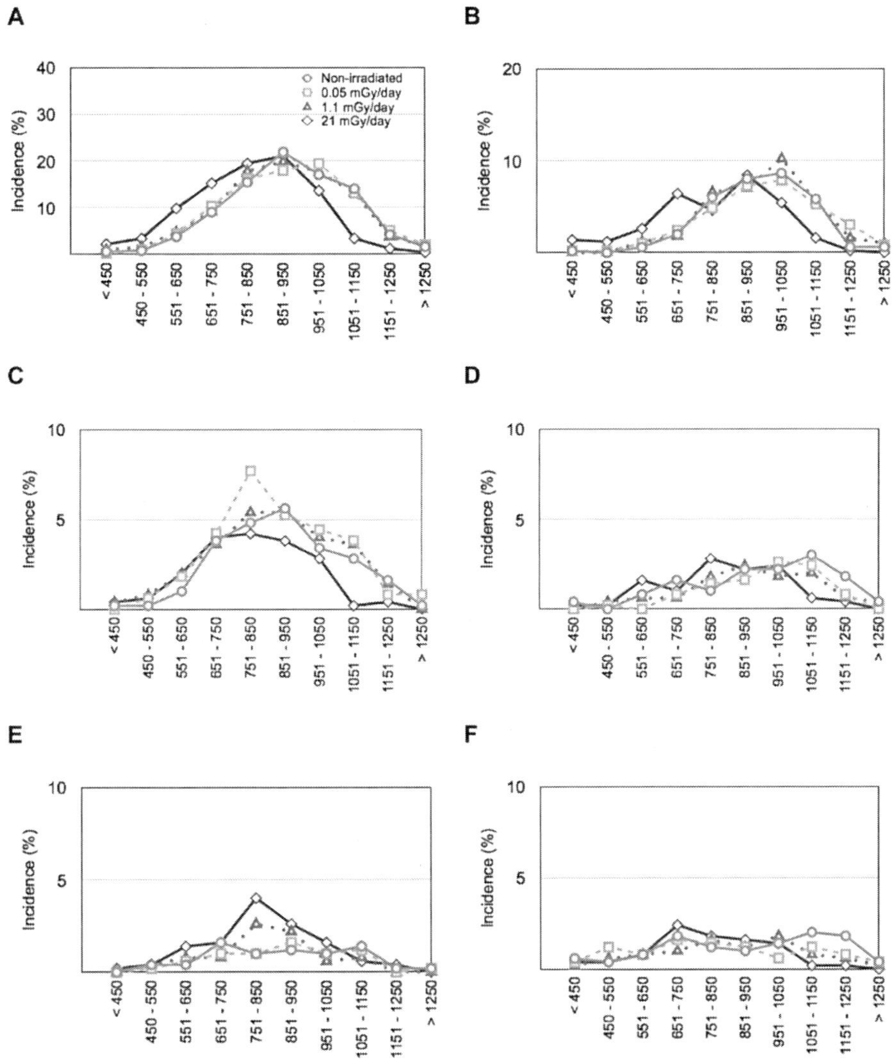

Figure 8.5.1. Temporal distribution of the different major causes of death (COD) in male B6C3F1 mice exposed to low dose-rates of gamma rays: (A) all neoplasms, (B) malignant lymphoma, (C) liver neoplasms, (D) lung neoplasms, (E) hemangiosarcoma, and, (F) soft tissue neoplasms.

Temporal Distribution

Graphic representations of the temporal distributions for neoplasms that caused majority of the deaths show no increase in incidences. There is, however a shift of the line towards the left further confirming that neoplasms caused early death in the irradiated mice (Figures 8.5-1A, 8-5-2A).

For male mice exposed to 21 mGy/day, the shift was due to a combination of several neoplasms, namely, malignant lymphomas (Figure 8.5-1B), liver neoplasms (Figure 8.5-1C) and hemangiosarcomas (Figure 8.5-1E). In female mice exposed to 21 mGy/day however, the shift to the left was mostly due to malignant lymphoma (Figure 8.5-2B) and hemangiosarcoma (Figure 8.5-2E).

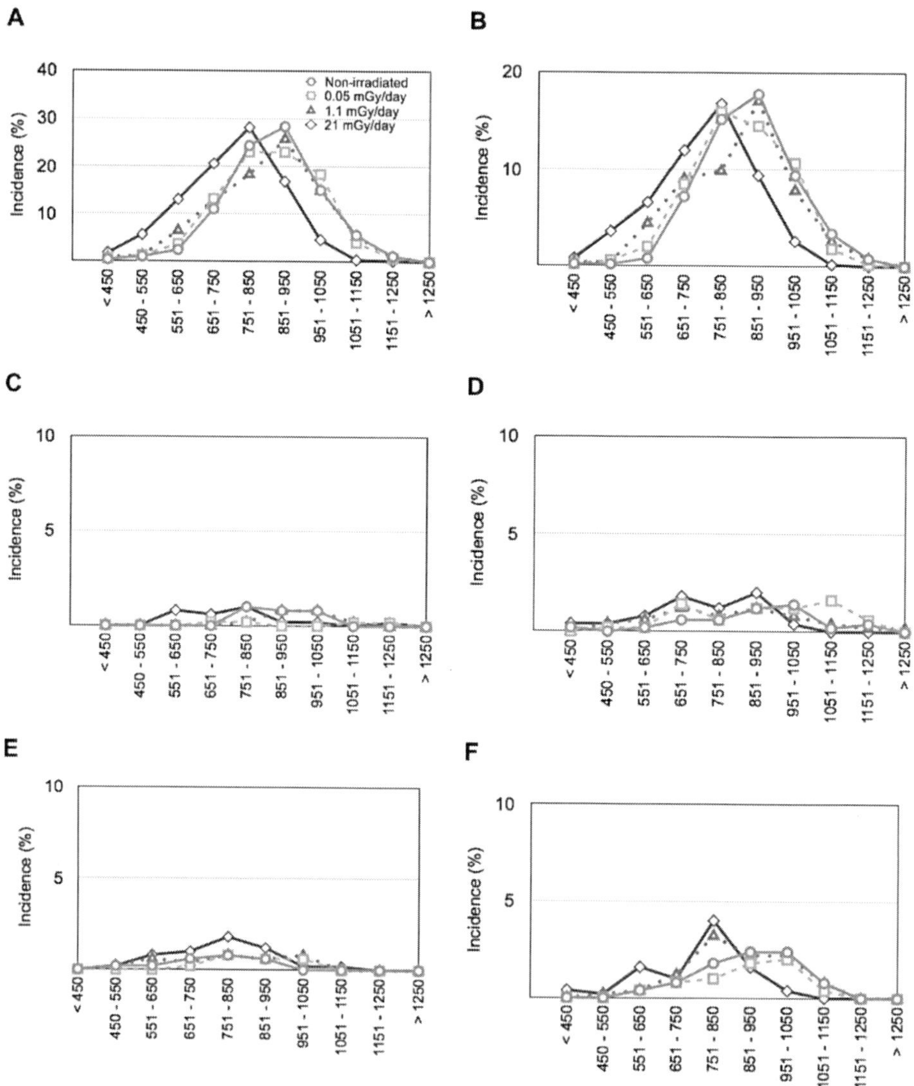

Figure 8.5.2. Temporal distribution of the different major causes of death (COD) in female B6C3F1 mice exposed to low dose-rates of gamma rays: (A) all neoplasms, (B) malignant lymphoma, (C) liver neoplasms, (D) lung neoplasms, (E) hemangiosarcoma, and, (F) soft tissue neoplasms.

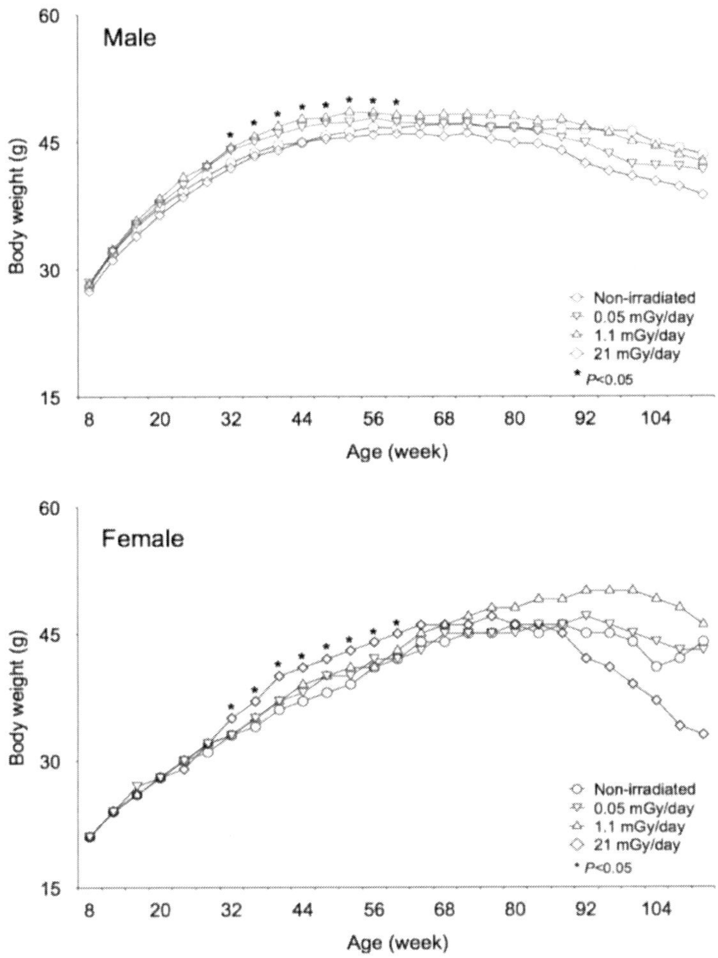

Figure 8.6. Body weights of B6C3F1 mice chronically exposed to low dose rates of gamma rays for 400 days. *[Radiat. Res. (2007) 167: 417-437].*

Body Weights

Male mice irradiated at 1.1 mGy/day showed significant weight gain during the irradiation period from age 32 weeks up to 60 weeks of age (Figure 8.6) [9]. Female mice irradiated at 21 mGy/day showed a statistically significant increase ($P<0.05$) in weight gain during the same irradiation period as the male mice (Figure 8.6). The weight gain during the irradiation period could be due to one or a combination of several factors including increased feed intake, decreased metabolic rate, decreased activity and increased fat deposition (proliferation of adipose cells).

II. Adiposity

To further elucidate the physiology of the striking increase in the bodyweights of female mice during chronic exposure to low dose-rates of gamma rays, a total 343 SPF B6C3F1 female mice were used under the same husbandry (SPF) conditions as the life span study.

In the first experiment (Expt. 1), a total of 325 mice (165 non-irradiated controls and 160 irradiated at 21 mGy/day to a total of 6 or 8 Gy), were used and housed in groups of 5. A second experiment (Expt. 2) to study feed consumption utilized 18 individually housed mice (9 non-irradiated and 9 irradiated at 21 mGy/day to a total of 4.4 Gy).

Body Weights

Bodyweights of group-housed mice (Expt. 1), measured at 7 (pre-irradiation), 37, 44 and 51 weeks of age, were significantly heavier in the irradiated mice from the 29th week (37 weeks of age) of the irradiation period up to the 43rd week (51 weeks of age), with weight gain gradually easing off at around the 44th week [16].

In Expt. 2, bodyweights and feed consumption were measured daily, 5 days a week (Monday to Friday), from 8 to 40 weeks of age. Based on the data from Expt. 1, the individually housed mice (Expt. 2) were sacrificed at 40 weeks of age before body weight gain decreases [16].

Adipose Tissue Deposits

The amounts of subcutaneous inguinal and abdominal (perigonadal and mesenteric) fat deposits dissected from randomly selected mice sacrificed at 51 (Expt. 1) and 40 (Expt. 2) weeks of age were significantly increased in irradiated mice, regardless of housing conditions. Histomorphometry (only in Expt. 2) of frozen sections of perigonadal fat show that the adipocytes of irradiated mice were significantly larger in size than those of the non-irradiated controls (Figure 8.7).

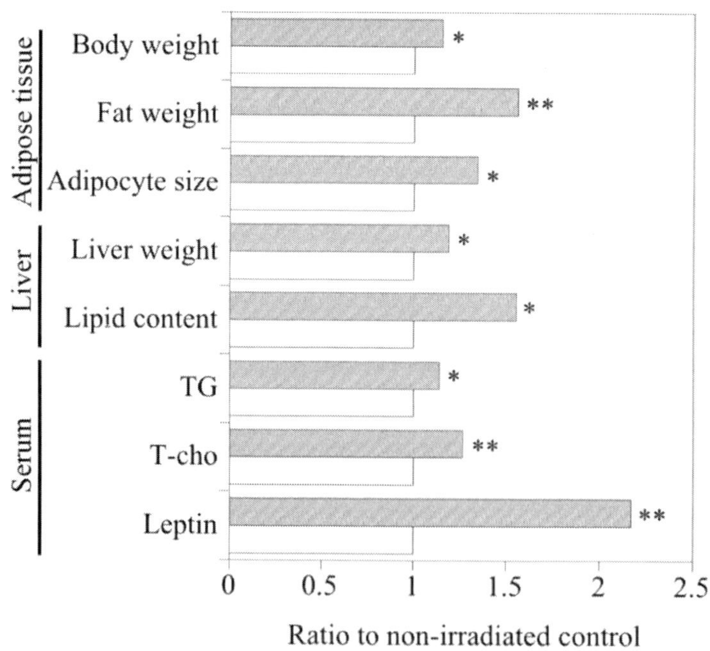

Figure 8.7. Parameters related to lipid metabolism in female B6C3F1 mice chronically exposed to low dose rates of gamma rays expressed as ratios in relation to the non-irradiated controls (white bar = non-irradiated; gray bar = 21 mGy/day). * = P<0.05, ** = P<0.01.

Liver, Serum Chemistry, Adipocytokines

Prior to sacrifice, sera from blood collected via cardiac puncture were separated and stored frozen at -80°C until analyzed. Total cholesterol and HDL cholesterol levels in serum were significantly increased in both group-housed and individually housed irradiated mice whereas serum triglycerides were significantly increased only in the individually housed (Expt. 2) irradiated mice.

Using commercially available ELISA kits, serum levels of insulin, leptin, and adiponectin were measured [16]. Although leptin levels were significantly increased in irradiated mice in both experiments, adiponectin levels were significantly increased only in irradiated mice that were individually housed (Expt. 2). The ratio adipocytokines adiponectin to leptin in the sera of irradiated mice was significantly higher than the non-irradiated controls in both experiments. Serum glucose and insulin levels were significantly increased in group-housed (Expt. 1) irradiated mice only (Figure 8.7).

Only liver weights, liver triglyceride content and total cholesterol of group-housed irradiated mice were significantly increased. Total lipids in the liver, extracted according to Folch et al. [17], of individually housed irradiated mice (Expt. 2) were significantly increased [16].

Feed Consumption

There was no significant increase in the food consumption of individually housed mice (Expt. 2) and calculation for feed efficiency ratios showed significant increases in mice exposed to irradiation at the 16th to 20th (25-29 weeks of age) and at the 21st to 25th (30-34 weeks of age) weeks from the start of irradiation [16].

CONCLUSION

In humans, high doses of radiation have been associated with increased cancer incidences including leukemia, breast, bladder, colon, liver, lung, esophagus, ovarian, multiple myeloma and stomach cancers. Due to the very long latent period of cancer, it is very hard to distinguish those that may be attributable to radiation exposure from those that occur spontaneously or from those resulting from exposure to other carcinogens. Cancer etiology is complex and is further modified by factors such as sex, age at the time of exposure, genetic predilection, lifestyle and environment factors (chemical and physical hazards).

Results from animal studies have not been consistent, but they remain the only source of information on the shape of the dose-response relationships for cancer in both high and low LET radiations. The application of a single model to all neoplasms is impossible [18] as there are different mechanisms of tumorigenesis as there are host responses, many of which are biological factors that are still undetermined. When an animal is exposed to radiation, the effect on the animal as a whole results either from one organ or from the interaction of several organs, and will depend upon many factors such as dose, species, sex, age, to name a few [19].

At low dose and low dose-rate exposures, the irradiated animals initially appear normal and identical to the non-irradiated ones. As the animals grow older however, the irradiated animals have been shown to age sooner and die earlier with higher incidences of some neoplasms [19].

Results of the life span study show that despite the significant over-all shortening of life spans in the 21 mGy/day groups, there was no significant increase in the incidence of neoplastic and non-neoplastic lesions that caused majority of the deaths. The length of the lifespan depended on the cause of death. Other than the shift of the time axis in the 21 mGy/day group [8] suggestive of precocious aging, we have no evidence as to whether radiation exposure at this dose was either due to a shortening of the latency period for the development of neoplasms, or, due to an increase in the proliferative capacity of the neoplastic cells, or both. Experiments to further clarify cancer latency at low dose-rate exposures are currently in progress.

There was no evidence indicating hormesis or adaptive response to the low dose-rates of radiation used since there was no lengthening of the lifespan that could be attributed to decreased tumor incidence. Since the incidence of some tumors increased while others decreased, there was no consistent relationship between radiation exposure and incidence of neoplasia.

The increase in bodyweight of female mice chronically exposed to gamma rays at a low dose-rate of 21 mGy/day is due to increased fat deposition or adiposity with no comparable increase in feed consumption. Non-cancer effects of radiation exposure, such as cardiovascular disease [20~21] and obesity [22~26], particularly in childhood cancer survivors who have undergone radiation therapy [27] have gained increased attention in the recent years.

The physiological changes in irradiated individuals that lead to fat deposition and consequent body weight gain are influenced by many factors including radiation regimen (radiation type, dose and dose-rate, anatomy of the irradiated body part), age, gender (hormones), genetic background, environment (nutrition, pollution), behavior and metabolism (calorie intake, insulin resistance, immunity, etc.).

Gut microbiota has also been reported to be a critical factor involved in energy metabolism and its role in obesity has been well studied in both mice [28, 29] and humans [30] particularly in inflammatory conditions such as endotoxemia and dysbiosis. Mice with normal intestinal microbiota exposed to acute high dose-rate (HDR) gamma rays exhibit enhanced intestinal radiosensitivity compared to germ-free mice [31], suggesting that microbial organisms may mediate tissue radiosensitivity.

The mechanism by which obesity occurs in female B6C3F1 mice exposed to continuous low dose-rate gamma-irradiation is not clear. New experimental data (unpublished) from female B6C3F1 mice chronically irradiated with gamma rays at 21 mGy/day show that significant body weight gain occur simultaneously with loss of PCNA positive ovarian follicles (ovarian atrophy) and anestrus (menopause) suggesting that radiation-induced premature menopause (via oocyte killing) play an important role in body weight gain.

These findings need to be considered in the estimation of cancer risks, where obesity is known to increase it, but also of non-cancer risk, such as cardiovascular diseases, for low dose-rate radiation exposures.

The absence of a statistically significant effect in mice exposed to 0.05 mGy/day in the life span study however does not imply that there is no radiation effect since the probability at which an effect observed is statistically significant depends on the magnitude of the effect and sample size.

In order to relate the health effects of radiation to the results of animal experiments such as these, further research on the mechanisms of radiation carcinogenesis using molecular biology techniques, as well cross sectional studies are required.

This work was performed under contract with the Aomori Prefectural Government, Japan.

REFERENCES

[1] Evaluation of the linear-non-threshold dose-response model for ionizing radiation (NCRP Report No. 136) (2001) Maryland: *National Council on Radiation Protection and Measurements.*

[2] International Committee on Radiation Protection (ICRP) 2005 Low-dose Extrapolation of Radiation-related Cancer Risk. *ICRP Publication* 99. Ann. ICRP 35 (4).

[3] Health Risks from Exposure to Low Levels of Ionizing Radiation (BEIR VII Phase 2) (2006) Washington, DC: The National Academies Press.

[4] United Nations Scientific Committee on the Effects of Atomic Radiation (UNSCEAR) 2010 Report of the United Nations Scientific Committee on the Effects of Atomic Radiation. United Nations, New York.

[5] United Nations Scientific Committee on the Effects of Atomic Radiation (UNSCEAR) (1993) Sources and effects of ionizing radiation, Report to the general assembly, with Annexes. United Nations, New York.

[6] Rao GN, Birnbaum LS, Collins JJ, Tennant W, Skow LC (1988) Mouse strains for chemical carcinogenicity studies: Overview of a workshop. *Fundamentals of Applied Toxicology* 10: 385-394.

[7] Rao GN, Pierporsch WW, Crawford DD, Edmondson J, Haseman JK (1989) Influence of viral infections on body weight, survival and tumor prevalence in B6C3F1 mice in carcinogenicity studies. *Fundamentals of Applied Toxicology* 13: 156-164.

[8] Tanaka S, Tanaka IB III, Sasagawa S, Ichinohe K, Takabatake T, et al. (2003) No lengthening of life span in mice continuously exposed to gamma rays at very low dose rates. *Radiation Research* 160: 376-379.

[9] Tanaka IB III, Tanaka S, Ichinohe, K, Matsushita S, Matsumoto T, et al. (2007) Cause of death and neoplasia in mice continuously exposed to very low dose rates of gamma rates. *Radiation Research* 167: 417-437.

[10] Mohr U, editor (2001) International classification of rodent tumors: The mouse. Springer-Verlag, Berlin, Heidelberg.

[11] Maronpot RR, editor (1999) Pathology of the mouse: Reference and atlas. Cache River Press, Vienna, Ill.

[12] Ettlin RA, Stirnimann O, Prentice DE (1994) Causes of death in rodent toxicity and carcinogenicity studies. *Toxicologic Pathology* 23: 165-178.

[13] Fry RJM, Garcia AG, Allen KH, Sallese A, Staffeldt E, et al. (1976) Effects of pituitary isografts on radiation carcinogenesis in mammary and harderian glands of mice. In: Biological and environmental effects of low-level radiation, vol. 1. Vienna: IAEA (Publication STI/PUB/409). pp. 213-237

[14] Lipman RD (1997) Pathology of aging rodents: Inbred and hybrid models. *Experimental Gerontology* 32: 215-228.
[15] Haseman JK, Young E, Eustis SL, Hailey JR. (1997) Bodyweight-tumor incidence correlation in long-term rodent carcinogenicity studies. *Toxicologic Pathology* 25: 256-263.
[16] Nakamura S, Tanaka IB III, Tanaka S, Nakaya K, Sakata N, et al. (2010) Adiposity in female B6C3F1 mice continuously irradiated with low-dose-rate gamma rays. *Radiation Research* 173: 333-341.
[17] Folch J, Lee M, Sloane Stanley GH (1957) A simple method for the isolation and purification of total lipids from animal tissues. *Journal of Biological Chemistry* 226: 497-509.
[18] Covelli V, Smith H (1989) Animal Studies: Introduction. In: Baverstock KF, Stather JW, editors. Low Dose Radiation: Biological Bases for Risk Assessment. London: Taylor & Francis. pp. 179-190.
[19] Casarett AP (1968) *Radiation Biology*. New Jersey: Prentice-Hall, Inc. p. 266.
[20] Preston DL, Shimizu Y, Pierce DA, Suyama A, Mabuchi K (2012). Studies of mortality of atomic bomb survivors. Report 13: Solid cancer and noncancer disease mortality: 1950-1997. *Radiation Research* 178: 146-172.
[21] Shimizu Y, Kodama K, Nishi N, Kasagi F, Suyama A, et. al. (2010) Radiation exposure and circulatory disease risk: Hiroshima and Nagasaki atomic bomb survivor data, 1950-2003. *British Journal of Medicine* 340: b5349.
[22] Odame I, Reilly JJ, Gibson BE, Donaldson MD (1994) Patterns of obesity in boys and girls after treatment for acute lymphoblastic leukaemia. *Archives of Disease in Childhood* 71: 147-149.
[23] Lustig RH, Post SR, Srivannaboon K, Rose SR, Danish RK, et al. (2003) Risk factors for the development of obesity in children surviving brain tumors. *Journal of Clinical Endocrinology and Metabolism* 88: 611-616.
[24] Gurney JG, Ness KK, Stovall M, Wolden S, Punyko JA, et al. (2003) Final height and body mass index among adult survivors of childhood brain cancer: childhood cancer survivor study. *Journal of Clinical Endocrinology and Metabolism* 88: 4731-4739.
[25] Murphy AJ, Wells JC, Williams JE, Fewtrell MS, Davies PS, et al. (2006) Body composition in children in remission from acute lymphoblastic leukemia. *American Journal of Clinical Nutrition* 83: 70-74.
[26] Janiszewski PM, Oeffinger KC, Church TS, Dunn AL, Eshelman DA, et al. (2007) Abdominal obesity, liver fat, and muscle composition in survivors of childhood acute lymphoblastic leukemia. *Journal of Clinical Endocrinology and Metabolism* 92: 3816-3821.
[27] Tukenova M, Guibout C, Oberlin O, Doyon F, Mousannif A, et al. (2010) The role of cancer treatment in long-term overall and cardiovascular mortality after childhood cancer. *Journal of Clinical Oncology* 28: 1308-15.
[28] Bäckhed F, Ding H, Wang T, Hooper LV, Koh, GY, et al. (2004). The gut microbiota as an environmental factor that regulates fat storage. *Proceedings of the National Academy of Sciences of the United States of America* 101: 15718-15723.
[29] Bäckhed F, Manchester JK, Semenkovich CF, Gordon JI (2007) Mechanisms underlying the resistance to diet-induced obesity in germ-free mice. *Proceedings of the National Academy of Sciences of the United States of America* 104: 979-84.

[30] Sanz Y, Santacruz A, De Palma G (2008) Insights into the roles of gut microbes in obesity. Interdisciplinary Perspectives on Infectious Diseases, vol. 2008, Article ID 829101, 9 pages. doi:10.1155/2008/829101.

[31] Crawford PA, Gordon JI (2005) Microbial regulation of intestinal radiosensitivity. *Proceedings of the National Academy of Sciences of the United States of America* 102: 13254-13259.

In: Fukushima Nuclear Accident
Editor: Shizuyo Sutou
ISBN: 978-1-63463-748-0

Chapter 9

DOSE-EFFECT RELATIONSHIP BETWEEN RADIATION AND CANCER INDUCED BY REPETITIVE OR CONTINUOUS IRRADIATION

Akira Ootsuyama*

Division of Radiation Biology and Health, School of Medicine,
University of Occupational and Environmental Health, Japan, Fukuoka, Japan

ABSTRACT

We focused on the carcinogenic effects caused by repetitive or continuous radiation at a low dose-range, which have received attention in recent years, and we scrutinized the experimental methods used to evaluate radiation-induced carcinogenesis. For this study, we selected a mouse species with as few specific autochthonous cancers as possible. Skin cancer was selected for analysis as the target cancer induced by radiation, because it is a rare autochthonous cancer in mice. Beta rays were selected as the radiation mode for inducing cancer. The advantage of using beta rays is their low penetration power into the tissues, thus protecting other organs such as the digestive and hematogenic organs, from the effects of radiation. The benefit of our method of irradiating the mice with beta rays is that only the skin cancer has to be monitored, and it is possible to perform long-term experiments because of little influence of the beta rays on organs other than skin. The skin on the back of the mice were exposed to beta rays repetitively three times a week until the occurrence of cancer or death, and the dose per exposure was set in the range of 0.5 Gy to 11.8 Gy. For the dose range of 2.5 Gy to 11.8 Gy per exposure, the length of the latency period and the carcinogenic rate (100%) were almost the same in each experimental group. When the dose per exposure was reduced to 1-1.5 Gy, an extension of the latency period was seen, but the carcinogenic rate remained at 100%. Furthermore, when the dose per exposure was reduced to 0.75 Gy, not only was the latency period increased but also the carcinogenic rate was decreased. When the dose per exposure was 0.5 Gy, skin cancer did not occur at all, even when irradiation was continued until the death of the last mouse in this experimental group. The mice in the 0.5 Gy exposure

* Correspondence to: Akira Ootsuyama, Tel: +81-93-691-7433, Fax: +81-93-692-0559, E-mail: aootsuya@med.uoeh-u.ac.jp.

group died from the same causes of death and showed the same lifespans as those of the control group. The results showed that this exposure dose did not cause cancer, even when the mice were repetitively or continuously exposed to radiation throughout their lifespan. We refer to this as the non-carcinogenic dose, as determined by the experiment, or the threshold-like dose.

INTRODUCTION

Carcinogenesis caused by radiation exposure has been a concern since humans began using radiation, and the mechanisms of radiation carcinogenesis have been a subject of considerable interest and study for nearly a century. The risk of radiation-induced cancer has been inferred from epidemiological studies of the A-bomb survivors and studies of survivors of radiation exposure accidents as well as considerable amount of data in animal experiments. However, there is little consensus regarding the epidemiological risk by low-dose radiation, because the carcinogenic mechanisms are not yet clear.

1. Anxiety of Fukushima Citizens Regarding Radiation Effects After the Fukushima Daiichi Nuclear Power Plant (FDNPP) Accident

A questionnaire regarding anxiety issues was administered after the FDNPP disaster to residents of a radiologically contaminated district and its periphery in Fukushima in 2011 and 2013 [1]. The fear of carcinogenesis by chronic daily radiation exposure was rated as the highest among the various concerns, even though two years had passed since the disaster. Anxieties regarding the radiation effects to subsequent generations and radiological contamination of food and water were also high. As scientists in the radiation biology field, we aim to alleviate their concerns, which appear to be due to a lack of knowledge regarding radiation.

Traditionally, the A-bomb survivor data have been used to to estimate the effects of radiation doses near100 mSv through linear extrapolation of higher dose data. However, the carcinogenic mechanism in the A-bomb survivors having acute radiation exposure would likely differ from that following repetitive or continuous radiation exposure, which is the concern of Fukushima citizens. Therefore, understanding the carcinogenic mechanism following repetitive or continuous radiation exposure are now required. In this chapter, we address the dose-effect relationship of carcinogenesis induced by repetitive or continuous radiation exposure using experimental animal models, in contrast to the existing data obtained from epidemiological studies and from carcinogenesis induced by single, high-dose radiation exposure.

2. Issues with Radiation-Induced Carcinogenesis Experiments Using Animal Models

There are many potential factors that can affect the results of radiation-induced carcinogenesis experiments in animal models such as the radiation method, e.g., radiation type and exposure method (whole or partial body; single, repetitive, or continuous exposure; dose and dose rate), the experimental animal, e.g., species, age, sex, hair growth cycle, and genetic modifications, and the analysis endpoint, e.g., occurrence of cancer or cancer-related death. Hence, despite the considerable amount of results from previous animal experiments, there continues to be uncertainty regarding radiation-induced carcinogenesis. In the present study, we have taken many of these factors into consideration in choosing the various experimental conditions in order to enable a reliable assessment of radiation-induced carcinogenesis.

In such experiments using animal models, there are many cases in which a specific autochthonous cancer is prone to occur because of the genetic background of the animal model species, and these cancers may be increasingly observed in radiation-induced cancer experiments. Therefore, in order to evaluate only carcinogenesis induced by the radiation itself, it is necessary to exclude autochthonous cancers (i.e., the non-radiation-induced cancers) from the endpoint analysis. However, it can be very difficult to perform the experiment in such a manner. Because autochthonous cancers generally have a short latency period, there are many cases in which they mask the occurrence of the target cancer during the experimental period.

3. An Advantageous Procedure to Avoid Autochthonous Cancer Issues

In the radiation carcinogenic experiments described here, the mouse species prone to autochthonous cancer were excluded, and skin cancer, which is rare in the experimental mouse model, was selected as the radiation-induced cancer for analysis. In addition, to avoid radiation effects in organs other than the skin, we selected beta rays to irradiate only the skin. With these considerations, even in the experiments using long-term repetitive or continuous radiation exposure, radiation-induced cancers would be expected to occur without shortening the lifespan of the animals.

Results and Discussions

Using the experimental method that we produced after extensive consideration of various experimental conditions, we discuss the results from the analysis of the relationship between the radiation dose effect and cancer induced by repetitive irradiation. Furthermore, we showed how a defect in a tumor suppressor gene (p53) affects the relationship between dose effect and radiation-induced carcinogenesis. The defect in p53 also has an effect on the induction of autochthonous cancers specific to the genetic background of that species. However, the mouse species-specific autochthonous cancer evaluated here was a thymic lymphoma and differed from the skin cancer, the target of this investigation.

1. Radiation-Induced Cancer and Dose Effect Using Repetitive Irradiation

To analyze only the target cancer induced by radiation, we used 7-week-old ICR female mice, in which there are thought to be lower incidences of species-specific autochthonous cancers. The autochthonous cancers that developed in these mice comprised mainly lung cancers, seen in approximately 20% of female mice. Malignant lymphoma was seen in approximately 10% of females [2]. In addition, Yamate et al. did not report skin cancer as an autochthonous tumor in the ICR mouse model of their study [2]. We used a ^{90}Sr-^{90}Y disk source (diameter of 2 cm, dose rate of 2.24 ~ 1.79 Gy/m) as the radiation source. We shaved the hair off the backs of the mice and stuck the disk to the skin, and this was performed three times a week until the occurrence of skin cancer or death. The mice were divided into 10 groups (0.5, 0.75, 1.0, 1.5, 2.5, 3.0, 3.5, 4.7, 8.0, and 11.8 Gy dose per exposure per mouse) and compared with a control group. Each experimental group included 15 to 31 mice. The irradiation was started at 7 weeks after birth. The carcinogenic rate was calculated using the Kaplan-Meier method [3]. The cancers that developed in these experimental groups were determined pathologically as squamous cell carcinoma, basal cell carcinoma, fibrosarcoma or osteosarcoma. Osteosarcoma is not a skin cancer, but it may have resulted from radiation exposure to the spine, which is a rare occurrence [4-8]. In addition, in a radiation-induced skin carcinogenesis experiment using a rat model [9, 10], cancer developed in hair follicle cells, but we did not detect such a cancer in our mouse model, despite mice and rats both being rodents. Because the radiation sensitivity of hair follicle cells in mice is higher than that in rats, the hair follicle cells may die before they develop cancer, thus preventing such adnexal cancers from developing in mice [11]. We previously performed a radiation-induced skin carcinogenesis experiment using a high-dose single exposure of beta rays. Fifty ICR female mice were irradiated with 27 Gy, and the most severe acute injuries manifested in skin approximately 20 days after irradiation and then gradually formed scars, but no tumors developed until 23 months after irradiation [12]. This indicates that a large single dose of irradiation, such as that provided by an A-bomb, could induce a lower incidence of skin cancer. Because the skin damage by high-dose radiation might rip the ground for the carcinogenesis off.

Even though the high-dose range group (2.5 Gy ~ 11.8 Gy per exposure) includes widely different doses, the changes in both the latency period and the carcinogenic rate were very small. In the irradiation groups, epilation occurred in the irradiated region of the skin within several months after the start of irradiation, and in the high dose exposure group, erythema, wet dermatitis, erosion, ulcer and crust formation also occurred. The carcinogenic rate did not change when we reduced the dose from the 2.5 Gy ~ 11.8 Gy range to 1.5 Gy per exposure, but the latency period was extended significantly. In this experimental group, epilation began approximately 10 months after the start of irradiation, and erythema, dry dermatitis and crust formation were observed in the irradiated skin area, but no ulcers developed in any of the mice. When we again reduced the dose from 1.5 Gy to 0.75 Gy per exposure, the carcinogenic rate decreased, and the latency period was prolonged. When the dose was further reduced to 0.5 Gy per exposure, we detected no skin cancer development during the lifespan of any of the mice. Furthermore, we had to shave the hair from the irradiated region several times, because even epilation did not occur in the 0.5 Gy exposure group. The radiation sensitivity of the hair follicle cells is higher than that of the surrounding squamous cells [13], but the absence of epilation implies that this radiation dose was not sufficient to cause this typical non cancer-related effect in skin tissue irradiated repetitively with 0.5 Gy. These results indicate that mice irradiated repetitively with the low radiation doses during their lifespan do not develop skin cancer. We refer to this dose as the threshold-like dose. (Figure 1)

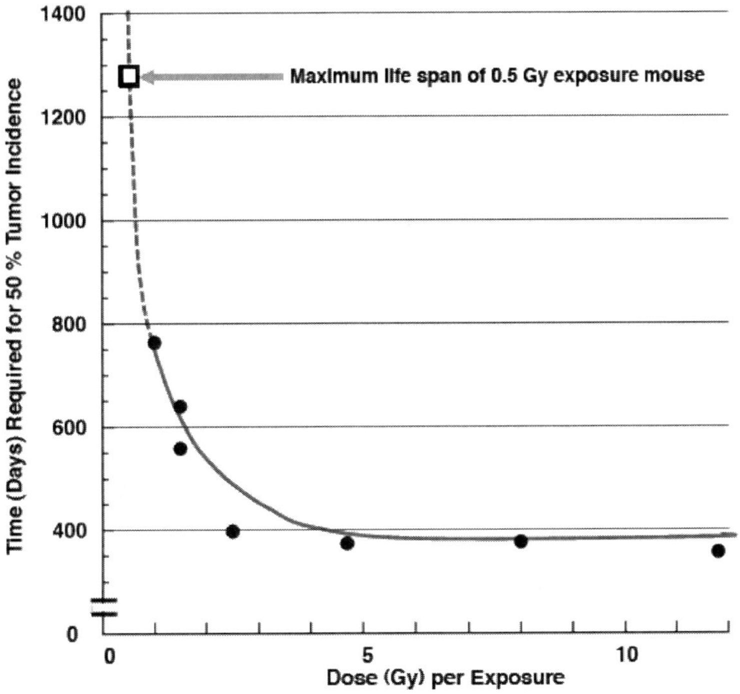

Figure 1. The relationship between the time required until 50% cancer incidence and the exposure per dose.

●; indicate the day at which 50% cancer incidence occurred in the 11.8, 8.0, 4.7, 2.5, 1.5, 1.0 Gy exposure groups. □; indicate the day at which death of the last mouse occurred in the 0.5 Gy exposure group. When the exposure per dose decreased, the solid line indicating the time required until 50% cancer incidence increased sharply. Because radiation-induced skin cancer was not seen in the 0.5 Gy group, the estimated 50% cancer incidence time is represented by a dashed line for exposure doses below 1.5 Gy. The threshold-like dose of radiation-induced carcinogenesis is suggested by this figure under these experimental conditions.

The 0.5-Gy dose was used as the minimum dose in this experiment, but it is actually not a so-called low dose. When we calculated the total exposure dose of the mouse which lived the longest in the 0.5 Gy exposure group, it was approximately 257 Gy. Even though the mice received a very large total dose of radiation, the observation that neither epilation nor skin cancer developed during the life span of the mice is important knowledge. In other words, in the case of repetitive or continuous irradiation, the radiation influence by the total dose is not as important. Similar results have been reported in previous animal experiments [14] but, surprisingly, have not been regarded as significant.

In the experimental groups irradiated with doses ranging from 2.5 Gy ~ 11.8 Gy, we did not see a marked difference in either the latency period or the carcinogenic rate, which was puzzling. The dose-effect relationships between the cell mutation rate and the cell survival rate in experiments using cultured cells [15-17] might be important to solve this puzzle. In the experimental group irradiated with 2.5 Gy ~ 11.8 Gy doses, the rate of cell mutation was high and the rate of cell survival was low in the high dose groups (e.g., 11.8 Gy group). On the contrary, the rate of cell mutation was low and the rate of cell survival was high in the low dose groups (e.g., 2.5 Gy group). As a result, it is thought that the balance between cell mutation and survival rates was the influential factor that resulted in the same latency periods and carcinogenic rates between the different radiation groups. Therefore, there might be an optimal carcinogenic dose in the 2.5 Gy ~ 11.8 Gy range for radiation-induced skin cancer. The non-cancer lesions, such as dermatitis, erosion and ulcer, in the irradiated region, which were present before the development of skin cancer in the mice of each experimental group, were thought to derive from the lethal (non-survival) region. On the other hand, it might be that the carcinogenic rate decreased and the latency period increased because both ratios of lethal region and mutational region continued to decrease when the single irradiation dose was reduced below this dose range. Furthermore, the repair mechanism for radiation-induced injuries may still be effective in the presence of a low dose, and radioresistance may develop as an adaptive response because of the repetitive exposure used in these experiments. It is thought that not only the prolonged latency period but also the decreased carcinogenic rate are induced rapidly by these compounded effects. In addition, severe dermatitis and ulcers would not have occurred because the ratio of the lethal region decreased sharply in the lower exposure dose groups. In the 0.5 Gy exposure group, this dose could not effectively kill the hair follicle cells despite their high sensitivity to radiation; thus the repair mechanism is almost completely functional in the skin tissue after irradiation. This might be why the 0.5-Gy exposure dose did not cause skin cancer within the life span of the mice.

Of course, the radiation induced carcinogenic process is not as simple as this explanation but rather seems to be a multistep process, and it is necessary to also consider epigenetic changes and/or genetic instability. When we consider the skin as an organ, the tissue consists of many cells in various stages of differentiation and of the cell cycle, and thus the radiation sensitivity among them is not uniform. Therefore, the results of the cell culture experiment cannot be compared directly to the results of this experiment, but the ratios of the lethal and mutational regions induced by irradiation might affect the next stages in the carcinogenic process.

2. Radiation-Induced Cancer and the Dose-Effect Relationship under the Influence of Defective Cancer Suppressor Genes

Generally speaking, since individuals may have a genetic background which increases their susceptibility to an autochthonous cancer, and since the conditions of exposure vary in each situation, we cannot perfectly apply the results of our experiment to the typical model of radiation-induced carcinogenesis in humans. In an experiment that investigated the carcinogenic effect of a low dose exposure, irradiated mice developed autochthonous cancers specific to their genetic background and exhibited shorter latency periods of the radiation-induced cancer in comparison with the non-irradiated group [18, 19]. This shortened latency period seemed to be caused by the accelerated induction of the autochthonous cancer. The cancer-susceptible genetic background was shown to have a large influence on the latency period of the radiation induced carcinogenesis.

The p53 gene is generally known to inhibit carcinogenesis via broad mechanisms. To investigate the does-effect relationship between the radiation dose and carcinogenesis according to different genetic backgrounds, we used mice with a defective p53 gene under the same irradiation conditions described above. Approximately 31% of the radiation-induced cancers obtained from the experiment using the ICR mice had some form of mutation in the p53 gene, which might have had a significant influence on the radiation-induced skin cancer [20]. The p53 gene knock out (KO) mouse (p53-/-) that we used frequently develop thymic lymphomas as a specific autochthonous cancer, because they inherit the genetic characteristics of C57BL/6N mice; their 50% survival rate is approximately 6 months [21]. For the radiation-induced carcinogenesis experiment, we used the p53-/- KO mice, F1 mice (p53+/-) obtained from crossing the p53-/- KO mice with C57BL/6N mice (p53+/+), and p53 wild-type mice (p53+/+) as the control. It is believed that because of their very short life span, the p53-/- KO mice are at risk of death prior to the development of cancer. Because our goal was to investigate the radiation influence under the presence versus absence of the p53 gene, we decided to use the F1 mice (p53+/-) because of their 50% survival rate at approximately 500 days.

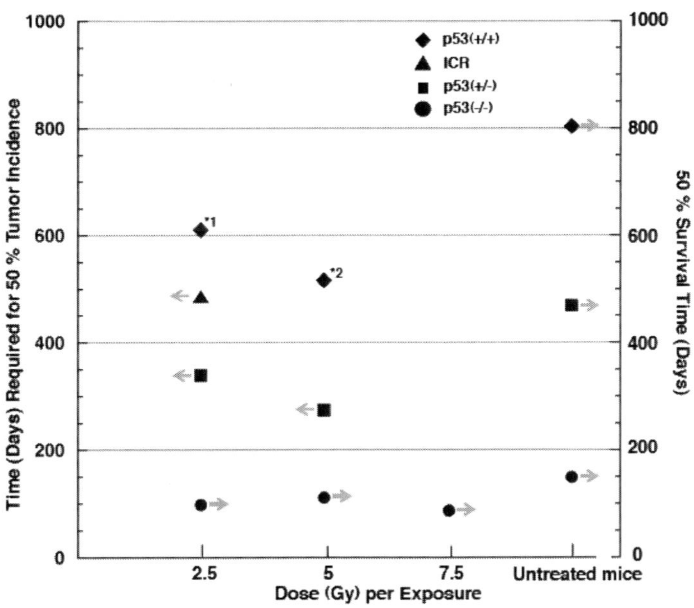

Figure 2. The relationship between the time required until 50% cancer incidence and the exposure per dose according to the p53 gene allelotype in mice.

Small arrows indicate the y-axis to which that value refers. •; indicate the day at which 50% survival was reached in each radiation exposure group of the p53 knockout (-/-) mice; radiation-induced skin cancer was not seen in these mice. ■,▲; indicate the day at which 50% cancer incidence was reached in the F1 (p 53 +/-) and ICR mouse radiation exposure groups. ◆; indicate the day at which the highest cancer incidence was reached in the p53 wild-type (+/+) mouse exposure groups (because the cancer incidences did not reach 50% in these groups, the plotted values indicate the highest cancer incidence times). *1; the cancer incidence was 25% (2/8 mice). *2; the cancer incidence was 18% (6/33 mice).

We used a ^{90}Sr-^{90}Y disk source (diameter of 2 cm, dose rate of 2.24 ~ 1.79 Gy/m) as the irradiation source. We shaved the back hair off the mice and stuck the disk to the skin three times a week until the occurrence of skin cancer or death. There were three experimental groups (2.5, 5.0, 7.5 Gy dose per single exposure) plus a control group. Each experimental group included 8-46 mice. The irradiation was started at 7 weeks after birth. The carcinogenic rate was calculated using the Kaplan-Meier method. The cancers that developed in these experimental groups were determined pathologically as squamous cell carcinoma, basal cell carcinoma, fibrosarcoma or osteosarcoma.

Using p53-/- KO mice, the experimental groups were exposed to either 2.5, 5.0 or 7.5 Gy radiation dose per single exposure. In this carcinogenic experiment, we did not see cancer occurrence in any of the exposure groups, and the survival curves for each group were similar to that of the non-irradiated group (p53-/-) (Figure 2).

Using p53 wild-type mice (p53+/+), the experimental groups were exposed to either 2.5 or 5.0 Gy radiation. A shorter latency period and higher carcinogenic rate were seen in these mice after exposure to 5.0 Gy, in comparison with p53 wild-type mice (p53+/+) exposed to 2.5 Gy. A longer latency period and lower carcinogenic rate were evident in the 2.5 and 5.0 Gy groups in comparison with the 2.5 Gy ICR mouse group. (Figure 2)

In the F1 mice (p53+/-), the 5.0 Gy exposure group showed a shortened latency period and increased carcinogenic rate compared with the 2.5 Gy exposure group. (Figure 2)

We had originally expected a shortened latency period in the p53-/- KO mouse group compared with the other mice. However, there was no development of skin cancer in this group. The cancer-prone genetic background was effective in shortening the latency period of mouse-specific autochthonous cancers. Furthermore, it did not seem to have any effect on the radiation-induced skin cancer. The defective p53 gene was assumed to contribute significantly to the development of autochthonous thymic lymphoma in the p53-/- KO mouse. The autochthonous thymic lymphoma may have already begun to develop during the fetal stage of these mice. Even though we began irradiation in the 7th week after birth, the autochthonous thymic lymphoma developed before the effects caused by p53 gene loss on radiation-induced skin carcinogenesis occurred. The thymic lymphoma seemed to have masked the radiation-induced skin cancer. Thus, the survival curves of all the p53-/- KO mouse groups might have appeared the same as that of the non-irradiated group, because they were unrelated to the radiation-induced carcinogenesis.

The potential reasons for this are shown by the experiment using the F1 mice (p53+/-), of which the 2.5 Gy and 5.0 Gy irradiation groups showed a shortened latency period and increased carcinogenic rate compared with the 2.5 Gy and 5.0 Gy irradiation groups of the p53 wild-type mice (p53+/+), respectively. Furthermore, the 2.5 Gy irradiation group of the F1 mice (p53+/-) showed a shortened latency period in comparison with the 2.5 Gy ICR mouse group. The 2.5 Gy and 5.0 Gy irradiation groups of the p53 wild-type mice (p53+/+) showed a prolonged latency period and decreased skin carcinogenic rate in comparison with the 2.5 Gy ICR mouse group. Even though the p53 wild-type mouse has the same p53 genetic background as the ICR mouse, it is thought that the C57BL/6N background mouse has lower sensitivity to radiation with respect to skin carcinogenesis in comparison with the ICR mouse. In the case of the p53-/- KO mouse, the mouse-specific autochthonous cancer had an advantage over the radiation-induced skin cancer. The defective p53 gene status in the F1 mouse (p53+/-) apparently contributed to the radiation-induced skin cancer.

In regard to the p53 gene status of the skin cancers that developed in the experiments using p53 wild-type (p53+/+), F1 (p53+/-) and ICR mice, we analyzed the differences in p53 gene mutations and allelotypes [22] using PCR (polymerase chain reaction) and SSCP (single-strand conformation polymorphism) methods [23]. Surprisingly, a switch to the p53 null allelotype was seen in 60% of the skin cancers that developed in the F_1 mice (p53+/-).

This result indicates that the wild-type p53 gene is lost during the carcinogenic process after the start of irradiation. Also, in the p53 wild-type mice (p53+/+), the allelotype in three of the five skin cancers changed to the p53 heterozygous status (+/-), and mutations were found in all remaining p53 genes. In the ICR mouse group, while there was no change in allelotype, p53 gene mutations were found in three of the four skin cancers. However, it is unknown whether these changes are the cause or the result of the skin cancer.

Thymic lymphoma, which is an autochthonous cancer, occurs earlier in p53 gene KO mice (p53-/-) regardless of whether they undergo irradiation. It appeared that radiation was not involved in the developmental mechanics of the radiation-induced skin cancer in this study. However, the latency period of skin cancer was shorter in F1 mice (p53+/-) in comparison with p53 wild-type mice (p53+/+); a loss of the wild-type p53 gene might be caused by repetitive irradiation, and it is suggested that the p53 null (-/-) status contributed to the developmental mechanics of radiation-induced skin cancer. In the skin cancers of the p53 wild-type mice (p53+/+), loss of one of the normal p53 genes may have resulted from radiation.

In addition, the mutations may have occurred in the remaining p53 genes due to radiation exposure. However, the order of events may have been reversed in this situation. Regardless, because both alleles of the p53 gene were expected to be affected, the latency period should be longer than that of the F1 mouse (p53+/-).

3. Anxiety in Fukushima Citizens Regarding Cancer Induced by Repetitive Irradiation Exposure

From the results of these experiments, there is a chance that cancer does not occur when the radiation dose per exposure decreases, even if the radiation exposure is long term; however, the tendency of carcinogenesis after radiation exposure depends on the genetic background. The carcinogenic risk induced by repetitive or continuous radiation exposure in humans seems to be considerably lower than that suggested from the results of these animal experiments, possibly because the experimental animal model used already has a strong cancer-prone genetic background compared with humans. However, the estimated range of the radiation-induced carcinogenesis risk might be considerably broad considering the non-uniform genetic background of humans. The citizens of Fukushima who remain in the radiological pollution district or along the periphery of the pollution district may be exposed to repetitive or continuous radiation. If their exposure level can be limited to within 20 mSv/year of the upper limit (when the radiological pollution district is in the restoration stage. ICRP Pub.109), it does not seem necessary to particularly fear a carcinogenic risk.

Conclusion

To evaluate radiation-induced carcinogenesis by repetitive or continuous radiation at a low dose range, beta rays were exposed three times a week to the back skin of mice until the occurrence of skin cancer or death. Because beta rays do not penetrate beyond the skin, the radiation influence on other organs decreases. And spontaneous skin cancer is rare in experimental mice. The dose per exposure was set at a range of 0.5 Gy ~ 11.8 Gy. The length of the latency period and the carcinogenic rate (100%) were almost the same in high-dose (2.5, 4.7, 8.0, and 11.8 Gy) groups. The latency period was extended, but the carcinogenic rate was maintained at 100% in mid-dose (1.0 and 1.5 Gy) groups. In a 0.75 Gy group, the latency period was extended and the carcinogenic rate decreased. When the dose per exposure was 0.5 Gy, skin cancer did not occur at all until the death of the last mouse. The mice of this group died of the same causes of death and showed the same life spans as those of the control group. We refer to this non-carcinogenic dose (0.5 Gy) as the threshold-like dose. In view of this, it appears plausible that in a similar manner a threshold dose would exist for humans which would be much above the anticipated cumulative radiation dose the returning Fukushima residents would face.

References

[1] Kohzaki M, Ootsuyama A, Moritake M, Abe T, Kubo T, et al. (2015) What have we learned from a questionnaire survey of citizens and doctors both inside and outside of Fukushima? –Survey comparison between 2011 and 2013–. *Journal of Radiology and Protection* 35: N1-N17. (in press)

[2] Yamate J, Tajima M, Kudou S (1986) Comparison of Age-Related Changes between Long-Lived CRJ: CD-1 (ICR) and CRJ:B6C3F1 Mice. *Japanese Society of Veterinary Science* 48: 273-284.

[3] Hoel DG, Walburg Jr. HE, Branch B (1972) Statistical analysis of survival experiments. *Journal of National Cancer Institute* 49: 361-372.

[4] Ootsuyama A , Tanooka H (1988) One hundred percent tumor induction in mouse skin after repeated beta irradiation in a limited dose range. *Radiation Research* 115: 488-494.

[5] Ootsuyama A, Tanooka H (1989) Induction of osteosarcomas in mouse lumbar vertebrae by repeated external beta-irradiation. *Cancer Research* 49: 1562-1564.

[6] Ootsuyama A , Tanooka H (1991) Threshold-like dose of local beta irradiation repeated throughout the life span of mice for induction of skin and bone tumors. *Radiation Research* 125: 98-101.

[7] Ootsuyama A , Tanooka H (1993) Zero tumor incidence in mice after repeated lifetime exposures to 0.5 Gy of beta radiation. *Radiation Research* 134: 244-246.

[8] Tanooka H , Ootsuyama A (1993) Threshold-like dose response of mouse skin cancer induction by repeated beta irradiation and its relevance to radiation-induced human skin cancer. *Recent Results Cancer Research* 128: 231-241.

[9] Burns FJ, Albert RE, Sinclair IP, Bennett P (1973) The effect of fractionation on tumor induction and hair follicle damage in rat skin. *Radiation Research* 53: 235-240.

[10] Burns FJ, Albert RE, Sinclair IP, Vanderlaan M (1975) The effect of 24-hour fractionation interval on the induction of rat skin tumors by electron radiation. *Radiation Research* 62: 478-487.
[11] Albert RE, Burns FJ, Bennett P (1972) Radiation-induced hair-follicle damage and tumor formation in mouse and rat skin. *Journal of the National Cancer Institute* 49: 1131-1137.
[12] Hoshino H, Tanooka H (1975) Interval effect of beta-irradiation and subsequent 4-nitroquinoline 1-oxide painting on skin tumor induction in mice. *Cancer Research* 35: 3663-3666.
[13] Ootsuyama A, Tanooka H (1986) Unscheduled DNA synthesis after beta-irradiation of mouse skin in situ. *Mutation Research* 166: 183-5.
[14] Tanooka H (2011) Meta-analysis of non-tumour doses for radiation-induced cancer on the basis of dose-rate. *International Journal of Radiation Biology* 87: 645-652.
[15] Borek C, Hall EJ (1973) Transformation of mammalian cells in vitro by low doses of X-rays. *Nature* 243: 450-453.
[16] Watanabe M, Suzuki K (1991) Expression dynamics of transforming phenotypes in X-irradiated Syrian golden hamster embryo cells. *Mutation Research* 249: 71-80.
[17] Suzuki K, Watanabe M (1991) Morphological transformation induced by X-rays in syrian/golden hamster embryo cells. *Alternatives to animal testing and experimentation* 1: 74-78.
[18] Tanaka S, Tanaka IB 3rd, Sasagawa S, Ichinohe K, Takabatake T, et al. (2003) No lengthening of life span in mice continuously exposed to gamma rays at very low dose rates. *Radiation Research* 160: 376-379.
[19] Tanaka IB 3rd, Tanaka S, Ichinohe K, Matsushita S, Matsumoto T, et al. (2007) Cause of death and neoplasia in mice continuously exposed to very low dose rates of gamma rays. *Radiation Research* 167: 417-437.
[20] Ootsuyama A, Nagao M, Makino H, Ochiai A, Yamauchi Y, et al. (1994) Frequent p53 mutation in mouse tumors induced by repeated beta irradiation. *Molecular Carcinogenesis* 11: 236-242.
[21] Gondo Y, Nakamura K, Nakao K, Sasaoka T, Ito K et al. (1994) Gene replacement of the p53 gene with the lacZ gene in mouse embryonic stem cells and mice by using two steps of homologous recombination. *Biochemical and biophysical research communications* 202: 830-837.
[22] Hulla JE, French JE, Dunnick JK (2001) Chromosome 11 allelotypes reflect a mechanism of chemical carcinogenesis in heterozygous p53-deficient mice. *Carcinogenesis* 2: 89-98.
[23] Orita M, Iwahana H, Kanazawa H, Hayashi K, Sekiya T (1989) Detection of polymorphisms of human DNA by gel electrophoresis as single-strand conformation polymorphisms. *Proceedings of the National Academy of Sciences of the United States of America* 86: 2766-2770.

In: Fukushima Nuclear Accident
Editor: Shizuyo Sutou
ISBN: 978-1-63463-748-0

Chapter 10

THE EFFECTS OF DOSE RATE IN RADIATION RELATED CANCER RISK – A BRIEF REVIEW OF EPIDEMIOLOGICAL STUDIES IN HIGH BACKGROUND RADIATION AREAS AND OTHERS

Suminori Akiba*
Professor of Epidemiology and Preventive Medicine, Kagoshima University Graduate School of Medical and Dental Sciences, Kagoshima, Japan

ABSTRACT

This chapter makes a brief review of the association between cancer risk and low-dose-rate exposure to low-LET ionizing radiation. Important epidemiological studies for evaluating the cancer risk associated with protracted exposure include those on residents in natural high background radiation (HBR) areas and in the man-made HBR areas. Among epidemiological studies in natural HBR areas, most important ones are those conducted in Karunagappally in Kerala State, India, and Yangjiang in Guangdong Province, China. One of the most important studies for man-made HBR is the Techa River study. The results obtained from those studies suggest that, even if solid cancer risk is increased by low-dose-rate exposure to low-LET ionizing radiation, the excess relative risk for solid cancer, estimated assuming a linear dose-response, is lower in magnitude for low-dose-rate exposure than for high-dose-rate exposure of atomic bomb survivors. That may also be true for leukemia.

INTRODUCTION

The effect of dose rates on the magnitude of cancer risk after exposure to ionizing radiation is yet unclear. In the case of exposure to high LET (Linear Energy Transfer) ionizing radiation , such as radon and its progeny inhalation into the lung, shorter durations of

* Corresponding to: Suminori Akiba, Tel: +81-99-275-5295, Fax:+81-99-275-5299, E-mail: akiba@m.kufm.kagoshima-u.ac.jp.

exposure to alpha particles is considered to be associated with lower lung cancer risks per dose [1,2]. However, in the case of low-LET radiation, such as X-rays and gamma rays, the dose-rate effect on cancer risk is unclear. The UNSCEAR 1993 report gave the definitions of low and high dose rates as the dose rates below 6 mGy/h and 6 Gy/h or larger, respectively [3].

The Life Span Study (LSS) of atomic-bomb survivors is by far the most important epidemiological study on the cancer risk associated with acute exposure to low-LET ionizing radiation. In a few years after the bombing, leukemia risk started to increase and reached a peak in the early 1950s [4]. On the other hand, solid cancer risk started to increase 5-10 years after the bombing, and the risk persisted for several decades. The difference between leukemia and solid cancer risks was not limited to the time pattern of risk, but was also observed in their dose-response relationships. The excess relative risk (ERR) of leukemia increases in a linear-quadratic manner with an increase of radiation dose, and the ERR per dose in the low-dose range is lower than that for the higher dose range [4,5]. On the other hand, the ERR/Gy of solid cancer risk is considered to increase linearly with radiation dose [5].

This chapter presents a brief overview of epidemiological studies on the late health effects associated with low-dose-rate exposure to low-LET radiation such as gamma-rays and X-rays. Mainly whole-body exposure will be discussed.

Questions remain concerning cancer risk after low-dose-rate exposure. Most epidemiological studies on low-dose-rate exposure have relatively low cumulative doses (<100 mGy), making it difficult to evaluate the magnitude of radiation-related risk per dose. It should be noted that such evaluation requires a long-term follow-up of a large population. Note also that there are many risk factors, such as smoking, which are much more strongly associated with cancer risk compared to low-dose radiation exposure. As a result, even a weak association of radiation with such a factor can confound the relationship between radiation exposure and cancer risk in epidemiological studies.

Important epidemiological studies for evaluating the cancer risk associated with relatively low-dose-rate exposure include those on residents in natural and man-made high background radiation (HBR) areas. Important epidemiological studies for natural HBR are those conducted in Karunagappally in Kerala State, India, and Yangjiang in Guangdong Province, China. The most important study for man-made HBR is the Techa River study.

1. Solid Cancer Risk among Residents in Karunagappally, India

Solid cancer is a collective term for cancers excluding leukemia and lymphoma. In this paper, the terms "solid cancers" and "cancers excluding leukemia" will be used interchangeably. The coastal areas of Karunagappally Taluk (an administrative division) are known for natural HBR from the thorium-containing monazite sand. This taluk consists of 12 panchayats (an administrative unit of rural area). Among them, Chavara and Neendakara panchayats have median outdoor radiation doses of more than 4 mGy/y. In certain locations on the coast, ambient radiation levels are as high as 70 mGy/y. A cohort of approximately 380,000 residents in Karunagappally was established in the 1990s to examine the potential health effects of HBR. Recently, Dr. RRK Nair of the Regional Cancer Center (RCC), Trivandrum and his colleagues reported the results of statistical analysis examining cancer risk in relation to individual cumulative radiation dose, adjusting for potential confounding factors, including tobacco use and drinking [6]. The strength of Karunagappally cohort study includes the following points: i) relatively large cumulative doses (161 mGy on average); ii) a large population size; iii) adjustment for potential confounding factors such as tobacco use and socioeconomic status in statistical analysis; and iv) a high-quality cancer registry, described in Cancer Incidence in Five Continents published by the International Agency for Research on Cancer (IARC) [7].

1.1. Cancer Case Ascertainment

Cancer cases among the cohort were ascertained by the Natural Background Radiation Cancer Registry (NBRR) of Karunagappally, which was established in 1990 [8]. The registry reports have been presented in "Cancer Incidence in Five Continents" vol. VII [9], vol. VIII [10] and vol. IX [11]. Every month, registry workers visit the RCC, Trivandrum (Director: Dr. Paul Seabstian), the comprehensive cancer center in the state of Kerala, which is located 70 km south to Karunagappally. The monthly visit identifies more than half of cancer cases registered in the Karunagappallly Cancer Registry [12]. Cancer cases are also ascertained by monthly clinics to provide follow-up care for local cancer patients. In addition, registry workers visit primary health centers with cancer screening facilities 3-4 times a year. Also registered are the cancer cases identified by screening camps, which are held twice a year on average in all panchayats in the taluk. Additional efforts to identify cancer patients include annual visits to the following medical facilities: i) Trivandrum Medical College Hospital in Trivandrum, ii) major pathological laboratories in Karunagappally taluk and its neighboring areas, and in Trivandrum, and iii) all the hospitals and medical practitioners in Karunagappally taluk. Death reports were obtained from the death registers kept in the Vital Statistics Division of the panchayats. House visits to the deceased were also made to supplement information on cause of death.

1.2. Cancer Incidence

The most common cancer among men was lung cancer. The most important risk factor of lung cancer in the study area is bidi smoking. Bidi is made of 0.15–0.25 g of sun dried flaked tobacco rolled into a conical shape in a dried rectangular piece of Temburni leaf (*Diospyros melanoxylon*) and a thread securing the roll [12]. Cigarette smoking is not common because of relatively high prices. Tobacco chewing is common not only among men but also among women. This habit is the major cause of oral and pharyngeal cancers in the study area. Among women, cancers of the uterus and breast are the two most common malignancies. Table 1 compares the distribution of major cancer sites in Karunagappally with that in Hiroshima. In Hiroshima, oropharyngeal cancers are less common while lower digestive tract cancers and liver cancer are more frequent.

Table 1. Cancer incidence (/100,000) in Karunagappally and Japan

	Karunagapally, India (1993-1997)*		Hiroshima, Japan (1991-1995)**	
	Men	Women	Men	Women
Oral cavity and pharynx	27	6	6	2
Esophagus	6	3	12	2
Stomach	5	2	86	34
Colon	1	1	59	28
Rectum	2	2	27	12
Liver	4	0.3	43	13
Gallbladder	1	0.2	7	6
Pancreas	2	1	9	5
Larynx	4	0	4	0.2
Trachea, bronchus, lung	19	4	40	12
Breast	0.1	15	0.1	37
Uterus		16		18
All sites, including skin	101	75	364	212

Incidence rates were adjusted for age using the Segi world population .
*Nair MK et al., 2002 [12]
** Usui et al., 2002 [13]

1.3. Base-Line Survey

In 1990, the RCC (then Director: Dr. MK Nair) established a cohort of Karunagappally residents with the support of the Department of Atomic Energy, Government of India. In 1992, Mr. Gangadharan of the RCC, biostatistician and the lead organizer of field work, visited the Radiation Effects Research Foundation in Hiroshima, and asked for advice on their epidemiological studies from Drs. Itsuzo Shigematsu and Jack Schull. Since 1999, the Health Research Foundation (HRF) in Kyoto (Chairman: Dr. Junji Konishi; former Chairmen: Drs. Kanji Torizuka and Tsutomu Sugahara) has helped continue this follow-up study.

During the period 1990-1997, NBRR workers conducted house-to-house visits to collect information on socioeconomic status and lifestyles, using a 6-page standardized questionnaire. This survey collected information on 359,619 subjects in 71,674 households, which correspond to 93% and 94% of the population and the households in Karunagappally, respectively, counted by the 1990 census.

Among male adults (aged 30-84 years), 73% and 18% are Hindu and Moslem, respectively. Fishermen and farmers accounted for less than 20% of the cohort. Illiterates were only 6%. More than 60% of study subjects answered to have smoked bidis in the past or to smoke bidis currently. Approximately 50% were past and current alcohol drinkers [12]. The most popular alcohol beverages in this population are locally produced alcoholic beverages, including toddy (palm wine) and arrack (a strong liquor made from coconut flowers, sugarcane or other locally grown plants). Among adult women (aged 30-84 years), 72% and 20% are Hindu and Moslem, respectively. Among the female cohort, less than 5% were fishermen and farmers, and 17% were illiterate. Tobacco chewing was experienced by

24% of the cohort members, and, among those aged less than 70 at the time of interview, more than 50% were past or current tobacco chewers [12]. Tobacco smoking and alcohol drinking are rare among women [14].

1.4. Dosimetry

In the base-line survey, NBRR office workers conducted one indoor and one outdoor measurement of gamma radiation levels in approximately 72,000 houses, using sodium iodide (NaI) detector-based scintillometers and plastic scintillometers, designed and supplied by the Bhabha Atomic Research Center. Indoor measurements were conducted in the center of the main room with maximum occupancy.

No questions regarding house occupancy were included in the questionnaire used in the baseline survey. In 2000, NBRR office workers conducted an interview survey of approximately 8,000 residents to obtain information on the fraction of time spent indoors and outdoors [8]. The surveyed subjects were residents in 2% of all houses, which were randomly selected from the entire taluk. The results obtained from the survey have shown that house occupancy factor varies from 0.5 to 0.89 depending on sex and age. Those values are consistent with the value of 0.8 cited by the UNSCEAR Report 2000 [15].

The annual dose for each individual was calculated using the formula:

$$\text{Annual dose (mGy)} = \{[\text{Indoor dose/y} - 0.227] \times OF_{indoor} + [\text{outdoor dose/y (mean) of ward or panchayat} - 0.252] \times OF_{outdoor}\} \times CF,$$

where OF is an age- and sex-specific house occupancy factor and CF is a conversion factor for air kerma to organ-specific absorbed dose presented in the ICRP 74 report [16]. The CF of ^{232}Th used in the present study was 0.782 for the colon and 0.791 for red bone marrow. Our recent studies use conversion factors presented in the ICRP 116 report [17]. The CFs of children aged 1-14 years and infants aged less than 1 year were increased by 10% and 30%, respectively. In the formula, 0.227 and 0.252 mGy/y, which correspond to the air kerma values for the cosmic ray component of the measured indoor and outdoor doses, respectively, [6] were subtracted in order to estimate radiation doses from terrestrial radiation. The internal doses from ingestion and inhalation of radionuclides were not considered in the cumulative dose calculation.

The individual cumulative dose was calculated by adding up the annual dose in a time-dependent manner. When calculating cumulative doses, information on migration was taken into account. In the estimation of ERR per dose for all cancers excluding leukemia, colon dose was used for risk analysis since the LSS used it [18].

1.5. Cancer Risk Estimation

Assuming a linear dose-response, the ERR of all cancers excluding leukemia in relation to cumulative dose 10 years prior to the time at risk was estimated using the following Poisson regression model:

H_0 (sex, attained age, follow-up interval, bidi smoking, education, occupation) [1 + ß D], where D is cumulative radiation dose calculated in a time-dependent manner, and ß, the coefficient of D, describes the relative change in rates associated with dose. H_0 is the baseline cancer rate, depending on sex, attained age, follow-up intervals and other covariates (socio-demographic factors and bidi smoking). In leukemia risk estimation, cumulative doses were lagged by 2 years. The entry into the cohort was the date of the base-line survey (date of interview). The end of follow-up was the date of death for those deceased, the end of follow-up or the date attaining age of 85, whichever occurred first.

From the above analysis, the ERR of all cancers excluding leukemia was estimated to be -0.13/Gy (95% CI: -0.58, 0.46). In site-specific analysis, no cancer site was significantly related to cumulative radiation dose. Leukemia was not significantly related to natural HBR, either [6].

2. Solid Cancer Risk among Residents in Yangjiang, China

Yangjiang in Guandong Province, China, is also known for natural HBR. Most of the inhabitants have lived in the HBR area for six or more generations [19]. The annual doses of external radiation from natural sources, including thorium, in the HBR and the control areas were estimated to be 2.10 and 0.77 mGy on average, respectively [20,21].

In the 1970s, Chinese researchers led by Dr. Wei Luxin of Laboratory of Industrial Hygiene (LIH) of the Chinese Government in Beijing conducted epidemiological surveys in Yangjiang, China, and their early results were published in the Science journal in 1980 [19]. The health survey during the period between 1972 and 1975 did not demonstrate any detrimental effects of HBR. They concluded that the study population was not large enough to detect any significant effects in such a low-dose range of ionizing radiation or there is a practical threshold dose. In the 1980s, Dr. Wang Zuoyuang of LIH and Dr. John Boice of the US National cancer Institute conducted a collaborative work in Yangjiang to examine the prevalence of thyroid nodularity among middle aged women since thyroid nodules are considered to be the most sensitive endpoint to radiation exposure, and are related to subsequent thyroid carcinoma risk. The prevalence of thyroid nodules in HBR areas was found to be similar to that in the control area [22]. In this study, chromosome aberrations in lymphocytes were also examined. The frequency of unstable-type chromosome aberration was significantly elevated among residents in the HBR areas than in the control areas.

In the 1990s, the HRF started a collaborative work with the research group led by Dr. Wei Luxin, including Drs. Tao Zufan, Zha Yongru and Sun Quanfu. As a part of this collaborative study, chromosome aberrations were examined by Chinese and Japanese cytologists. They found a positive correlation between natural radiation levels and the frequency of unstable-type chromosome aberration in lymphocytes [23,24].

The most recent mortality study analyzed the data obtained from the follow-up of a cohort of approximately 32,000 men and women aged 30-74 living in Yangjiang during the period 1979-1998 [25]. In the risk analysis of cancers excluding leukemia, cumulative external radiation dose lagged by 10 years was estimated for each individual based on hamlet-specific indoor and outdoor doses, and sex- and age-specific house occupancy factors. Cumulative radiation doses from natural radiation, excluding cosmic ray, among HBR and control area residents were 84.8 mGy and 21.6 mGy, on average, respectively. The mortality follow-up study accumulated approximately 737,000 person-years at risk, and ascertained 6,005 deaths, including 956 cancer deaths (15 deaths of leukemia). Since cancer treatment is usually prohibitively expensive for local farmers, which is the predominant occupation of cohort members, cancer was usually fatal in the study area; therefore, cancer mortality rates could be considered similar to cancer incidence rates for most of cancers, including breast cancer. The estimated ERR/Gy of cancers excluding leukemia was -1.01 (95% CI: -2.53, 0.95). In site-specific analysis, liver-cancer mortality was inversely related to the cumulative dose (P=0.002). Since mortality of liver diseases other than cancer tended to increase with cumulative radiation dose (P=0.061), misdiagnosis between liver cancer and non-cancer liver diseases, such as liver cirrhosis, was suspected. For cancers excluding liver cancer and leukemia, the ERR/Gy was 0.19 (95% CI: -1.87, 3.04).

3. The Study of Residents Along The Techa River

The Techa River study is one of the most important studies of residents in man-made HBR areas. This river, which runs through areas east of the Ural Mountains in central Russia, was contaminated by radionuclides from the Mayak Production Association Plant (Mayak PA). During the period between 1949 and 1956, this plant discharged 10^{17} Bq of radionuclides into the river [26]. The maximum release was during the period 1950-52. These releases contaminated the riverbed with the long-lived radionuclides Sr-90, Cs-137 and other radionuclides. Estimated releases of Sr-90 and Cs-137 were 11.6 PBq and 12.2 PBq, respectively [27]. Flood plain soils along the Techa River were also contaminated during spring floods caused by melting snow. In some areas along the river shoreline, flood swamps measured 300 m to 2 km in width [28]. Inhabitants used the floodlands for various purposes, including cattle breeding, and were exposed to radiation [29]. During the period 1949-1951, the population of Metlino, which is the town most closely located (7 km downstream) to the nuclear waste releasing site, received an average dose of about 2 Gy from the external exposure [26, 27]. An extraordinarily powerful flood in April to May in 1951 caused heavy contamination in the river basin [28, 29]. In August 1951, when the largest amount of radioactive waste was released into the environment, the dose rates from external exposure in some gardens of this settlement is estimated to have reached 30-1000 microR/s (0.016 – 0.53 mGy/m) [30].

In 1967, the Urals Research Center for Radiation Medicine (URCRM) initiated a study of the long-term health effects of residents, and established the Techa River Cohort (TRC) [26, 31]. TRC includes approximately 27,000 residents.

To the original cohort, approximately 5,000 migrants who moved to the study areas before 1960 were added. Added cohort members were restricted to those born before 1949. The extended cohort (ETRC) consists of approximately 18,000 people in the 25 villages in Chelyabinsk Oblast (7-148km from the release point) and approximately 11,000 people in the 16 riverside villages in Kurgan Oblast (155-237 km from the release point). Dr. Marina Degteva and her colleagues described the significant features of the dosimetry for this cohort as follows [32]:

I. internal doses from Sr-90 were calculated for each individual on the basis of representative doses for the village where the individual had the main exposure;
II. internal doses from Sr-89 and Cs-137 were calculated using Sr-90 dose and their relative concentrations in river water for the reference site; and
III. external doses were estimated from measured dose rates along the river bank and presumed age-specific life styles (concerning time spent on the river bank, *etc.*).

In the most recent study, 2,303 solid cancer deaths were ascertained during the period between 1950 and 2007. Poisson regression analysis of solid cancer gave an ERR/Gy of 0.61 (95% CI: 0.04, 1.27). Its wide confidence interval precludes us from drawing any definitive conclusions on the effects of dose rates on solid cancer risk per dose. This study has several limitations. Most residents in the highly contaminated areas moved to nearby major cities (Chelyabinsk, Kurgan, and Ekaterinburug) or other areas within Chelyabinsk, Kurgan and Sverdlovsk oblasts in the 1950s [33].

It should be noted that relocation tended to start earlier in areas with higher radiation doses [34]. Although emigration is likely to have a large impact on lifestyles, no adjustment for those factors was made in radiation-related cancer risk analysis. Another limitation of this study is the presence of not a small number of loss to follow-up, largely due to migration.

4. Comparison of Solid Cancer Risk between Low and High-Dose-Rate-Exposures

It is of interest to compare the ERRs per dose of solid cancer risk observed in HBR studies with that obtained by the LSS of atomic bomb survivors since that is a comparison between low and high-dose-rate exposures. Pooled data analysis is a possible approach to compare risk estimated by the LSS with those obtained from the Karunagappally, Yangjiang and Techa River studies. One of the problems involved in such an approach is the difficulty in determining the age at exposure in HBR studies.

Another problem is the lack of information on lifestyles, including SES and smoking in the studies of atomic bomb survivors, Yangjiang residents and Techa River basin residents. A possible approach to overcome those problems is to create a table with dose-category specific observed and expected numbers of solid cancer cases for each study, (incident cases of all cancers excluding leukemia for the Karunagappally study, and the deaths from all cancers excluding leukemia in the Yangjiang study), and to conduct Poisson regression analyses of grouped data, using the pooled tables. Such an analysis showed that the estimate of ERR per dose among atomic-bomb survivors was significantly larger than that observed in the Karunagappally study [36]. The addition of Yangjiang study data did not change results evidently. Apparently, the Yangjiang study does not have a statistical power to detect a cancer risk difference from that of the LSS.

A comparison of the Karunagappally and Techa River studies showed no significant difference in their ERR per dose. In this analysis, no adjustment for attained age was made since the average attained age of the LSS (=59.2 years) was similar to those of the studies in Karunagappaly and Yangjiang. In the LSS, the average age at the time of the bombing was 28.0 years, and was approximately half of the average attained age of the Karunagappally and

Yangjiang studies. Therefore, no adjustment for age at exposure was conducted, either. In other words, it was assumed that the average ages at exposure in the HBR area studies were approximately half the average attained ages.

5. Nuclear Worker Studies

The most extensive study on cancer risk among nuclear workers is the IARC 15-country study. In this study, the mortality follow-up data of nuclear workers in 15 countries were pooled and analyzed. The standard analysis of this study focused on external exposure to low-LET radiation. The estimated ERR for all cancers excluding leukemia (5,024 deaths) was 0.97/Gy (95% CI: 0.14, 1.97) [37, 38]. In this analysis, a stratification on the duration of employment (<10 years vs 10+ years) was made. Without such stratification, the estimate was 0.31/Gy. Both estimates are much larger than the estimate of -0.07/Gy (90% CI: -0.4, 0.3), which was obtained from the IARC 3-country study [39], which pooled the nuclear worker data of the US, the UK and Canada. The IARC 15-country study is known to have problems [40, 41]. One of them is a large ERR estimate obtained from the Canadian data, which was 6.65 (90% CI: 2.56, 13.0) per sievert. This anomalously large estimate stems from the data of Atomic Energy of Canada Limited (AECL) workers, which accounts for approximately 75% of the Canadian workers in the data used in the standard analysis. It is reported that some dosimetry information, including dates of hire, was lost when dosimetry data were transferred from the AECL worker records to the National Dose Registry [42]. However, underlying mechanisms resulted in the large ERR estimate is still unclear.

Dr. Peter Jacob and his colleagues conducted a meta-analysis of cancer risk using the data from seven cancer-mortality studies (five nuclear worker studies, a Chernobyl emergency worker study, and a Techa River study). The ERR estimate obtained from their meta-analysis was approximately 0.6/Gy. This value is a factor of 1.21 (90% CI: 0.51, 1.90) larger than that for the corresponding estimate obtained from the LSS [43]. Among the studies included in their meta-analysis, Chernobyl clean-up workers had a relatively large ERR per unit dose for solid cancer deaths, which was 2.04/Gy.

Interestingly, the estimate obtained from a cancer incidence study of those clean-up workers was 0.33/Gy. The cause of this large difference between cancer mortality and incidence is unclear. However, generally, cancer incidence studies have more accurate information on cancer diagnosis than mortality studies do. The mean cumulative dose of clean-up workers in the cancer incidence study was 130 mGy [44] while the overall average cumulative recorded dose in the IARC-15 country study was as low as 19.4 mGy [37].

Since the meta-analysis conducted by Dr. Peter Jacob and his colleagues was not limited to nuclear worker studies, Drs. Akiba and Mizuno conducted a meta-analysis focusing on those workers. Their meta-analysis included all the major studies of nuclear workers in Canada, France, the UK and the US that had reported an estimate of ERR per unit dose for all cancers excluding leukemia or solid cancer. The ERR estimate was 0.14/Gy (95% CI: -0.12, 0.41) [45]. This is not a statistically significant excess. It is of note that the estimated value is similar to what was observed in the studies in Yangjiang and Taiwan, where the dose rates were relatively low (Table 2).

An important study that was not included in their meta-analysis was the Mayak study [46], which has a very high average dose (=0.81Gy). This study was not included in the IARC 15-country study, either. A problem of this study is the internal exposure to alpha emitting plutonium. Shilnikova and her colleagues reported the mortality of solid cancers excluding cancers of the lung, liver and skeleton since internal exposure in those three organs cannot be ignored. The ERR estimate was 0.08/Gy (90%CI: 0.03, 0.14) [46].

Table 2. ERR/Gy for solid cancer in major epidemiological studies

Studies			ERR/Gy
Atomic bomb survivors	Incidence Preston et al., 2007 [18]	Age ATB=23.9 Attained age=57.7	0.712 (SE=0.051)
Techa River study residents	Incidence Krestinina et al., 2007 [33]		1.0, 95%CI: 0.3, 1.9
	Mortality Schonfeld et al., 2013 [26]		0.61, 95%CI: 0.04, 1.27
Karunagappally study, residents	Incidence Nair et al., 2009 [6]		-0.13*, 95%CI: -0.58, 0.46
Yangjiang study residents	Mortality Tao et al., 2012 [25]		0.19*, 95%CI: -1.87, 3.04
IARC 3-country study, nuclear workers	Mortality Cardis et al., 1995 [39]		-0.07, 90% CI:-0.4, 0.3
IARC 15-country study, nuclear workers	Mortality, Cardis et al., 2005 [37]	Stratification on duration of employment	
		Yes	0.97, 90% CI: 0.14, 1.97
	mortality, Cardis et al., 2007 [38]	No	0.31, 90% CI: -0.23, 0.93
Meta-analysis of nuclear worker data	Akiba and Mizuno, 2012 [45]		0.14 95%CI: -0.12, 0.41
Mayak workers nuclear workers	mortality Shilnikova et al., 2003 [46]		0.08** 90%CI: 0.03, 0.14

* All cancers excluding leukemia.

** All cancers excluding cancers of the lung, liver and bone.

6. LEUKEMIA RISK

For non-CLL leukemia nuclear workers, the IARC 3-country and 15-country studies [36, 38] reported ERRs at 100 mGy of 0.22 (90% CI: 0.01, 0.57) and 0.19 (90% CI: <0, 0.85), respectively (Table 3). Recently, Daniels and his colleagues reported the results of a pooled analysis of US nuclear worker data. They got an ERR estimate of 0.09 (95% CI:−0.17, 0.65) at 100 mGy [47].

Populations included in their analysis consisted of workers who were employed at the Hanford Site, the Savannah River Site, the Oak Ridge National Laboratory, the Los Alamos National Laboratory, including the Zia Company employees in Los Alamos, the Idaho National Laboratory, and the Portsmouth Naval Shipyard.

Table 3 shows ERRs per dose obtained from other major studies. The studies of Canadian nuclear workers were not included since the validity of their data has been questioned.

Here, a meta-analysis of nuclear worker mortality studies was conducted. From this analysis, IARC 3-country and 15-country studies were excluded. The summary estimate of ERR at 100 mSv was 0.14 (95%CI: -0.11, 0.39) as shown in Table 3. In the studies included in this meta-analysis, radiation exposure was mainly whole-body external exposure.

Table 3. ERR at 100 mGy for non-CLL leukemia in major mortality studies of nuclear workers

Studies	Mean dose	Leukemia deaths	ERR at 100 mGy	Authors
IARC 3-country study, pooled analysis	40.2	119	0.218 90%CI: 0.01, 0.57	Cardis et al., 1995 [39]
IARC 15-country study, pooled analysis	19.4	196	0.193 90% CI: <0, 0.714	Cardis et al., 2005 [38]
US nuclear workers pooled case-control study	26.5	264	0.09 95%CI: -0.17, 0.65	Daniels et al., 2013 [47]
US Rocky Flats facility workers, 1952-1983	41	6	-0.72 90% CI: <0, 4.2	Gilbert et al., 1993 [48]
U.S. Nuclear Power Plant workers, 1979-1997	25.7	26	0.567 95%CI: -0.256, 3.04	Howe et al., 2004 [49]
US Rocketdyne workers, 1948-2008	13.5	159	0.06 95% CI: -0.50, 1.23	Boice et al., 2011 [50]
U.K. National Registry for Radiation Workers (NRRW), -2001	24.9	198	0.171 90%CI: 0.006, 0.429	Muirhead et al., 2009 [51,52]
French combined cohort	22.5	60	0.396 90%CI: <0, 1.682	Metz-Flamant et al., 2014 [53]
Japanese nuclear workers 1991-2002	12.2	80*	−0.193 95%CI: -0.612, 0.857	Akiba and Mizuno, 2012 [45]
Meta-analysis of the studies listed above**		793	0.14 95%CI: -0.11, 0.39	Present study

Mean doses are in mGy or mSv.

* A CLL case is included.

** The method of meta-analysis used in this study was the same as that used in Akba and Mizuno [45].

The Mayak and US mound studies, which were excluded from this meta-analysis, reported somewhat smaller estimates of ERR per dose (Table 4). The workers in those studies were exposed to internal exposure to plutonium and other radionuclides [46, 54].

The pooled analyses of the IARC or the meta-analysis of the current study shown in Table 3 did not include the studies of Chernobyl liquidators, either. Two case–control studies of Chernobyl liquidators (clean-up workers) reported increased risks of leukemia. The ERR of non-CLL leukemia per 100 mGy was 0.273 (95% CI: <0, 1.35) for Ukrainian clean-up

Table 4. ERR for non-CLL leukemia in major epidemiological studies of workers

Studies	Mean dose	Cases	ERR at 100 mGy	Remarks and authors
Nuclear workers				
Meta-analysis of nuclear worker data in Table 3	12-41	793 deaths	0.14 95%CI: -0.11, 0.39	present study
Mayak workers, mortality -1997	810	66 deaths	0.099 90%CI: 0.045, 0.212	Shilnikova et al., 2003 [46]
US mound workers exposed to plutonium and others, 1944-1979.		21 deaths	0.04 95%CI:-0.37, 0.71	Boice et al., 2014 [54]
Chernobyl liquidators				
Chernobyl clean-up workers, incident cases in Russia (1993-1998), Belarus (1993-2000) and Baltic countries (1990-1998)	14.7 (cases), 12.8 (controls)	19	0.50 90%CI: -0.38, 5.70	A case-control study nested in a cohort Kesminiene et al., 2008 [56]
Ukrainian Chernobyl workers, incidence, 1986-2006	132.3 (cases), 81.8 (controls)	52	0.221, 95%CI: 0.005, 0.761	Zablotska et al., 2013 [57]
Chernobyl clean-up workers, incident cases in Russia				Ivanov et al., 2012 [58]
1986–1997	108	51	0.50, 90%CI: 0.06, 1.45	
1998–2007		60	-0.16, 90%CI:-0.26, 0.06	

Mean doses are in mGy or in mSv.
CLL: chronic lymphocytic leukemia

Table 5. ERR for non-CLL leukemia in major epidemiological studies of residents

Studies	Mean dose	Cases	ERR at 100 mGy	Remarks and authors
Atomic bomb survivors **cohort study,** incidence			0.39-0.4	age at the time of the bombing >=20 Hsu et al., 2013 [4]
HBR studies				
Techa River basin residents, incidence, 1953-2007	420	72	0.22 95% CI: 0.08, 0.54	Krestinina et al., 2013 [59]
Residents of Co-60 contaminated building in Taiwan, incidence	48	6	0.19, 90% CI: 0.01, 0.31	Hwang et al., 2008 [60]
Karunagappally residents, incidence, 1990-2005	161	20	0.372 95% CI: <0, 33.68	Nair et al., 2009 [6]
Meta-analysis of various studies, including workers			0.19 95%CI: 0.07, 0.32	Daniels & Schubauer-Berrigan, 2013 [61]

Mean doses are in mGy or in mSv.
CLL: chronic lymphocytic leukemia.

workers in the study reported by Romanenko and his colleagues [55], and 0.5 (90% CI: -0.38, 5.70) in a study of workers from Belarus, Russia and the Baltic countries conducted by Kesminiene and her colleagues [56]. Recent Ukrainian and Russian clean-up worker studies reported ERRs at 100mGy of 0.221 (95%CI: 0.005, 0.761) for non-CLL [57] and 0.498 (90%CI: 0.059–1.447) for the period 1986-1997 [58], respectively.

The LSS of atomic-bomb survivors showed a linear-quadratic dose-response relationship between leukemia incidence and radiation dose. The ERR per dose in relatively low-dose ranges was lower than those in medium-high doses. A recent cancer mortality study reported ERR estimates of 0.15 at 100 mGy and 3.1 at 1 Gy [Ozasa et al].

Since an ERR of 0.15 at 100 mGy corresponds to 1.5 at 1 Gy, the ERR per dose in the dose range of 100 mGy was approximately half of that at 1 Gy. In the most recent LSS of cancer incidence, the ERR/Gy for leukemia other than chronic lymphocytic leukemia (CLL) or adult T-cell leukemia (ATL) was 4.7 (95%CI: 3.3, 6.5). For those with ages ATB (at the time of the bombing) of 0-19, 20-39 and 40+, the estimates were 6.5 (95%CI: 4.0, 10.3), 3.9 (95%CI: 2.3, 6.1) and 4.0 (95%CI: 2.1, 6.9), respectively [4]. CLL and ATL are usually excluded from radiation-related leukemia risk analysis since they are not considered to be caused by radiation exposure.

In order to make comparisons with those studies, it would be a good idea to restrict atomic bomb survivors to those with age ATB of 20 years or older since most of nuclear workers are 18 years or older. Such an age ATB restriction in the atomic-bomb survivor study will make the estimates to be 3.9-4.0 per gray, which are 0.39-0.40 at 100 mGy (Table 5).

Tables 4 and 5 summarize ERR at 100 mGy reported by major epidemiological studies. Among the man-made HBR studies, the Techa River study obtained an estimate of 0.22 per 100 mGy [59], which is slightly larger than those obtained from those of nuclear workers. In this study, Sr-90 was one of the major sources of bone marrow exposure. A cohort study of residents in Co-60 contaminated buildings in Taiwan obtained an estimate similar to nuclear worker studies [60]. Recently, Daniels and Schubauer-Berrigan conducted a meta-analysis of various studies, including those of nuclear workers and residents, and obtained an ERR estimate of 0.19 (95% CI: 0.07, 0.32) at 100 mGy [60]. Their analysis included the following studies:

I. the cohort study of residents in Co-60 contaminated building residents in Taiwan reported by Hwang et al. in 2008 [60];
II. the Techa River study reported by Krestinina et al. in 2005 [62];
III. the Ukrainian Chernobyl worker study reported by Romanenko et al. in 2008 [55];
IV. the case-control study nested in cohorts of U.S. nuclear weapons facility workers [63] in all the facilities (excluding the Idaho National Laboratory) included in the pooled analysis of Daniels et al. [47] described above;
V. a study on nuclear workers at the Idaho National Laboratory;
VI. the study on workers at the Rocky Flats nuclear weapons facility reported by Gilbert et al. in 1993 [48];
VII. the Rocketdyne worker study reported by Boice et al. in 2006 [64];
VIII. the IARC 15-coutry study [37];
IX. the case-control study nested in a cohort of Chernobyl liquidators reported by Kesminiene et al. in 2008 [56]; and
X. the Karunagappally study [6].

Tables 4 and 5 include all of these studies, with some of them having updated data.

Taken together, the estimate of ERR per dose for leukemia in medium to high-dose ranges was approximately two-fold larger than that for a low-dose range, and was larger than those obtained from the studies of low-dose-rate exposure. Exceptions are Chernobyl liquidator studies reported by Dr. Ausra Kesminiene and her colleagues [56] on Russian and Baltic country workers, and by Dr. Victor Ivanov and his colleagues on Russian workers [58], which had an ERR/Gy of 0.5. However, their confidence intervals are wide. It should also be noted that, in the comparisons between the LSS (acute exposure to high-dose-rate radiation) and studies of protracted low-dose-rate exposure, including nuclear workers, the time since exposure, which strongly affects ERR per unit dose, was not taken into account. Age at exposure is not completely adjusted, either.

Conclusion

In the case of solid cancer risk, even if low-dose-rate exposure is associated with its increase, its magnitude, estimated assuming a linear dose-response, may be lower than that from high-dose-rate exposures of atomic-bomb survivors. That may also be true for leukemia.

References

[1] Lubin JH, Boice JD Jr, Edling C, Hornung RW, Howe GR, et al. (1995) Lung cancer in radon-exposed miners and estimation of risk from indoor exposure. *Journal of National Cancer Institute* 87: 817-827.

[2] Grosche B, Kreuzer M, Kreisheimer M, Schnelzer M, Tschense A (2006) Lung cancer risk among German male uranium miners: a cohort study, 1946-1998. *British Journal of Cancer* 95: 1280-1287.

[3] UNSCEAR (1993) UNSCEAR 1993 Report to the General Assembly, with Scientific Annexes. Sources and Effects of Ionizing Radiation. Annex F: Influence of dose and dose-rate on stochastic effects of radiation. United Nations, New York: United Nations Scientific Committee on the Effects of Atomic Radiation.

[4] Hsu WL, Preston DL, Soda M, Sugiyama H, Funamoto S, et al. (2013) The incidence of leukemia, lymphoma and multiple myeloma among atomic bomb survivors: 1950-2001. *Radiation Research* 179: 361-382.

[5] Ozasa K, Shimizu Y, Suyama A, Kasagi F, Soda M, et al. (2012) Studies of the mortality of atomic bomb survivors, Report 14, 1950-2003: an overview of cancer and noncancer diseases. *Radiation Research* 177: 229-243.

[6] Nair RRK, Rajan B, Akiba S, Jayalekshmi P, Nair MK et al. (2009) Background radiation and cancer incidence in Kerala, India- Karunagappally study. *Health Physics* 96: 55-66.

[7] Boice JD Jr, Hendry JH, Nakamura N, Niwa O, Nakamura S, Yoshida K (2010) Low-dose-rate epidemiology of high background radiation areas. *Radiation Research* 173: 849-854.

[8] Nair M, Gangadharan P, Jayalakshmi P, Nair RRK, Mani KS (2004) Natural background radiation cancer registry. Trivandrum: Regional Cancer Center, Trivandrum; *Technical Report* 1990-1999.
[9] Nair MK, Amma S, Mani KS (1997) India, Karunagappally 1990-1994, In: Parkin DM, Whelan SL, Ferlay J, Raymond L, Young J editors. Cancer Incidence in Five Continents, vol VII. Lyon: International Agency for Research on Cancer. IARC Publication No. 143.
[10] Nair MK, Gangadharan P, Jayalakshmi P, Mani KS (2002) Cancer incidence in Karunagappally 1993-1997, Kerala, India. In: Parkin DM, Whelan SL, Ferlay J, Teppo L, Thomas DB editors. Cancer Incidence in Five Continents, vol VIII. Lyon: International Agency for Research on Cancer; IARC Publication No. 155.
[11] Jayalekshmi P, Rajan B. 2009. Cancer incidence in Karunagapally 1998-2002, Kerala, India. In: Curado MP, Edwards B, Shin HR, Ferlay J, Heanue M, Boyle P, Storm H. editors. Cancer Incidence in Five Continents, vol IX. In: Lyon: International Agency for Research on Cancer; IARC Publication No. 160.
[12] Jayalekshmy PA, Akiba S, Nair MK, Gangadharan P, Rajan B et al. (2008) Bidi smoking and lung cancer incidence among males in Karunagappally cohort in Kerala, India. *International Journal of Cancer* 123: 1390-1397.
[13] Usui S, Inoue J, Koyama K. Cancer incidence in Hiroshima, Japan, 1991-1995. Lyon: International Agency for Research on Cancer; IARC Publication. 2002
[14] Jayalekshmi PA, Gangadharan P, Akiba S, Nair RRK, Tsuji M et al. (2009) Tobacco chewing and female oral cavity cancer risk in Karunagappally cohort, India. *British Journal of Cancer* 100: 848-852.
[15] United Nations Scientific Committee on the Effects of Atomic Radiation. 2000 Report to the General Assembly with scientific annexes. Vol I. Annex A: Dose assessment methodologies. New York: United Nations; 2000. p. 28.
[16] International Commission on Radiological Protection (1996) ICRP Publication 74.Conversion coefficients for use in radiological protection against external radiation.; *Annals of the ICRP* 26: 3-4.
[17] Petoussi-Henss N, Bolch WE, Eckerman KF, Endo A, Hertel N et al. (2010) ICRP Publication 116. Conversion coefficients for radiological protection quantities for external radiation exposures. *Annals of ICRP* 40: 1-257.
[18] Preston DL, Ron E, Tokuoka S, Funamoto S, Nishi N, et al. (2007) Solid cancer incidence in atomic bomb survivors: 1958-1998. *Radiation Research* 168: 1-64.
[19] High Background Radiation Research Group (1980) Health survey in high background radiation areas in China: High Background Radiation Research Group, China. *Science* 209: 877-880.
[20] Yuan YL, Shen H, Sun QF, Wei LX (1999) Estimation of individual doses from external exposures and dose-group classification of cohort members in high background radiation area in Yangjiang, China. *Chinese Journal of Radiology and Medical Protection* 19: 99-103.
[21] Morishima H, Koga T, Tatsumi K, Nakai S, Sugahara T et al. (2000) Dose measurement, its distribution and individual external dose assessments of inhabitants in the high background radiation areas in China. *Journal of Radiation Research* 41 suppl: 9-23.

[22] Wang ZY, Boice JD Jr, Wei LX, Beebe GW, Zha YR, et al. (1990) Thyroid nodularity and chromosome aberrations among women in areas of high background radiation in China. *Journal of National Cancer Institute* 82: 478-85.

[23] Jiang T, Hayata I, Wang C, Nakai S, Yao S et al. (2000) Dose-effect relationship of dicentric and ring chromosome in lymphocytes of individuals living in the high background radiation areas in China. *Journal of Radiation Research* 41 suppl: 63-68.

[24] Hayata I, Wang C, Zhang W, Chen D, Minamihisamatsu M et al. (2004) Effect of high level natural radiation on chromosomes of residents in southern China. *Cytogenetic and Genome Research* 104: 237-239.

[25] Tao Z, Akiba S, Zha Y, Sun Q, Zou J, et al. (2012) Cancer and non-cancer mortality among Inhabitants in the high background radiation area of Yangjiang, China (1979-1998). *Health Physics* 102: 173-181.

[26] Schonfeld SJ, Krestinina LY, Epifanova S, Degteva MO, Akleyev AV et al. (2013) Solid cancer mortality in the Techa river cohort (1950-2007). *Radiation Research* 179: 183-189.

[27] Shutov VN, Travnikova IG, Bruk GYa, Golikov VY, Balonov MI et al. (2002) Current contamination by 137Cs and 90Sr of the inhabited part of the Techa river basin in the Urals. *Journal of Environmetal Radioactivity* 61: 91-109.

[28] http://users.physics.harvard.edu/~wilson/publications/pp747/techa_cor.htm

[29] http://www-ns.iaea.org/downloads/rw/projects/emras/emras-aquatic-techa.pdf

[30] Mokrov YG (2004) External radiation exposure of residents living close to the Mayak facility: main sources, dose estimates, and comparison with earlier assessments. *Radiation Environment and Biophysics* 43: 127-39.

[31] Kossenko MM, Thomas TL, Akleyev AV, Krestinina LY, Startsev NV et al. (2005) The Techa River Cohort: Study Design and Follow-up Methods. *Radiation Research* 164: 591-601.

[32] Degteva MO, Kozheurov VP, Tolstykh EI, Vorobiova MI, Anspaugh LR et al. (2000) The Techa River Dosimetry System: Dose Reconstruction for Population Risk Analysis. In: IRPA 10 Proceedings of 10th International Congress of the International Radiation Protection Association on Harmonization of radiation, human life and the ecosystem; May 14-19; Hiroshima, Japan; Tokyo: Japan Health Physics Society; CD-ROM; Paper No. T-19 (1)-4.

[33] Krestinina LY, Davis F, Ostroumova EV, Epifanova SB, Degteva MO et al. (2007) Solid cancer incidence and low-dose-rate radiation exposures in the Techa River cohort: 1956–2002. *International Journal of Epidemiology* 36: 1038–1046.

[34] Kossenko MM, Degteva MO (1994). Cancer mortality and radiation risk evaluation for the Techa River population. *Science of Total Environment* 142: 73-89.

[35] Akiba S, Nair RRK, Jayalekshmi P, Sebastian P (2014) Solid cancer risk in high level natural radiation areas of India and China. *Bhabha Atomic Research Institute Bulletin, in press.*

[36] Akiba S (2013). Cancer risk associated with low-dose and low-dose-rate ionizing radiation exposure. *Genes and Environment* 35:80-87.

[37] Cardis E, Vrijheid M, Blettner M, Gilbert E, Hakama M et al. (2005) Risk of cancer after low doses of ionizing radiation: retrospective cohort study in 15 countries. *British Medical Journal* 331: 77-82.

[38] Cardis E, Vrijheid M, Blettner M, Gilbert E, Hakama M et al. (2007) The 15-Country Collaborative Study of Cancer Risk among Radiation Workers in the Nuclear Industry: estimates of radiation-related cancer risks. *Radiation Research* 167: 396-416.

[39] Cardis E, Gilbert ES, Carpenter L, Howe G, Kato I et al. (1995) Effects of low doses and low dose rates of external ionizing radiation: cancer mortality among nuclear industry workers in three countries. *Radiation Research* 142: 117-32.

[40] Shigematsu I (2005) More information is needed for verifying low dose radiation effects. *British Medical Journal* 331: 77–81.

[41] Wakeford R (2005) Cancer risk among nuclear workers. *Journal of Radiological Protection* 25: 225-228.

[42] Ashmore JP, Gentner NE, Osborne RV (2010) Incomplete data on the Canadian cohort may have affected the results of the study by the International Agency for Research on Cancer on the radiogenic cancer risk among nuclear industry workers in 15 countries. *Journal of Radiological Protection* 30: 121-129.

[43] Jacob P, Rühm W, Walsh L, Blettner M, Hammer G et al. (2009) Is cancer risk of radiation workers larger than expected? *Occupational and Environmental Medicine* 66: 789-796.

[44] Ivanov V, Ilyin L, Gorski A, Tukov A, Naumenko R (2004) Radiation and epidemiological analysis for solid cancer incidence among nuclear workers who participated in recovery operations following the accident at the Chernobyl NPP. *Journal of Radiation Research* 45: 41-44.

[45] Akiba S, Mizuno S (2012) The third analysis of cancer mortality among Japanese nuclear workers, 1991-2002: estimation of excess relative risk per radiation dose. *Journal of Radiological Protection* 32: 73-83.

[46] Shilnikova, NS., Preston DL, Ron E, Gilbert ES, Vassilenko EK et al. (2003) Cancer mortality risk among workers at the Mayak nuclear complex. *Radiation Research* 159: 787-798.

[47] Daniels RD, Bertke S, Waters KM, Schubauer-Berigan MK (2013) Risk of leukaemia mortality from exposure to ionising radiation in US nuclear workers: a pooled case-control study. *Occupational and Environmental Medicine* 70: 41-48.

[48] Gilbert ES, Cragle DL, Wiggs LD (1993) Updated analyses of combined mortality data for workers at the Hanford Site, Oak Ridge National Laboratory, and Rocky Flats Weapons Plant. *Radiation Research* 136: 408-421.

[49] Howe GR, Zablotska LB, Fix JJ, Egel J, Buchanan J (2004) Analysis of the mortality experience amongst U.S. nuclear power industry workers after chronic low-dose exposure to ionizing radiation. *Radiation Research* 162: 517-526.

[50] Boice JD Jr, Cohen SS, Mumma MT, Ellis ED, Eckerman KF, et al. (2011) Updated mortality analysis of radiation workers at Rocketdyne (Atomics International), 1948-2008. *Radiation Research* 176: 244-258.

[51] Muirhead CR, O'Hagan JA, Haylock RG, Phillipson MA, Willcock T et al. (2009) Mortality and cancer incidence following occupational radiation exposure: third analysis of the National Registry for Radiation Workers. *British Journal of Cancer* 100: 206-212.

[52] Muirhead CR, O'Hagan JA, Haylock RGE, Phillipson MA, Willcock T, et al. (2009) HPA-RPD-062 - Third analysis of the national registry for radiation workers:

Occupational exposure to ionizing radiation in relation to mortality and cancer incidence. Chilton: Health Protection Agency.
[53] Metz-Flamant C, Laurent O, Samson E, Caër-Lorho S, Acker A et al. (2013) Mortality associated with chronic external radiation exposure in the French combined cohort of nuclear workers. *Occupational and Environmental Medicine* 70: 630-638.
[54] Boice JD Jr, Cohen SS, Mumma MT, Ellis ED, Cragle DL et al. (2014) Mortality among mound workers exposed to polonium-210 and other sources of radiation, 1944-1979. *Radiation Research* 181: 208-28.
[55] Romanenko AY, Finch SC, Hatch M, Lubin JH, Bebeshko VG et al. (2008) The Ukrainian-American study of leukemia and related disorders among Chornobyl cleanup workers from Ukraine: III. Radiation risks. *Radiation Research* 170: 711-720.
[56] Kesminiene A, Evrard AS, Ivanov VK, Malakhova IV, Kurtinaitis J, et al. (2008) Risk of hematological malignancies among Chernobyl liquidators. *Radiation Research* 170: 721-35.
[57] Zablotska LB, Bazyka D, Lubin JH, Gudzenko N, Little MP et al. (2013) Radiation and the risk of chronic lymphocytic and other leukemias among Chornobyl cleanup workers. *Environmental Health Perspectives* 121:59-65.
[58] Ivanov VK, Tsyb AF, Khait SE, Kashcheev VV, Chekin SY et al. (2012) Leukemia incidence in the Russian cohort of Chernobyl emergency workers. *Radiation Environment and Biophysics* 51: 143-149.
[59] Krestinina LY, Davis FG, Schonfeld S, Preston DL, Degteva M et al. (2013) Leukaemia incidence in the Techa River Cohort: 1953-2007. *British Journal of Cancer* 109: 2886-2893.
[60] Hwang SL, Hwang JS, Yang YT, Hsieh WA, Chang TC et al. (2008) Estimates of relative risks for cancers in a population after prolonged low-dose-rate radiation exposure: a follow-up assessment from 1983 to 2005. *Radiation Research* 170: 143-148.
[61] Daniels RD, Schubauer-Berigan MK (2011) A meta-analysis of leukaemia risk from protracted exposure to low-dose gamma radiation. *Occupational and Environmental Medicine* 68: 457-464.
[62] Krestinina LY, Preston DL, Ostroumova EV, Degteva MO, Ron E et al. (2005) Protracted radiation exposure and cancer mortality in the Techa River Cohort. *Radiation Research* 164: 602-611.
[63] Schubauer-Berigan MK, Daniels RD, Fleming DA, Markey AM, Couchet JR et al.(2007) Risk of chronic myeloid and acute leukemia mortality after exposure to ionizing radiation among workers at four U.S. nuclear weapons facilities and a nuclear naval shipyard. *Radiation Research* 167: 222-232.
[64] Boice JD Jr, Marano DE, Cohen SS, Mumma MT, Blot WJ et al. (2006) Mortality among Rocketdyne workers who tested rocket engines, 1948-1999. *Radiation Research* 2006;166: 98-11.

In: Fukushima Nuclear Accident
Editor: Shizuyo Sutou
ISBN: 978-1-63463-748-0

Chapter 11

COMPARATIVE ANALYSES OF LEAVES CONTAMINATED WITH ^{137}Cs COLLECTED IN CHERNOBYL AND FUKUSHIMA

Hiroo Nakajima*
Department Radiation Biology and Medical Genetics,
Graduate School of Medicine,
Osaka University, Osaka, Japan

ABSTRACT

The accident at the Fukushima No. 1 Nuclear Power Plant caused radioactive material to spread over a wide area, dispersing over not only soil and buildings, but also, given Fukushima's abundant natural environment, over plant leaves in fields, mountains, and agricultural land. These plants are consumed as food, and animals that have eaten them will have been exposed to radiation. Therefore, investigating the radioactivity of leaves will enable us to ascertain the contamination level in the plants and animals in that area. The leaf surface is an extremely small area compared to the area of Fukushima Prefecture, and the distribution of the radioactive material on the leaves is uneven, as it is in Fukushima Prefecture.

In this chapter, the author introduces a method of monitoring the radiation dose in leaves contaminated by the Chernobyl and Fukushima nuclear accidents. This method measures the difference in radioactive contamination status between the leaves and the change in status over time in an easy and quick manner using image analysis of individual leaves. This differs from the conventional method of measuring the Bq per kg, which requires a large amount of sample material.

---

* Correspondence to: Hiroo Nakajima, Tel: +81-6-879-3815, Fax: +81-6-6879-3819, E-mail: nakajima@radbio.med.osaka-u.ac.jp.

INTRODUCTION

The unprecedented Great East Japan Earthquake and Fukushima No. 1 Nuclear Power Plant accident occurred on March 11, 2011. Following these incidents, because of the hydrogen explosion of the reactor building, radioactive materials such as water-soluble highly radioactive iodine (^{131}I) and radioactive cesium (^{134}Cs and ^{137}Cs) were spread.

The distribution of radioactive fallout from a nuclear accident differs according to weather and topography. Many scientists, students, and volunteers created a map to show this distribution [1, 2]. The method, at the time, was to take samples of surface soil with fixed dimensions of 5-cm depth and 5-cm diameter, mix the sample until it was of even consistency, then use a germanium semiconductor detector to measure radioactivity in the sample. Then, the radiation level of each soil sample point was detected for each nuclide in Bq per kg to evaluate the distribution and degree of contamination [1, 2]. This was the best method for ascertaining the topographical contamination status across a wide area, and it was effective for producing the map. Moreover, the method of measuring Bq per kg is generally used to inspect for contamination of food, such as agricultural and marine products, and water. However, although this method enables us to measure the average contamination status per standard volume or weight, it cannot be used to detect the dose distribution on the surface of individual objects.

Radioactive cesium, one of the other materials released during the nuclear accident, is an alkali metal of the same type as potassium. Therefore, it may contaminate animals and plants in an ecosystem not only by surface contamination but also internally. In this case, measuring the Bq per kg cannot distinguish between surface contamination and *in vivo* contamination of living organisms.

To predict the amount of radioactive material absorbed by animals in an ecosystem and the subsequent changes in the amount accumulated within the body, it is important to understand the distribution of radioactivity in individual edible plants contaminated by radioactive fallout, as well as changes in this distribution over time. This is also important with respect to agricultural produce consumed by humans. Moreover, detecting the localization of radioactive material is important to identify the areas for decontamination. The importance of understanding the distribution and concentration of material in this manner is also evident in many everyday cases.

For example, rainwater of pH 5.6 or lower is known as acid rain. Why does acid rain cause problems by affecting marble and bronze statues or human hair when its acidity is less than that of cola drinks or orange juice (pH 2–4)? The answer is that acid rain is a solution containing small amounts of dissolved NOx, SOx, and HCl. These substances do not cause any problems at low concentrations; however, when water evaporates and the solution becomes more concentrated, SOx becomes concentrated sulfuric acid and NOx becomes concentrated nitric acid. Therefore, when materials at low concentration become highly concentrated through localization, they can have a different, much stronger, effect.

Figure 1 shows seven molecules evenly distributed per standard 1-liter volume (a large cube) (Figure 1a). Here, there is never more than one molecule in the small cube. However, if for some reason seven molecules are concentrated inside the small cube (Figure 1b), their concentration increases sevenfold in that area.

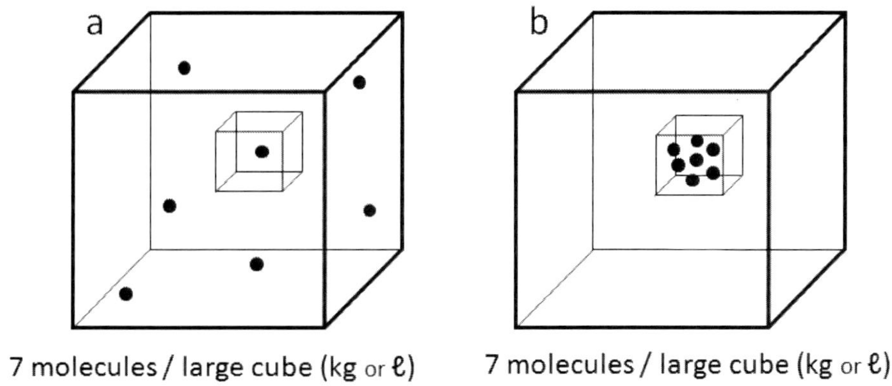

Figure 1. Density and distribution.

However, when measured by liter (large cube), the number of molecules in both Figure 2a and b is seven, and the measured concentration will be the same.

If the large cube represents an individual living organism, and the small cube represents a single cell within that organism, then if the situation in Figure 2b occurs, the biological reaction to that substance will be stronger than in the case of Figure 2a, and the biological effect will be significantly different.

Figure 2 (right panel) is an image taken using a transmission electron microscope, showing many nanoparticles localized within a living cell [3]. Figure 2 (left panel) shows an alveolar macrophage from a mouse that has inhaled airborne titanium dioxide nanoparticles. It is very difficult to detect tiny amounts of airborne nanoparticles within the body of a mouse following their inhalation. However, because alveolar macrophages engulf and collect many nanoparticles within the lung, these become concentrated inside the macrophages and can be easily detected as demonstrated in Figure 2 right [3]. In this manner, small amounts of substances within the body can become highly concentrated in certain cells. Therefore, in radioactive contamination, radioactive substances tend to localize within and on the surface of living organisms.

To ascertain this, we need to measure the radioactivity distribution within individual samples as well as measure the Bq per liter or kg. Radiation image analysis makes it possible to ascertain the radioactivity distribution on the surface of a sample. Since the Chernobyl accident, this method has been widely used in plants and animals [4-8]. However, a limitation of this method is that, while it can detect the strength of the radiation, it cannot determine the type of nuclide, and where the contamination is a mix of nuclides, it cannot quantify the dose of each contaminant. As such, it is not considered an important technique in environmental monitoring.

However, image analysis is a powerful method that has many advantages. It does not require a lot of sample material, allowing users to ascertain the distribution of radioactivity from a single leaf, and it enables a quick, easy, quantitative understanding of the changes in the overall contamination status over time after a nuclear accident.

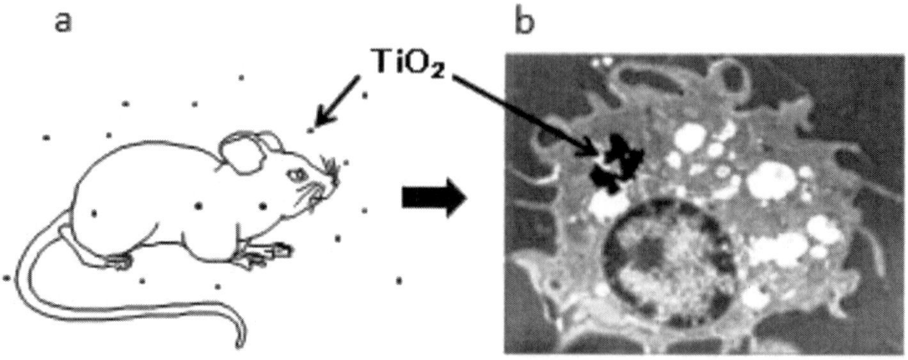

Figure 2. Localization of nanoparticles. (a) Inhalation of titanium dioxide nanoparticles. (b) Titanium dioxide nanoparticles concentrated in an alveolus macrophage. (Reprinted with permission from Elsevier [3]).

1. Images of Leaves Taken after the Fukushima No. 1 Nuclear Power Plant Accident

Figure 3 shows an image of a bamboo grass leaf taken on day 109 after the accident (June 29, 2011) alongside National Route 115 in Iitate village, some 42 km away from Fukushima No. 1 Nuclear Power Plant [9]. It was created by contacting the leaf to a BAS imaging plate (Fujifilm) and exposing it appropriately in a cool dark place. A scanner-type image analyzer (Typhoon FLA 7000, GE Healthcare) was used to form an image of the radiation strength and distribution on the leaf surface.

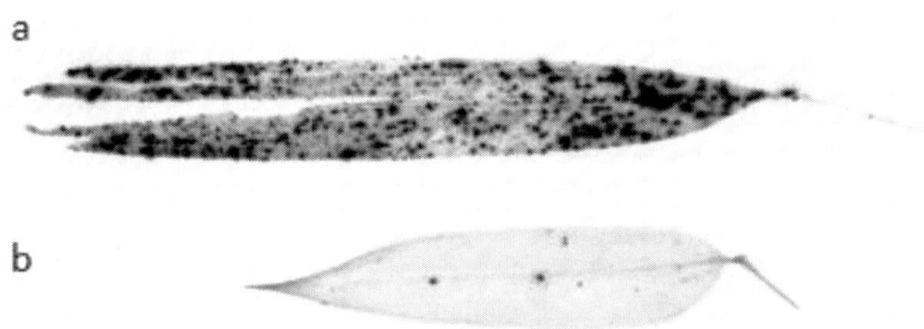

Figure 3. Distribution of radioactive cesium on bamboo leaves sampled at Iitate village located 42 km from the Fukushima No. 1 Nuclear Power Plant. Samples were taken 109 days after the Nuclear Power Plant accident. (a) old leaf. (b) young leaf. (Reprinted with permission from Wolters Kluwer Health [9]).

The upper leaf (Figure 3a) is an old leaf that had been growing since before the accident, and the lower leaf (Figure 3b) is a young leaf that sprouted after the accident. The leaves in the photos are rendered darker in areas where the radiation is stronger; areas emitting no radiation are left white and are not visualized. On the older leaf, we can clearly see that the radioactive fallout has formed innumerable dots that have adhered to the leaf surface. This shows that the radioactive material drifting in the air became concentrated in raindrops, which fell and adhered to the leaf surface, whereupon the raindrops dried, and the material remained until the leaf was gathered.

On the younger leaf, however, there are very few black dots on the image. This shows that the young leaf, which sprouted after the accident, was not exposed to radioactive fallout before it was gathered. In other words, there was no radioactive material falling during this period. However, the fine capillaries in the leaf and the surrounding areas are faintly colored, showing that a tiny amount of radioactive material has permeated evenly throughout the leaf. This shows that radioactive material collects on the surface of old leaves and moves throughout the interior of the leaf after being absorbed through the leaf surface as well as through the roots via the soil.

A bamboo grass leaf was collected 389 days after the accident at a point 10 km from the nuclear power station, and an image comparable to that shown in Figure 3 was obtained. This showed that once the contamination had settled in points, it remained until the leaf fell. This shows that old evergreen leaves contaminated by radioactive fallout retain high concentrations of contamination in the upper section of the tree until they fall, so that in several years' time, the falling of these leaves will cause the radiation level at the base of the tree to rise.

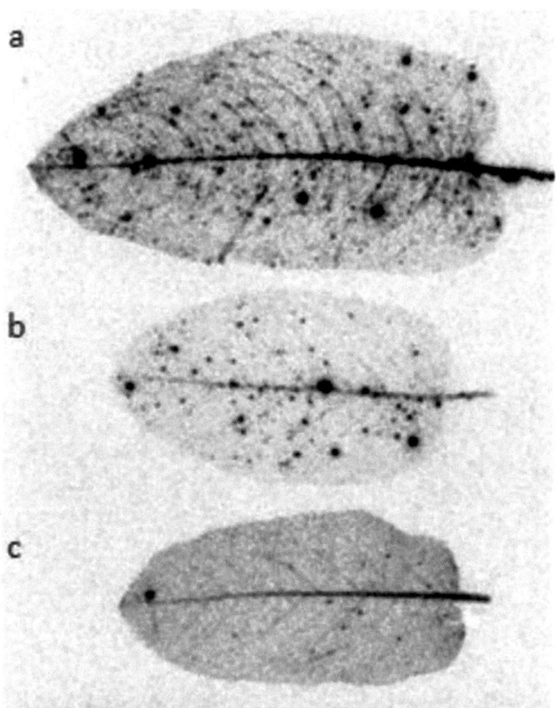

Figure 4. Distribution of radioactive cesium on sorrel leaves sampled at Fukushima city located 62 km from the Fukushima No. 1 Nuclear Power Plant. From the top panel, leaves were sampled at the same location on days 78 (a), 119 (b), and 133 (c). (Reprinted with permission from Wolters Kluwer Health [9]).

Figure 4 shows the contamination status of sorrel leaves collected from the front of JR Fukushima Station, an area 62 km from the nuclear power station, which was not designated as an evacuation zone [9]. In this image, the exposure time was lengthened to increase the sensitivity to detect weaker radiation than that detected on the bamboo leaves in Figure 3. The photos are of leaves taken from the same spot on (from top to bottom) day 78 (May 29), day 119 (July 9), and day 133 (July 23) after the accident. The contamination dose is low; however, the leaf from day 78 contains many black dots showing the adherence of radioactive fallout as observed on the bamboo grass leaf in Figure 3. Then, on days 119 and 133, the black dot contamination declines as the time since the accident increases. However, the leaf capillaries and their peripheral areas grow uniformly darker, enabling us to confirm that weak radiation is present throughout the entire leaf.

In both Figures 3 and 4, the main contamination is on the leaf surface due to the initial radioactive fallout after the accident; however, it is clear that over time the contamination status changes as contaminants accumulate inside the plant due to their absorption through the roots and leaves. Measurement of the radiation depicted in these images using a germanium semiconductor detector showed that most of the radioactivity was from ^{134}Cs and ^{137}Cs.

2. Comparison of Leaves from Fukushima and Chernobyl

Figure 5a shows an oak leaf collected from an area contaminated with high levels of radiation (Masani village, Belarus) in 1997, 11 years after the Chernobyl nuclear accident, and Figure 5b shows a bamboo grass leaf from Iitate village (the same leaf as shown in Figure 3), and Figure 5c shows a sorrel leaf from Fukushima City (the same leaf as shown in Figure 4) [9]. These photographs use the same image processing method to compare the radiation strength and distribution status.

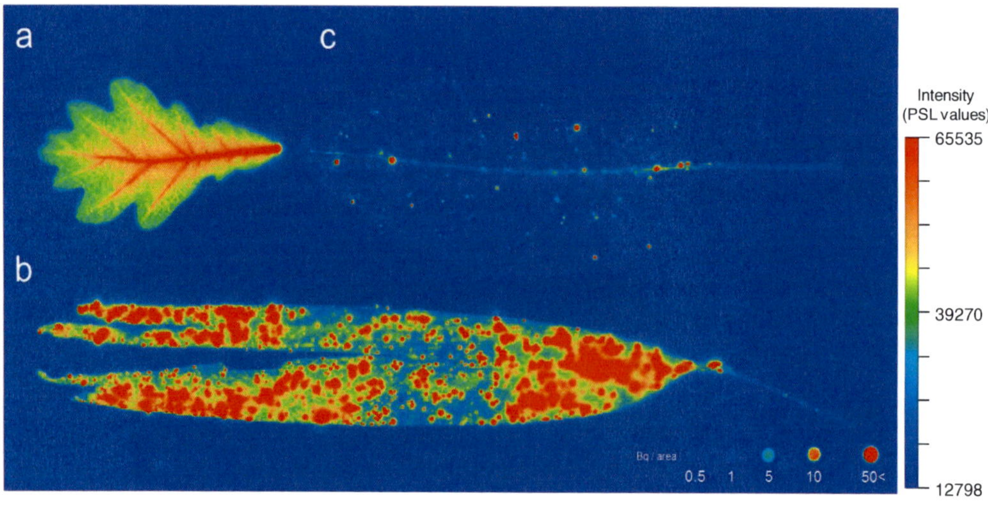

Figure 5. Comparison of an oak leaf (a) sampled in an area of high radioactivity (Masani village, Belarus) in 1997, 11 years after the Chernobyl Nuclear Power Plant accident; a bamboo leaf (b) from Iitate village; and a sorrel leaf (c) from Fukushima City. (Reprinted with permission from Wolters Kluwer Health [9]).

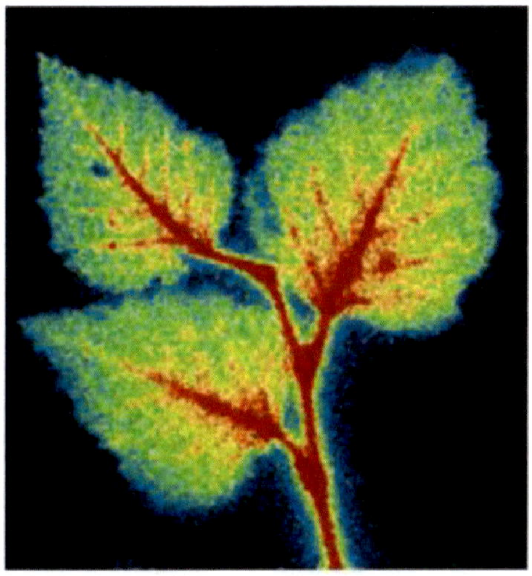

Figure 6. Distribution of radioactive cesium on birch leaves sampled at Masani village, Belarus in 1997.

The radiation strength is depicted by color, increasing in strength from blue, through green, yellow, and red. Comparing these three leaves, we can see clear differences in the contamination status in terms of radiation strength and distribution. In particular, the Chernobyl leaf, which was collected 11 years after the nuclear accident, shows no red dots of contamination at all, with the highest radiation appearing around the leaf capillaries and a gradation in intensity toward the tip. This indicates that all of the contamination is internal, occurring through absorption. This image of intensity gradation in the leaf is not unique—as shown in Figure 6 (birch leaves)—leaves sharing the same root system are affected in the same way. On the other hand, in the leaves from Fukushima, Figure 5b and Figure 5c, the radiation strength differs, but, as noted, both leaves show red dots of contamination. This difference may be due to the change in contamination status after 11 years, and suggests that the images of contaminated leaves from Fukushima will eventually change to be similar to those from Chernobyl.

3. Predictions from the Analysis of Leaf Radioactivity Images

3.1. Future of Plant Contamination Status in Fukushima

Further, it is likely that the radiation distribution in leaves growing in contaminated areas of Fukushima will change. Thus, rather than showing points of contamination on images, radiation will be clearly concentrated around the leaf capillaries, weakening in a gradient to the leaf periphery, as observed on the Chernobyl leaf. Moreover, over time, the contamination status is projected to change from that of the leaf in Figure 7a right, which is taken from a high-level contamination area (Masani), to that of the leaf in Figure 7a left, which is taken from a low-level contamination area (Babchin). In fact, in leaves collected from the same areas 8 years later, the radiation had decayed in the same way for both the high-level contamination area and the medium-level contamination area (Figure 7b). Figure 7c shows image analysis of frogs taken from the high-level contamination area of Masani (Figure 7c right) and the medium-level contamination area of Babchin (Figure 7c left). In these frogs, the radiation had decayed by 2005, the same as in the leaves (Figure 7d).

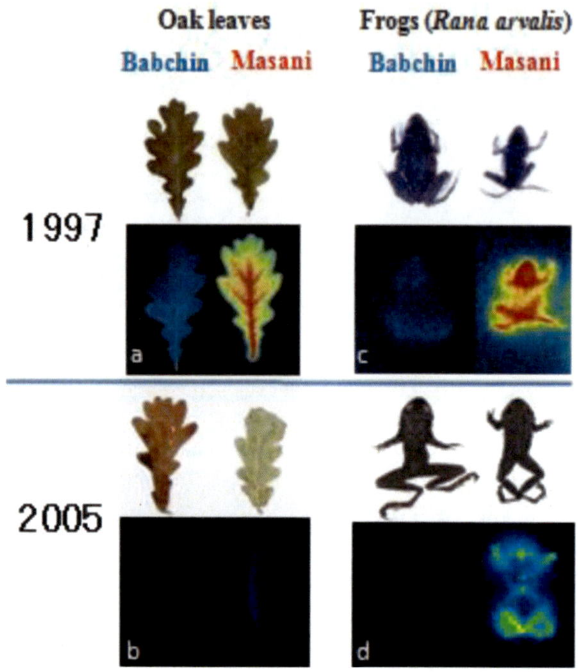

Figure 7. Comparison of the radiocesium content in 1997 and 2005 sampled at the same place in Belarus. (a, b) oak leaves, (c, d) frogs.

Moreover, comparisons of the ^{137}Cs concentration between 1997 and 2005 in different internal organs of mice, moles, frogs, and grasshoppers revealed that the radiation had decayed during that period, just as it had in the leaves. Taking the ^{137}Cs radioactivity (Bq/g) in the internal organs in 1997 as 100%, and then seeking the average residual rate of 137Cs in 2005, the average residual rate in 2005 was found to be 1.99% of the radioactivity observed in 1997 for most animal species taken from high- and medium-level contamination areas. Based on the rate of decrease (%) over the eight-year period between 1997 and 2005, the biological half-life during this period was calculated as approximately 1.44 years [10]. Ecosystems in Fukushima will likely follow a similar course of radioactivity decay in the future.

3.2. Changes in the Internal Contamination of Animals in Fukushima over Time

As can be seen in Figures 3, 4, and 5, young leaves that sprouted after the accident and leaves that were examined after some time had elapsed showed progressive internal contamination due to absorption. However, during the initial period after the nuclear accident, internal contamination in living organisms due to absorption was significantly lower than surface contamination. This might be because radioactive cesium fallout has a strong affinity to clay soil, making it difficult for plants to absorb. Additionally, the transfer coefficient to grains and vegetables was 0.01–0.026 [11,12]. (The transfer coefficient is defined as the concentration in edible crop parts (Bq/g raw weight) per unit concentration in the soil (Bq/g dry weight)). Since cesium has a strong affinity to the soil, the downward penetration of radioactive cesium into the deep soil layer is very slow. In soil samples taken from Fukushima in June 2011, contamination levels at a depth of 4 cm were less than one-tenth of those at the soil surface. A similar result was reported for soil from the Chernobyl area, where 95% of the radioactive cesium remained within a 5 cm depth even 10 years after the nuclear accident [13, 14].

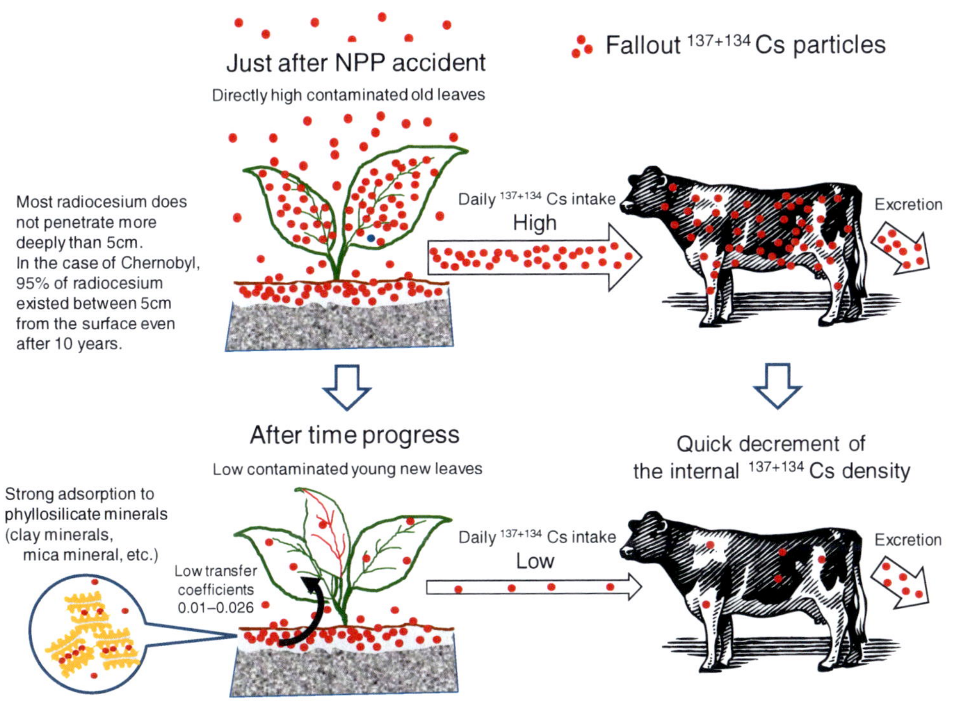

Figure 8. Changes in the internal contamination of wild animals in Fukushima.

As shown in Figure 8, it is thought that herbivorous domestic and wild animals in Fukushima consumed leaves with high-level surface contamination in the initial period after the nuclear accident and then consumed new leaves with low-level contamination thereafter. As a result, the time during which most radioactive cesium was internalized by wild animals was the time when they consumed plants with the most surface contamination due to radioactive fallout. Subsequently, the daily intake dose of cesium would have declined significantly because they ate new leaves that sprouted after the radioactive fallout had ceased. Moreover, half of the radioactive cesium that was released was ^{134}Cs, which has a physical half-life of two years. Therefore, also considering the physical decay of ^{134}Cs itself, after a certain period (8 years after the accident), the *in vivo* radioactive cesium present in animals would decrease rapidly even if they continued to consume wild plants.

In an experiment where mice were administered a single oral dose of ^{137}Cs, the ^{137}Cs declined by half in most of the internal organs within 2 days. In muscle tissue, however, the movement of ^{137}Cs was delayed, reaching its peak level on day 4, and then declining by half around day 13 [15]. This shows that ^{137}Cs does not accumulate internally in animals, and from this, we can predict that when the excreted amount exceeds the intake amount, the total internal contamination will gradually decrease. However, cesium that has adsorbed strongly to the soil gradually accumulates in shrubs and annual herbs with shallow roots, and when these plants die and form a compost layer, the ^{137}Cs becomes easier to absorb chemically and will be absorbed by larger plants. Therefore, with the passage of time, the plants in Fukushima would accumulate a certain level of radioactive cesium, similar to the leaves from Chernobyl (Figures 6 and 7). When animals consume these leaves at a constant daily rate, the gap between the intake and excreted amounts of radioactive cesium will narrow, and from year 10 after the accident, the amount of internal radioactive cesium in animals is expected to decrease more slowly. Ecological studies after the nuclear accident at Chernobyl showed two phases of radioactive cesium decline *in vivo*: a rapid decline phase (early phase) and a slow decline phase (late phase) [16, 17].

4. Application of the Image Analysis Method

4.1. Trees in Fukushima and Chernobyl

Figure 9 is an image of ^{137}Cs distribution in pine trees collected from the non-contaminated (Ribzky Belarus: a, b), moderately contaminated (Babchin Belarus: c, d), and highly contaminated areas (Masani, Belarus: e, f) in the eleventh year after the nuclear accident at Chernobyl [9].

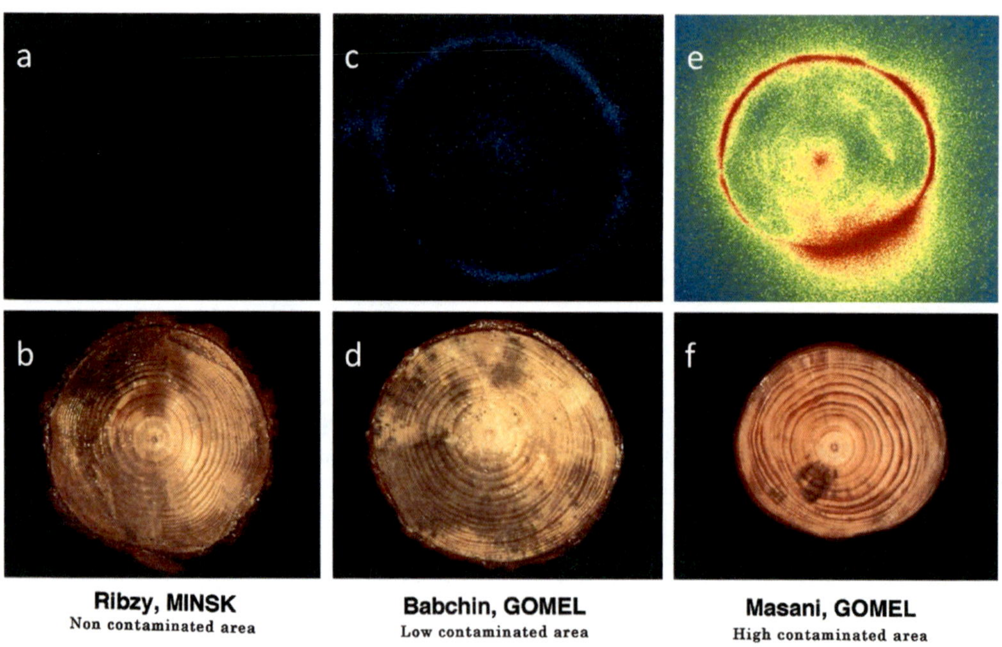

Figure 9. Comparison of pine trees sampled at areas of high contamination (Masani Belarus: a, b), moderate contamination (Babchin Belarus: c, d), and no contamination (Ribzky Belarus: e, f) in 1997, 11 y after the Chernobyl Nuclear Power Plant accident. (Reprinted with permission from Wolters Kluwer Health [9]).

The tree rings reveal the growth status of the tree each year, and therefore record the environmental conditions (sunlight or temperature) that influenced the tree's growth.

Unfortunately, there was no record of radioactive contamination from the Chernobyl nuclear accident in the tree ring corresponding to 1986. The bark shows the highest level of radioactivity, and the level of radioactivity decreases toward the center. The *in vivo* action of ^{137}Cs is similar to that of potassium and, therefore, is mostly present not in the 1986 tree ring, but in the tree bark section that was growing in 1997 when the tree was collected. It is thought that ^{137}Cs was taken up from the roots along with nutrients and water, and was present inside the ducts in the tree bark [5, 9].

Figure 10 shows images of the radioactive cesium distribution in tree ring cross sections from cedar trees collected from points 0 km (a), 5 km (b), and 10 km (c) from the Fukushima No. 1 Nuclear Power Plant on day 389 after the nuclear accident [9]. In Figure 10, all of the images show surface contamination of the trees, and it is clear that, by one year after the accident, the radioactive contamination had not yet progressed inside the tree or inside the tree bark through absorption from the roots.

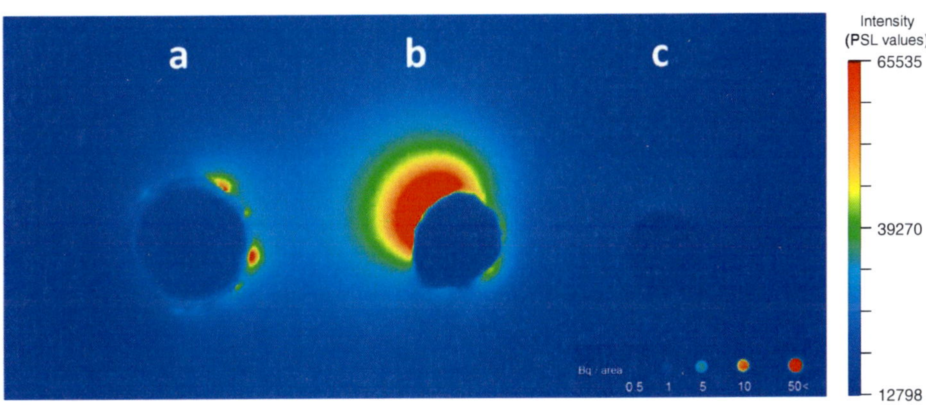

Figure 10. Comparison of cedar trees in high-contamination areas of Fukushima. Tree trunks were sampled at 0 km (a), 5 km (b), and 10 km (c) from Fukushima Nuclear Power Plant on day 389. (Reprinted with permission from Wolters Kluwer Health [9]).

There is a risk of building material contamination as a result of forest contamination. Therefore, it is possible that a large amount of timber could be discarded if the contamination level is considered high after measuring the radiation dose at the surface. However, ^{137}Cs is not incorporated into biological matter in the same way as ^{14}C, and as shown in Figure 9, it was not locked into the 1986 tree ring corresponding to the time of the Chernobyl accident. As discussed above, it has a similar *in vivo* action to potassium and is present mostly in the wood section where there are many ducts. Therefore, even if trees have grown in contaminated areas, it may be possible to decontaminate them by examining their contamination status through image analysis of the tree-ring cross sections (Figures 9 and 10) and by the comprehensive removal of the contaminated bark sections. Thus, image analysis may also be applied when considering which sections are to be removed.

Furthermore, as shown in Figure 11, technology also enables users to visualize the contamination status of discarded wood chips (Figure 11a) and contaminated soil before (Figure 11b top) and after (Figure 11b bottom) decontamination.

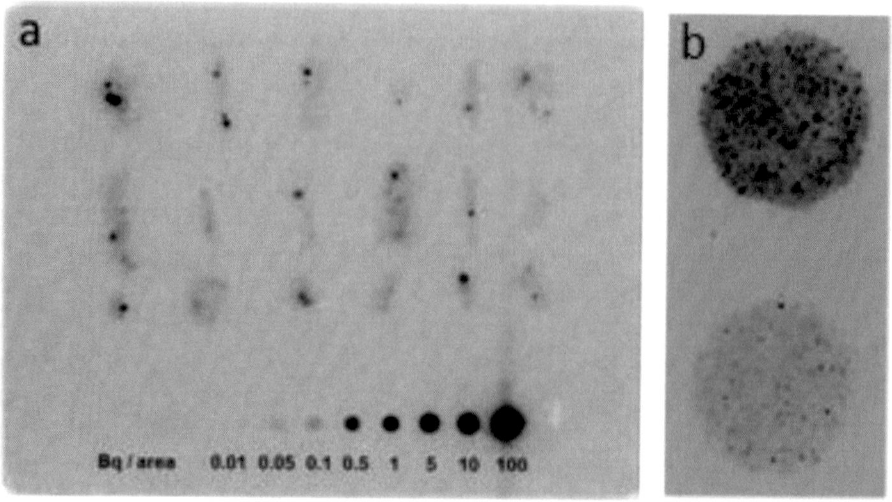

Figure 11. Visualized radioactivity of contaminated waste wood chips (a) and soil (c) sampled in Fukushima.

4.2. Selection of Plants for Decontamination

A previous study described the absorption of radioactive cesium by plants as a soil decontamination method. The candidate plants that have been shown to have high absorption efficiency for ^{137}Cs are those of the Amaranthaceae family, such as *Amaranthus retroflexus*, *Amaranthus lividus*, and *Achyranthes japonica*, and sunflowers from *Compositae* [6,18-19]. However, because root lengths and sizes are different in each plant, the plant species have to be selected carefully based on the contamination status to obtain better absorption efficiency of radioactive cesium in shallow areas. Furthermore, it is essential to avoid the destruction of the current ecosystem because of the introduction of these selected plants.

Thus, image analysis of leaves can be used, not only for evaluating the surface contamination of radioactive cesium, but also as a tool for assessing the leaves and identifying a suitable plant with good internal absorption from the plants that currently grow in the Fukushima area. Further, this method can be used to identify plants with distinct biological or chemical ability to accumulate radioactive cesium in the non-edible parts, as part of realizing efficient radioactive decontamination in combination with productive farming.

Figure 12 shows the image analysis of two types of grasses growing in the same place in the high-level contamination area within a 20-km radius of the nuclear accident site. Figure 12a shows the image analysis for short grass with shallow roots, and Figure 12b shows the image analysis for tall grass with deep roots.

The image analysis for the short grass shows that radioactive cesium has been absorbed not only in the roots, but also throughout the leaves. In the tall grass, the leaf shape is hardly rendered (white arrow: part of the stalk). This shows that even within the same area, radioactive cesium is efficiently taken in only by plants that spread their roots in the soil layer where cesium is present.

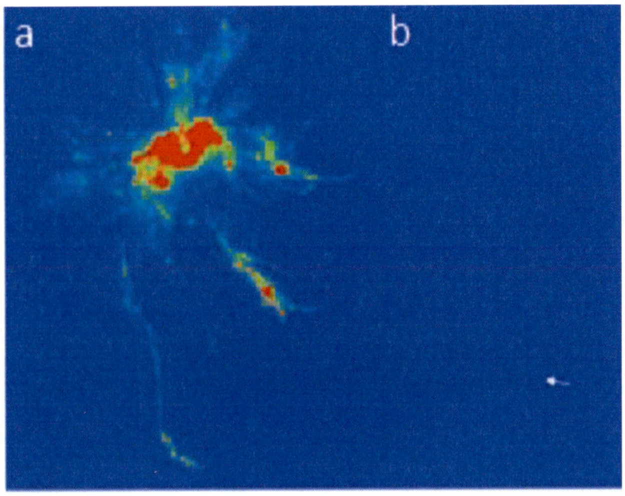

Figure 12. Comparison of the radioactivity of weeds growing in the same place. (a) Short height and shallow-rooted weed. (b) Tall height and deep-rooted weed.

5. Method of Obtaining and Analyzing Images

5.1. The Acquisition of the Image

Leaf samples taken from radioactive-contaminated areas were placed on a BAS imaging plate (20 cm × 25 cm or 20 cm × 40 cm, Fujifilm Corporation, Tokyo, Japan) covered by plastic wrap to prevent contamination, and were stored in cool, dark conditions for exposure periods of 1 to 7 days (depending on the strength of radioactivity) (Figure 13). Latent imaging was conducted using a scanner-type image analysis device (Typhoon FLA 7000, GE Healthcare) to visualize the distribution of radioactivity on leaf surfaces (Figure 14).

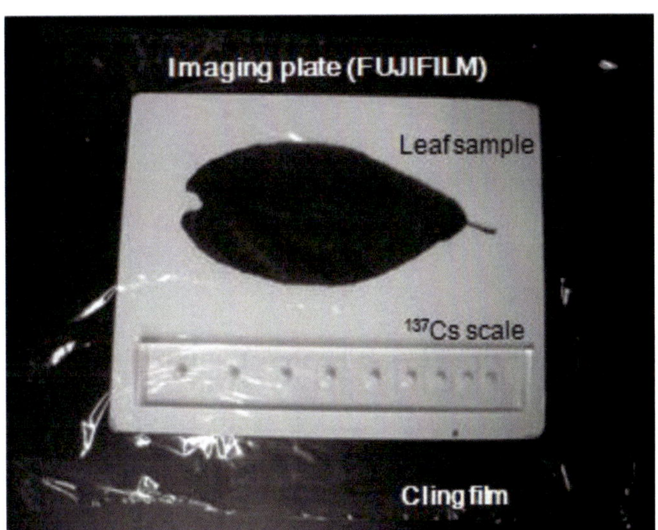

Figure 13. Acquisition of the image of a leaf and ^{137}Cs scale.

Figure 14. Scanner type image analyzer. Typhoon FLA 7000 (GE Healthcare).

5.2. The Quantification of Distributed Radioactivity on a Leaf

To quantify the radioactivity on a leaf, a concentration series of ^{137}Cs solutions (0.5, 1, 5, 10, and 50 Bq/5 μL) was prepared as a scale of radioactivity levels. Each solution was dropped onto water-absorbent paper (diameter 6 mm) and was exposed to the imaging plate containing leaf and wood samples (Figure 13).

5.3. Sensitivity of the Imaging Plate

Figure 15 shows the sensitivity of the imaging plate to ^{137}Cs radioactivity. Circular papers of 6-mm diameter (28.27 mm^2) contaminated with ^{137}Cs at respective strengths of 0.1, 1, 5, 10, 50, 100, 500, and 1000 Bq were exposed to the imaging plate for different lengths of time, namely 3, 6, 30, and 60 s; 10 min; and 24 h. The 0.1 and 1 Bq papers, which corresponded to a surface contamination level of 4 Bq/cm^2, were able to form an image with a 24-h exposure. At 500 Bq, the paper was able to form an image with a 3-s exposure. This shows that the imaging plate has higher exposure sensitivity than an autoradiograph.

5.4. Quantifying ^{137}Cs from Image Analysis

The intensity of radioactivity was assessed as the PSL (photo-stimulated luminescence) values at 50 μm/pixel by scanner-type image analysis device (Typhoon FLA 7000, GE Healthcare) (Figure 14). The range of PSL values was from 12,798 (background) to 65,535 (upper limit). In the images, areas of high radioactivity on leaf surfaces were represented as darker patches, whereas low-radioactivity areas were white. Individual spot (6-mm diameter) radioactivity on the leaves taken on day 109 after the accident in Iitate village (Figure 16c, the area of the open circle) was roughly estimated using the dose response curve of the PSL values of the prepared ^{137}Cs scale (Figure 16a, b). The dose range of measurable spots was 1.5–16.0 Bq (Figure 16d). It seemed that the sizes of fallout particles scattered by the first explosion at the Nuclear Power Plant were reflected in each radioactive spot.

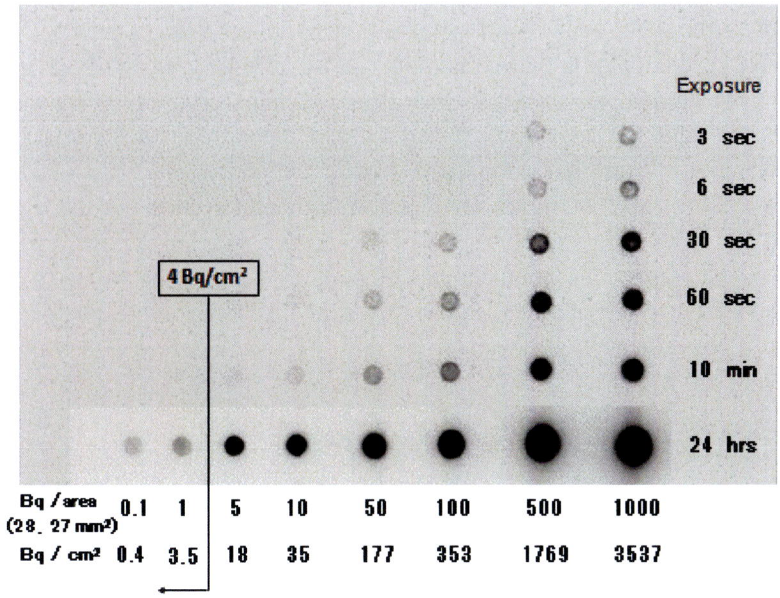

Figure 15. Exposure sensitivity of imaging plate (FUJIFILM) for ^{137}Cs radioactivity.

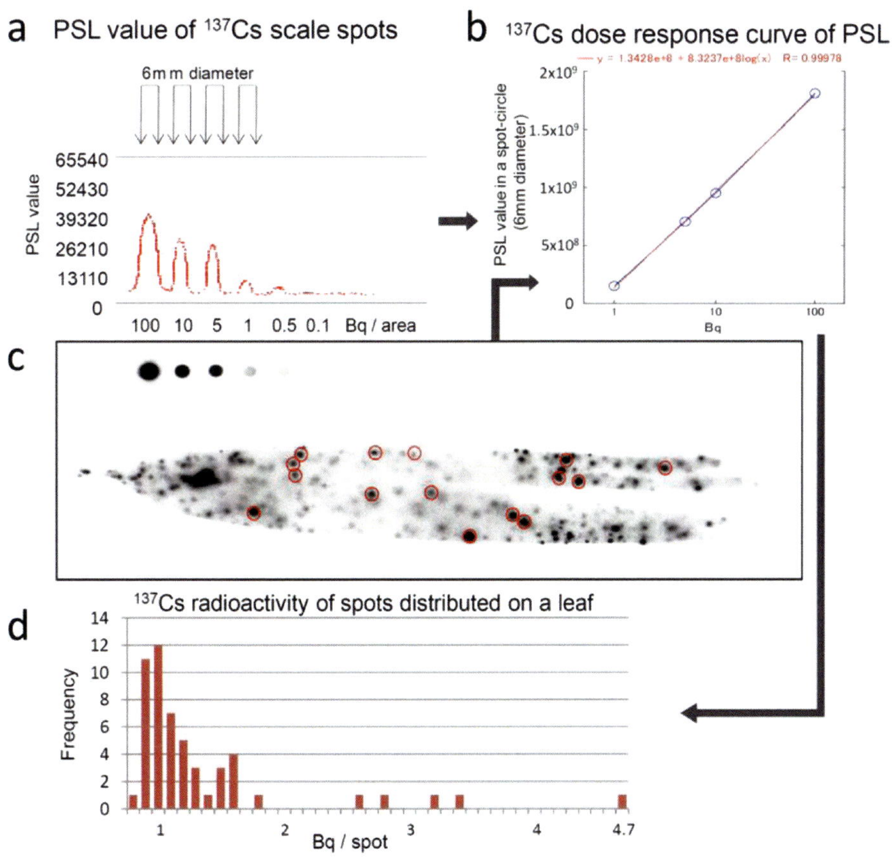

Figure 16. Quantifying ^{137}Cs from image analysis of a leaf taken on day 109 after the accident in Iitate village.

To visualize the semi-quantification of radioactivity on a leaf, the PSL values at 50 μm/pixel gained by the imaging plate were converted to a color gradient (blue, green, yellow, and red) using image analysis software (Win Roof, Mitani Corporation, Tokyo, Japan), and are shown on the right side in Figures 5 and 10. The color from blue to red represents the level of radioactivity from background to high. The series of ^{137}Cs radioactivity was visualized on a color scale as Bq per area (6 mm in diameter circle) at the right bottom of Figures 5 and 10.

To confirm the major radio-contaminants of samples, gamma spectrometry of leaves and trees were carried out by Standard Electrode Coaxial Ge Detectors GC3018 (CANBERRA Industries Inc. Connecticut, U.S.A.), and ^{137}Cs and ^{134}Cs were detected as major nuclides at about the ratio of 1:0.98 [9].

CONCLUSION

Plants are at the bottom of the food chain for both animals and humans. Therefore, it is important to ascertain their contamination status following a nuclear accident in order to minimize the spread of radioactive contamination to animals and people.

Using the radioactivity image analysis introduced in this chapter, it is easy to delineate changes in the contamination status of leaves (including surface dose distribution, leaf tissue internal and external dose distribution, *in vivo* action, and change in dose distribution over time) that cannot be determined by dose measurements made in kilogram units. Moreover, if the leaves of evergreen trees are used for sampling, it is possible to measure the accumulated dose of radioactivity over periods of more than one year (long term). In the case of deciduous trees and annual grasses, this method shows the short-term contamination status over 1-year periods. In particular, in low-level contamination areas, this method can increase detection sensitivity by measuring the accumulated dose in evergreen trees. Furthermore, image analysis using an imaging plate is excellent for the quantitative evaluation of radiation, offering detection sensitivity that is dozens to several hundred times that of photographic methods. Therefore, it can be used as a tool to monitor the radioactive contamination status of objects other than leaves.

Looking ahead, the focus in Fukushima seems likely to shift to determining how to decontaminate the confirmed contaminated areas, and how the actual contamination status will change over time. Conducting long-term image analyses of the leaves from trees growing in the same place over several years is a simple activity; however, it will enable people to make timely responses to the contamination status as it changes over time. As such, this image analysis method has the potential to help keep damage from radioactive contamination to a minimum.

REFERENCES

[1] Fujiwara M, Tanihata I, Saito K (2011) Distribution map of radiation dose in Fukushima [online]. Available at http://www.rcnp.osaka-u.ac.jp/dojo/. Accessed 3 September 2014.

[2] Ministry of Education, Culture, Sports, Science and Technology (MEXT) (2011) Distribution map of radiation dose around Fukushima Dai-ichi & Dai-Ni NPP [online]. Available at http://www.mext.go.jp/english/incident/1305901.htm. Accessed 3 September 2014.

[3] Nakajima H, Ozaki K, Hongyo T, Narama I, Todo T (2011) A rapid and easy method for the qualitative detection of intracellular deposition of inhaled nanoparticles. Nanomedicine: *Nanotechnology, Biology, and Medicine* 7: 881-888. doi:10.1016/j. nano.2011.02.004

[4] Bersina G, Brandt R, Vater P, Hinke K, Schu¨tze M (1995) Fission track autoradiography as a means to investigate plants for their contamination with natural and technogenic uranium. *Radiation Measurements* 24: 277-282.

[5] Nakajima H, Ryo H, Nomura T, Saito T, Yamaguchi Y, et al. (1998) Radionuclides carved on the annual rings of a tree near Chernobyl. *Health Physics* 74: 265-267.

[6] Soudek P, Valenová Š, Vavříková Z, Vaněk T (2006) 137Cs and 90Sr uptake by sunflower cultivated under hydroponic conditions. *Journal of Environmental Radioactivity* 88: 236-250.

[7] Sawidis T, Tsigaridas K, Tsikritzis L (2010) Cesium-137 monitoring using lichens from W. Macedonia, N. Greece. *Ecotoxicology and Environmental Safety* 73:1789-1796.

[8] Yamaguchi T, Sawano K, Furuhama K, Mori C, Yamada K (2012) An autoradiogram of skeletal muscle from a pig raised on a farm within 20 km of the Fukushima Daiichi nuclear power plant. *Journal of Veterinary Medical Science* 75: 93-94.

[9] Nakajima H, Fujiwara M, Tanihata I, Saito T, Matsuda N and Todo T (2014) Imaging plant leaves to determine changes in radioactive contamination status in Fukushima, Japan. *Health Physics*, 106: 565-570. doi:10.1097/HP.0000000000000020

[10] Nakajima H, Saito T, Yeliseeva KG, Ryo H, Hongyo T, et al. (2009) Ecological decrease, biological concentration and genome stress of radionuclides in plants and animals after Chernobyl catastrophe. *Proceeding of 10th International Conference on Environmental Mutagens (ICEM)*, p. 221

[11] Nisbet AF, Woodman RFM (2000) Soil-to-plant transfer factors for radiocaesium and strontium in agricultural systems. *Health Physics* 78: 279-288.

[12] International Atomic Energy Agency (1994) Handbook of parameter values for the prediction of radionuclide transfer in temperate environments. Vienna: IAEA; Technical Reports Series No.364.

[13] Nakajima H, Ryo H, Yamaguchi Y, Saito T, Yeliseeva KG, et al. (2000) Biological concentration of radionuclides in plants and animals after the Chernobyl catastrophe. In: Sato F, Yamada Y, Onodera J, eds. Biological effects of low dose radiation. Aomori, Japan: Institute for Environmental Sciences, pp. 199-205.

[14] Nakajima H, Saito T, Ryo H, Nomura T (2008) Ecological decrease and biological concentration of radionuclides in plants and animals after the Chernobyl catastrophe. In: Miura T, Kinoshita N, eds. Proceedings of the Eighth Workshop on Environmental Radioactivity. Tsukuba, Japan: High Energy Accelerator Research Organization (KEK) Proceedings 2007-16, pp. 113-118.

[15] Nakajima H (2011) How to convey the information of radiation impact correctly to the general public. *Radiation Biology Research Communications.* 46: 120-126.

[16] Jonsson B, Forseth T, Ugedal O (1999) Chernobyl radioactivity persists in fish. *Nature* 400: 147.

[17] Smith JT, Comans RNJ, Beresford NA, Wright SM, Howard BJ, et al. (2000) Chernobyl's legacy in food and water. *Nature* 405: 141.

[18] Dushenkov S, Mikheev A, Prokhnevsky A, Ruchko M, Sorochinsky B (1999) Phytoremediation of radiocesium-contaminated soil in the vicinity of Chernobyl, Ukraine. *Environmental Science and Technology* 33: 469-475.

[19] Fuhrmann M, Lasat M, Ebbs S, Cornish J, Kochian L (2003) Uptake and release of cesium-137 by five plant species as influenced by soil amendments in field experiments. *Journal of Environmental Quality* 32: 2272-2279.

In: Fukushima Nuclear Accident
Editor: Shizuyo Sutou

ISBN: 978-1-63463-748-0

Chapter 12

MONITORING OF RADIO-CONTAMINATION IN RETURNEES FROM EVACUATION ZONES IN FUKUSHIMA AND SPORADIC MEASUREMENTS OF RADIATION IN CONTAMINATED AREAS OF IBARAKI AND FUKUSHIMA PREFECTURES

Shizuyo Sutou*
School of Pharmacy, Shujitsu University, Okayama, Japan

ABSTRACT

The March 11 earthquake and tsunami crippled the Fukushima Daiichi Nuclear Power Plant (FDNPP) of Tokyo Electric Power Company (TEPCO). The blackout of the plant led to hydrogen explosion and wide areas were consequently contaminated with radioactive substances. People who lived in the evacuation zone that was set as the area within a 20 km radius from the power plant were forced to evacuate immediately. Fukushima Prefectural Disaster Headquarters (FPDH) and the Japanese Governmental Disaster Headquarters (JGDH) planned to let them return to their homes temporarily. They asked for help from related organizations, one of which is the Ministry of Education, Culture, Sports, Science and Technology (MEXT) of Japan. MEXT asked all Japanese universities and colleges to take part in screening radioactivity of the temporal returnees from the areas. I applied to the screening activities and measured radioactivity of the returnees at Bajikoen Off-site Center on July 14 – 17, 2011. This constituted only a tiny fraction of the temporal returning program, which started in May and lasted for several months. A total of 3,686 volunteers were recruited by MEXT for the program and 33,598 people out of 78,000 evacuees (43%) were screened during this period; none was found to be contaminated with radioactivity over 13,000 cpm, which was the criterion for decontamination. Making use of this opportunity of screening, I measured radioactivity in several areas in Fukushima and Ibaraki Prefecture. The results from these sporadic measurements are presented here.

* Correspondence to: Shizuyo Sutou, Tel & Fax: +81-86-271-8357, E-mail: sutou@shujitsu.ac.jp.

INTRODUCTION

On March 11, 2011, the earthquake and tsunami broke down the Fukushima Daiichi Nuclear Power Plant (FDNPP) and, as a result of hydrogen explosion, a wide range of areas has been contaminated with radioactivity. Based on the Act on Special Measures Concerning Nuclear Emergency Preparedness, the evacuation zone was set to the area within a 20 km radius from FDNPP.

People who lived in the zone were forced to evacuate at 20:30 on March 12, three h after the explosion. Fukushima Prefectural Disaster Headquarters (FPDH) planned to let them return to their home temporarily for two hours and asked help of the Japanese Governmental Nuclear Disaster Headquarters (JGDH), which then asked for help from relevant organizations and agencies for the temporary returning program. The Ministry of Education, Culture, Sports, Science and Technology (MEXT) was one of them and delivered a notice to all national, public, and private universities and colleges to dispatch a total of 30 volunteers a day for radioactive contamination screening of temporary returnees. This program started on May 10, 2011, but was conducted most actively in June and July, and continued intermittently in August and early September. Shujitsu University formed two volunteer teams each of which consisted of two faculty members.

The first team to which I belonged worked from July 13 to 18, 2011, and the second team worked from July 21 to 25, 2011. This chapter was compiled to record our voluntary activity that constituted some aspects of the large scale radioactivity screening program planned by FPDH and JGDH.

MATERIALS AND METHODS

A GM counter (model TGS-146B; Hitachi Aloka Medical, Ltd., Tokyo, Japan) was used to detect radioactivity. When radioactivity in higher air masses up to 5 m was measured, the detector muzzle and the body of the TGS-146B were connected with a 5.9-m-long coaxial cable cord. The muzzle was suspended using a string along a 5-m-long rod. Its position was adjustable. When radioactivity was counted with and without the extension cord in Iitate Village, Fukushima Prefecture, the mean % ± SD was 91.1% ± 5.2% ($n = 8$) of the control (without the extension cord).

The cord hampered the counting by less than 10%, but the effect was marginal. The present GM counter expresses radioactivity as counts per min (cpm). Effective doses were measured using an ion chamber scintillation counter (model ICS-321; Hitachi Aloka Medical, Ltd.), which is effective to detect both β-rays and γ-rays. This is equipped with a shielding cap to cut off β-rays.

The cap was used when needed. Effective doses are expressed as microsieverts per h (μSv/h). When radioactivity (cpm) and effective doses (μSv/h) were measured with TGS-146B and ICS-321 in Iitate Village, the mean cpm/μSv/h ± SD was 504.6 ± 102.1 ($n = 9$), respectively. Roughly speaking, 1 μSv/h corresponds to 500 cpm under the conditions in Fukushima.

RESULTS

1. Off-Site Centers

1.1. Locations of Four Centers

We arrived at the Fukushima Station on July 13, 2011, and lodged in a hotel near the station. The hotel had been damaged by the earthquake and was still under repair. It was just re-opened to guests from early July. Radioactivity around the hotel was arbitrarily measured. Radioactivity in my hotel room, on the road just in front of the hotel, and on the ground in the Cycle Park near the hotel was 60, 3,500, and 3,650 cpm, respectively. We attended the explanatory meeting held in a room of the Fukushima Prefectural Office. Of four off-site centers for the temporal returning program, Bajikoen (horse training center), the Kodo Gymnasium, the Kawauchi Gymnasium, and the Hirono Central Gymnasium, we were assigned to Bajikoen, which is located in Minamisoma City, 68 km east to Fukushima City, Fukushima Prefecture, for four days (Figure 1).

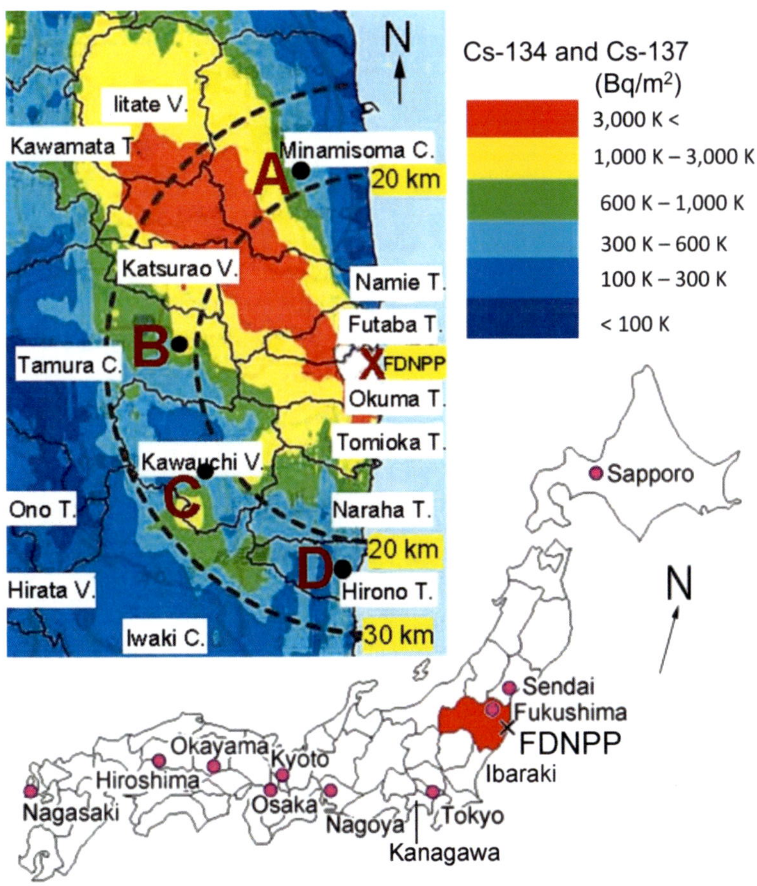

Figure 1. A map around the Fukushima Daiichi Nuclear Power Plant (FDNPP), shown by X, of Tokyo Electric Power Company (TEPCO) and off-site centers. The main Japanese archipelago is also shown. Four off-site centers are A, Bajikoen (horse training center); B, Kodo Gymnasium; D, the Kawauchi Gymnasium; and D, the Hirono Central Gymnasium. The contamination map was modified from the map presented by the Ministry of Education, Culture, Sports, Science and Technology (MEXT) [1].

1.2. Bajikoen off-Site Center

We used a rental car to commute between our hotel and the Bajikoen Off-site Center. When we measured radioactivity on the surface of asphalt pavement outside the Bajikoen building, it was 6,500 cpm or ca. 15 μSv/h. The layout of the Bajikoen Off-site Center is shown in Figure 2.

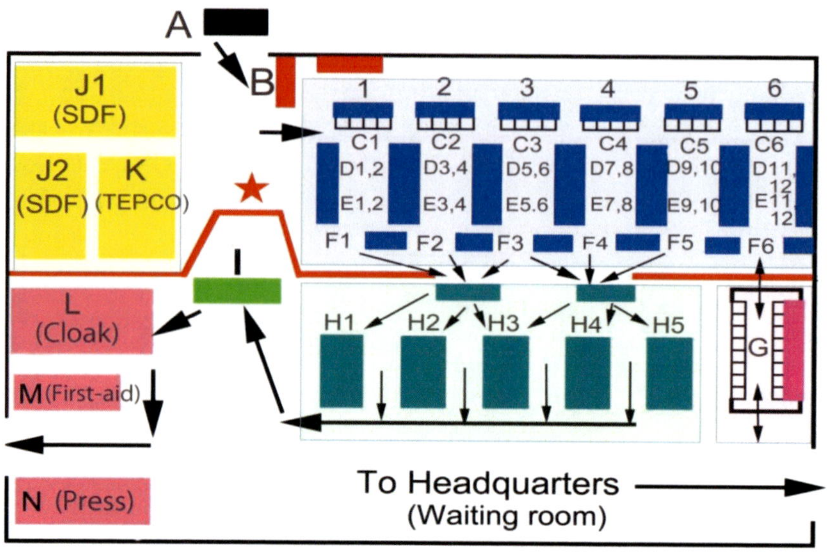

Figure 2. Layout of the Bajikoen Off-site Center (size: ca. 40 m x 60m). Filled squares represent tables and blank ones are chairs (all are not shown). A thick red line crossing the central area separated "hot" and "cold" areas. Immediately before screening, the leader (star mark) declared that the hot line became effective. People who tried to cross the hot line from the hot area to the cold area had to be screened each time. A shows a bus stop close to the entrance. B shows a reception table where two staff served returnees, who received a document that had been filled out before departure. There were 15 in-escorts who carried each returnee's baggage and escorted them from the bus to the reception to one of six screening booths. Returnees had their shoe-covers removed at C1–C6. A pair people at D examined a returnee head to toe for contamination, and another pair at E examined his or her baggage with GM counters. F shows relaying tables from screening to undressing. Six out-escorts were working in between the screening tables and relaying tables. They carried baggage and escorted returnees. G shows a resting area for volunteers who engaged in screening jobs. Six people were assigned to a booth, where five worked and one took a rest at 15 min intervals in turn. H1–H6 denote undressing tables. Five people were working at each table. At I is a recording table where effective doses measured using a dosimeter that had been hanging around a returnee's neck were recorded. J1-J2 are decontamination areas served by Self-Defense Forces. Other symbols show the following: K, TEPCO Survey Center; L, cloak area; M, first-aid; and N, a press corner.

2. Screening for Contamination

2.1. Practical Screening of Returnees

Screening of returnees was conducted as partly shown in Figure 3. To convey returnees, 16 buses were used on weekdays and 24 were employed on weekends at Bajikoen Off-site Center (Figure 3B). A bus could carry around 15 returnees who were accompanied by a leader, two attendants, and two radiation monitors (Figure 3C). Up to two persons per family were allowed to return. Returnees came to the off-site center at 9 o'clock. Several persons concerned made addresses and explanations to them (Figure 3D). A TEPCO person apologized for their giving tremendous troubles to the refugees, and around 20 TEPCO people bowed deeply and synchronously. After explanations, they put on a cap, a mask, gloves, a Tyvek suit, and shoe-covers (Figure 3E). Returnees left Bajikoen on buses for their homes after around 10:00 a.m. After seeing them off, the duties and procedures of the screening were explained for participants. The number of academic volunteers was not large. Many participants seemed to be dispatched from electric companies and energy-associated organizations or agencies on duty. There were experienced people and they contributed to conduct the screening smoothly. Returnees were allowed to stay in their homes for up to 2 h and take goods that could be contained in a plastic bag. They started to arrive at around 1:30 p.m.; their bus stopped at the entrance (Figure 2A). An escort, who carried the bag, led each returnee to the reception (Figure 2B, Figure 3F), where he or she received the inquiry sheet that had been filled before departure. Returnees were distributed to one of six screening booths (Figures 2C1-C6), to which the escort delivered his or her goods (Figure 3G).

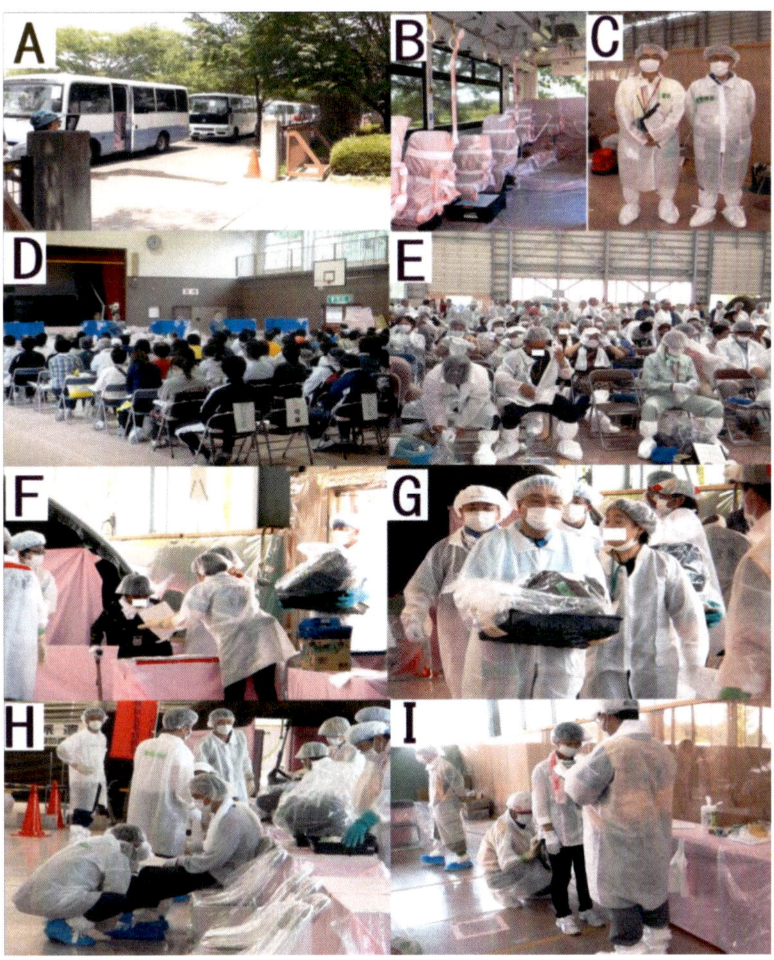

Figure 3. Screening of returnees. See the text for details.

One of five people worked at a booth removed the shoe-covers (Figure 3H), two examined the radioactivity of the goods, and two monitored the body from head to tip for contamination (Figure 3I). After the survey, one monitor put a check in the inquiry sheet as no problem. If more than 13,000 cpm were detected, goods or persons had to be subjected to decontamination treatments (Figures 2J,K). These cases were not found. Returnees were escorted to one of five undressing booths (Figure 2H). People at a booth helped undress them. They moved to the recording table (Figure 2I) where the identification card and a cumulative dosimeter hung from their neck were removed, and the date, name and a total effective dose was recorded. I worked at the recording table for a short period. Most cumulative dosimeters showed 2 to 4 μSv, and no one exceeded 10 μSv. They took their belongings at the cloakroom (Figure 2L) and, followed by an escort with goods, left Bajikoen not to their homes but to their refuges.

The Bajikoen Off-site Center was not air-conditioned. These were hot days and the ambient temperature was around 36°C. People in charge of screening were fully equipped like returnees (Figures 3E,G-I). Six people were assigned to a booth, where five worked and one took rest in the G area (Figure 2G) at 15 min intervals. A couple of returnees fell due to heatstroke and were taken to a hospital by an ambulance or a helicopter. We put our names and affiliations on our Tyvek suit (Figures 3F,H,I), otherwise returnees might take us as TEPCO employees and show hostility. When they learned that we were from universities, they sometimes expressed their thanks. When I worked putting on a black T-shirt, it became greyish after sweat evaporated.

The total number of returnees was 258 and 16 busses took them to their homes on the first day. As a result, a total of 338 people including attendants and others were monitored by us 36 examiners. Our duty was over at 3:30 p.m. Many other participants including medical doctors and volunteers nearby were working at Bajikoen; the number probably reached more than 150. The total number of returnees on the second day was 264 and 16 busses took them to their homes. A total of 344 people were monitored. Returnees favored weekends. The total number of returnees on the third day was 451 and 24 busses were used on the third day. They were divided into two groups. A total of 571 people were monitored on the third day. Monitoring of the second half was finished after 5:00 p.m. The total number of returnees was 464 and 24 buses were used on the fourth day. They were also divided into two groups. A total of 584 people were monitored. Monitoring was finished after 5:00 p.m. Thus a total of 1,837 people were monitored during the four days.

Another Shujitsu team worked from July 22 to 24 at the Kawauchi Gymnasium on July 22 and 23 and at the Hirono Central Gymnasium on July 24 (Figures 1C-D) [2]. The layouts of the off-site centers differed from that of Bajikoen (Figure 2) depending on the structures of the buildings. The screening procedures were almost the same as those of our team. Owing to the typhoon No. 6 (Ma-on), maximum ambient temperatures were 20°C on July 22 and 23, and 26°C on July 24; working conditions were not so bad as those our team had. Screening was conducted by people dispatched from nuclear power plants of electric companies, Shujitsu University, Ehime University, Oita University, and Tsukuba University. One of five booths was assigned to academic people.

2.2. Numbers of Organizations and Participants Engaged in Screening

Soon after the tragedy, off-site centers were set up and many volunteers were recruited. Table 1 shows data of volunteer activities, including radioactivity screening of returnees, screening of evacuees in refuges, medical care of workers in FDNPP, and contact and adjustment between organizations associated with radiation protection. Many employees of electric companies and associated organizations were working on duty. Taking into consideration that academic people were assigned to one or two screening booths of five or six (Figure 2C), many more people than the number listed in Table 1 were engaged in the temporal returning program.

Table 1. Tentative data of volunteer activities (March 12 – December 7, 2011)*

Organizations	No. universities	104
	No. independent administrative agencies	71
	Total	175
No. volunteers	Universities	931
	The Japan Atomic Energy Development Organization	2,448
	The National Institute of Radiological Sciences	251
	The Institute of Physical and Chemical Research	47
	Others	19
	Total	3,696

*Data by MEXT. Organizations associated with the Ministry of Health, Labour and Welfare (hospitals, pharmacies, research institutes etc.) are not listed. Some organizations associated with electric companies, power plants and others are not also listed.

Table 2. Tentative data of volunteer activities in off-site centers (March 12 – December 7, 2011)*

Off-site centers	No. of people screened
The Kodo Gymnasium	4,635
The Kodo Gymnasium/the Hirono Central Gymnasium	4,104
The Hirono Central Gymnasium	4,006
The Hirono Central Gymnasium/the Kawauchi Gymnasium	6,674
The Hirono Central Gymnasium/ the Kawauchi Gymnasium/Bajikoen	729
The Kawauchi Gymnasium	202
Bajikoen (horse training center)	13,248
Total	33,598

*Data by off-site centers. The population of the evacuation zone was 78,000 and the number of families was 27,843.

Table 2 shows tentative data of people treated within the four off-site centers. People (78,000) who lived in the evacuation zone were forced to leave their homes and less than half (33,600; 43%) could go home and stay there for 2 h (Table 2). They were allowed to fetch a bagful of goods. Fortunately, no one was contaminated heavily over 13,000 cpm. Most returnees whom we asked the situation of their homes answered that their gardens were full of high weeds.

A man replied that his house had collapsed due to aftershocks; he is a victim of both the tremor and radiation. Those who did not return home may include lots of double-punched people like him. Voluntary work of radioactivity screening was well organized and conducted smoothly, letting us feel that we Japanese could do things cooperatively and efficiently, once proper targets are established.

Bajikoen Off-site Center was used until July 17, 2012. Presently, six screening places are available in Fukushima Prefecture. The closest one to Bajikouen is the Fujihashi Screening Place located at Fujihashi, Namie Town.

2.3. Some Odd Landscapes

During off-duty periods, we measured radioactivity at various places in Fukushima Prefecture. It was odd to see that there were neither rice plants nor farmers in rice fields, which were covered with various kinds of weeds instead (Figure 4A). In some local communities along the way form the Off-site Center to our hotel, people were evacuated and there were no lights from the houses in the evening. Small local communities seemed to be more difficult to maintain when one or two members were evacuated. Close to the Pacific seashore, a whole community had been swept away by the tsunami and a barren grassland remained as if there was nothing from the start (Figure 4B). When we drove to another seaside city Minamisoma, tetrapods at the seawall had flown deep into lands (Figure 4C), and many fishing vessels carried by the tsunami were left in rice fields or beside roads several km deep inland (Figure 4D).

Figure 4. Some scars of radioactive contamination and tsunami. A, rice fields were abandoned and full of weeds; B, once there was a town at Kitaizumi, Hara-Ward, Minamisoma-City; C, tetrapods were carried from the seashore deep into lands at Kashima-Ward, Minamisoma-City; and D, fishing vessels were overlapped over one another several km from the seashore at Kashima-Ward.

3. SPORADIC MEASUREMENTS OF RADIATION IN CONTAMINATED AREAS

3.1. Radioactivity in Various Places in Fukushima, Ibaraki, and Kanagawa Prefectures

Table 3 shows excerpts from a lot of measurements of radioactivity in Fukushima, Ibaraki, and Kanagawa prefectures. Some data were reported previously [3]. Moriya City, where my house is located, is one of the hotspots in the Kanto district. It was not as contaminated as Fukushima areas, but was more contaminated than such areas as Joso City and Bando City, which are closer to FDNPP, than Moriya City. Deposits of debris consisting of sand and soil particles and fragments of leaves were found at scuppers of roadsides. Radioactive substances fell with precipitation and immediately became attached to the ground surface materials including the particles and fragments. These would have flowed with rainwater to the scuppers, but radioactive substances were apparently firmly fixed to them and resistant to being washed out by the water. Table 3 shows that the level of contamination differs greatly depending on the location despite close proximity to FDNPP. Table 3 also shows indoor radioactivity. Data varied to some degree, and some rooms in Fukushima Prefecture showed slightly higher radioactivity, part of which might be attributable to radiation from outside. As a whole, however, indoor radioactivity was markedly lower than that outdoors. Indoor data indicate that places that had not been exposed to precipitation were not contaminated.

Table 3. Measurements of radioactivity in some places in Moriya City and other areas

Locations	Distance (km from FDNPP)	Radioactivity (cpm)	Date
Asphalt pavement-1, Moriya City, Ibaraki Prefecture	180	2,520	Apr. 3, 2011
Scupper of a road side, Moriya City Ibaraki Prefecture	180	15,000	Apr. 3, 2011
Asphalt pavement, Joso City, Ibaraki Prefecture	170	800	Apr. 4, 2011
Asphalt pavement, Bando City, Ibaraki Prefecture	175	380	May 5, 2011
Asphalt pavement, Kamakura City, Kanagawa Prefecture	280	160	May 6, 2011
Park soil, Fukushima City, Fukushima Prefecture	60	3,650	July 14, 2011
Scupper near a park, Fukushima City, Fukushima Prefecture	60	10,500	July 14, 2011
Scupper beside a road, Iitate Village, Fukushima Prefecture	40	33,300	July 15, 2011
Room, Shujitsu University, Okayama City*	700	76	Mar. 30, 2011
Dining room, my house, Moriya City, Ibaraki Prefecture*	180	95	Apr. 3, 2011
Gymnasium, Shimotsuma Daiichi High School, Shimotsuma City, Ibaraki Prefecture*	160	60	Apr. 4, 2011
Hotel room, Fukushima City*	60	60	July 14, 2011
Dining room, house in Kawauchi Village, Fukushima Prefecture*	25	150	July 15, 2011
Living room, house in Minamisoma City, Fukushima Prefecture*	26	160	July 16, 2011

3.2. Contamination of Grounds and Electric Poles

3.2.1. Contamination Is Limited to Surface Soil

When the surface soil of the ground was removed differentially and radioactivity at different depths was measured in schools and parks, contamination was apparently limited to the very surface (Figure 5I). Soil samples from a middle school at Minamisoma City also showed surface-contamination (not shown). After about seven months, during which the area had heavy rains, especially in the rainy season and following typhoon No. 15, soil of the parks in Moriya City was measured on January 5, 2012 (Figures 5I,D-F). Still, the contamination was apparently confined to the upper 1–2 cm of surface soil. Data in Figure 5I show that attachment of radioactivity to soils is quite strong and difficult to wash out. These data strongly suggest that radioactive substances were trapped in rain or snow and immediately fixed to the surface before rainwater penetrated deep into the soil.

3.2.2. Radioactivity from the Ground Level to the Air 5 m High

Typical patterns of radioactivity from the ground level to the air 5 m high at Matsukaze Park in Moriya City and in elementary schools in Fukushima City are presented in Figure 5II. Radioactivity decreased rapidly with height up to 2 m high; then reached a plateau (Figures 5IIA,B,D). The distance of 1.5 – 2 m coincides well with the height limit of most β-rays from Cs-134 and Cs-137 [4]. When the β-ray shielding cap was attached to the muzzle of the GM counter, radioactivity at the place A showed a marked decrease from the start (Figure 5IIC), indicating that the initial decreases presented in Figure 5IIA,B,D are attributable to β-rays. Other data not presented in Figure 5II showed almost the same patterns, e.g., data at Ishigami Middle School, Minamisoma City, Fukushima Prefecture. Consequently, these data indicate that the basic pattern of radioactivity is the same even if places and intensities of radioactivity differ, and the mixed emission of β-rays and γ-rays extends to 2 m high; then only γ-rays reach the higher air at heights of 5 m, and theoretically, of 100 m [4].

When the radioactivity of the parking lot of an elementary school was measured, a typical pattern as shown in Figure 5IIB was obtained. When the upper 5 cm soils were removed and were wrapped in plastic sheets and embedded under the ground 0.5 m deep at a specified corner, radioactivity became a non-contamination level (Figure 5IIE). Places B and E are approximately 50 m apart. July 16, 2011, was a windy day, and if radioactive airborne particles were floating in the air, radioactivity over the ground would be higher than the ground level (Figure 5IIE). The same pattern to Figure 5IIB,E was also obtained at Nankodai Elementary School in Fukushima City. These findings provide us with very important hints to practically decontaminate the affected soils. Since radioactive substances were so firmly fixed to surface soils to be washed out by water (Figure 5I), wrapped and embedded soils would not ooze out and contaminate underground water for a decade or more.

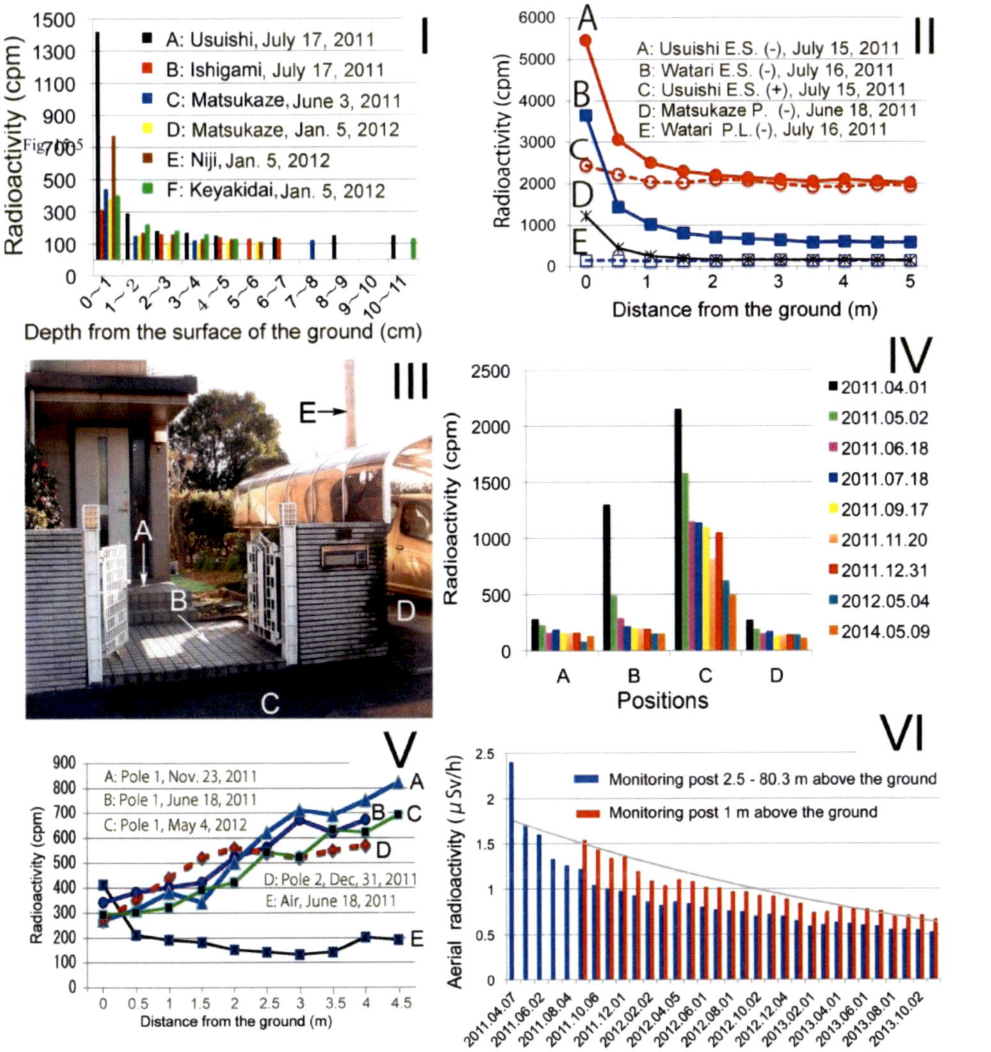

Figure 5. Sporadic measurements of radiation in contaminated areas. I. Radioactivity of soil surfaces in schools and parks. A, Usuishi Elementary School (Iitate Village, Fukushima Prefecture); B, Ishigami Middle School (Minamisoma City, Fukushima Prefecture); C and D, Matsukaze Park (Moriya City, Ibaraki Prefecture); E, Niji Park (Moriya City); and Keyakidai Park (Moriya City). II. Radioactivity from the ground to the air. A, Usuishi Elementary School (without a filter); B, Watari Elementary School (without a filter, Fukushima City, Fukushima Prefecture); C, Usuishi Elementary School (with a filter); D, Matsukaze Park (without a filter, Moriya City), and E, the parking lot of Watari Elementary School (without a filter). III. The entrance of my house at Moriya City. A, a tile under the porch roof; B, a tile inside the gate; C, a part of an asphalt road just outside the gate; D, the concrete floor of the garage; and E, one of two electric poles. IV. Radioactivity at the entrance of my house. Positions A, B, C, and D correspond to those of III. V. Contamination of electric poles. Pole 1 is shown in III, E and Pole 2 is located at the opposite side of Pole 1. VI. The decay of radioactivity in Fukushima City. The grey line is the decay curve of Cs-134 and Cs-137.

3.2.3. Differential Contamination Depending on Circumstances

Figure 5III shows the vicinity of the entrance of my house. When radioactivity of a tile under the porch roof (A), a tile inside the gate (B), a part of an asphalt road just outside the gate (C), and the concrete floor of the garage (D) were measured on April 1, 2011, radioactivity differed greatly (Figure 5IIIA-D). Figure 5III also shows the time course of the decay of radioactivity. As indoor areas were not contaminated (Table 3), places covered with a roof such as a porch and a garage were minimally contaminated (Figures 5IIIA,D). Open spaces were more heavily contaminated (Figures 5IIIB,C). Contamination must be affected by materials; an asphalt road (Figure 5IIIC) seems to absorb radioactive substances more easily than tile (Figure 5IIIB). Decay patterns depicted in Figure 5III indicate that radioactivity decreased rapidly at first; then decayed slowly, suggesting that rainwater washed away contaminated small particles that were detachable from contaminated surfaces, and that radioactive isotopes with short half-lives such as I-131, the half-life of which is 8 days, were involved in the initial rapid decay.

3.2.4. Contamination of Electric Poles

There are two electric poles of TEPCO in my house lot (Figure 5IIIE). To confirm that the higher in the air, the lower the radioactivity becomes, I put a ladder to one pole and started measuring radioactivity. As expected, radioactivity showed a gradual decrease at first, but then started to increase, leading me to suspect that airborne radioactive particles might be floating in the air. When the detector muzzle of the GM counter faced the surface of the electric pole, radioactivity started to increase (Figure 5IVB). When the muzzle was directed to the air opposite to the pole, radioactivity gradually decreased (Figure 5IVE). The result strongly suggests that radioactive substances were carried by rain or mist, which came from the upper air, and that the substances became instantaneously attached to the surface of the pole before the rainwater ran down. This resembles the rapid attachment of the radioactivity to surface soils (Figure 5I). The situation that upper parts were more contaminated than lower parts was confirmed 8 months (Figure 5IVA) and 14 months (Figure 5IVC) after the FDNPP accident and with another electric pole (Figure 5IVD). It is of considerable interest to read newspaper articles that upper parts of cider trees were more contaminated than lower parts (The Nihon Keizai Shimbun, June 7, 2011).

3.2.5. Decay of Radioactivity in Fukushima City

The Nikkei Shimbun started to report radioactivity of major cities in the air five days a week from March 24, 2011 except for Fukushima City. Radioactivity of γ-rays in Fukushima City appeared first on April 7, 2011 and that of the β- and γ-rays was reported first on August 23, 2011. Radioactivity at a monitoring post 1.8 – 34 m above the ground consists of γ-rays (Figure 5VI, blue) and radioactivity 1 m above the ground (Figure -5VI, red) is made up of β- and γ-rays.

All data of Fukushima City had been collected from The Nikkei Shimbun and data approximately twice a month are presented in Figure 5V. The gray line is a rough decay curve of Cs-134 and Cs-137 half lives of which are 30 and 2 years, respectively, and the ratio of Cs-134 to Cs-137 is approximately 1.0 in Bq basis and 2.7 in energy basis. Except for early days or a couple of months when I-131 was active, radioactivity most of which are emitted from Cs-134 and Cs-137 seems to be decaying according to the equation:

$$P(t) = 1/3.7 \text{ x } P(0)2^{-t/30} + 2.7/3.7 \text{ x } P(0)2^{-t/2},$$

where P(t) is a total ionizing dose at time t, t is time (years), and P(0) is the original dose.

DISCUSSION

FPDH has been conducting radioactivity screening from March 13, 2011 to the present [5]. By March 31, 2013, 264,978 persons had been screened, of whom 102 were found to be contaminated with radioactivity to a level greater than 100,000 cpm. These heavily contaminated people were found only among the 114,488 screened during the period from March 13 to 31, 2011. This means that no one has been heavily contaminated since April 1st, 2011 for about two years. Heavily contaminated people were subjected to whole-body decontamination treatment. Fortunately, FPDH reports that no hazardous effects have been detected among those people. The reason why contamination level of 100,000 cpm was detected is not clear.

Considering that 500 cpm is equivalent to 1 μSv/h, 100,000 cpm is equivalent to 200 μSv/h (0.2 mSv/h) or 4.8 mSv/d. Given that heavily contaminated people were exposed to radiation for a whole day or 4.8 mSv before decontamination, they are calculated to have received a half the nominal dose of one CT scan, i.e., 8.8-10 mSv per scan. No hazardous effects are expected among those people.

Contamination differed greatly depending on the places and specific locations within areas (Figure 1, Table 3). Very fortunately for Japanese residents, an east wind from March 11–14 carried large quantities of the radioactive substances over the Pacific. Unfortunately, a west wind on March 15 blew back toward populated areas [6]. A report by U.S.DOE (the Department of Energy of the U.S.A.) and NISA (the Nuclear and Industrial Safety Agency, the Ministry of Economy, Trade, and Industry, Government of Japan) indicates that the greatest distribution of contamination occurred on March 19, and that contamination patterns at March 23 and April 29 are almost identical [7]. This report suggests that 1) large-scale emissions of radioactive substances had ceased, and 2) contaminated areas were determined until March 23.

Rain or snow must trap radioactive substances and contaminate affected areas firmly and instantaneously as suggested by practically no-contamination indoors (Table 3) and covered areas (Figures 5IIIA,D; IVA,D), limited contamination to the surface soils (Figure 5I) and heavier contamination of upper parts of electric poles (Figures 5VA-D). Indeed, meteorological analyses concluded that the radioactive material transported by rain or snow from 5 PM, March 15, to 4 AM, March 16, was the major contributor to contamination in Fukushima Prefecture, whereas rains at 8 AM, March 21, and 6 AM, March 23, were the major sources in Ibaraki, Tochigi, Saitama, and Chiba Prefectures and in Tokyo Metropolis [8]. When we visited Usuishi Elementary School in Iitate Village, all students evacuated the school and a few teachers remained in the teachers' room. When I asked them if they had rain after March 11, a teacher replied that they had a snowfall of about 10 cm on March 15. This snow must have determined the basic contamination pattern shown in Figure 1.

CONCLUSION

The March 11, 2011, earthquake and tsunami crippled the FDNPP of TEPCO. People who lived in the area within a 20 km radius from FDNPP were forced to evacuate immediately.

FPDH and JGDH planned to let evacuees return to their homes temporarily. This temporary returning program was conducted at four off-site centers from May to September, mainly in June and July, 2011. Returnees were allowed to stay in their homes for two hours, and then they were screened for contamination together with their goods. To screen them, a total of 3,686 volunteers were recruited by MEXT. Together with many other participants, 33,598 returnees out of 78,000 evacuees (43%) were screened; none was found to be heavily contaminated. No hazardous effects have been detected among them.

References

[1] http://savechild.sub.jp/archives/17.html

[2] Sutou S, Egawa T, Mori H, Aso H (2013) The Fukushima Daiichi Nuclear Power Plant disaster: a report on volunteer activity for radioactivity screening of temporary returnees to the evacuation zone. *Genes and Environment* 35: 73-79.

[3] Sutou S (2013) Fukushima Daiichi Nuclear Power Plant disaster: generally applicable implications from sporadic and arbitrary measurements of radioactive contaminations in some areas of Ibaraki and Fukushima. *Genes and Environment* 35: 13.

[4] The Japan Isotope Association (2011) Isotope Handbook, 11th ed., Maruzen, Tokyo.

[5] http://www.pref.fukushima.lg.jp/sec/21045c/iryou-screeningkatsudou.html

[6] Brumfiel G (2011) Fallout forensics hike radiation toll. *Nature* 478: 435–436.

[7] http://energy.gov/situation-japan-updated-42312

[8] Kinoshita N, Sueki K, Sasa K, Kitagawa J, Ikarashi S, et al. (2011) Assessment of individual radionuclide distributions from the Fukushima nuclear accident covering central-east Japan. *Proceedings of National Academy of Sciences of the United States of America USA* 108: 19526-19529.

In: Fukushima Nuclear Accident
Editor: Shizuyo Sutou

ISBN: 978-1-63463-748-0

Chapter 13

COMMENTARY 1: BIOKINETICS OF ^{137}Cs AND ESTIMATION OF INTERNAL RADIATION DOSE; PHYSIOLOGICAL EFFECTS IN DESCENDANT MICE AFTER THE EVERY GENERATIONAL LOW DOSE-RATE INTERNAL ^{137}Cs RADIATION EXPOSURE, AS THE FUKUSHIMA SIMULATION EXPERIMENT

Hiroo Nakajima*
Department Radiation Biology and Medical Genetics,
Graduate School of Medicine, Osaka University, Yamada-oka, Osaka, Japan

ABSTRACT

To investigate the transgenerational effects of chronic low dose-rate internal radiation exposure after the Fukushima Nuclear Power Plant accident in Japan, every generations of mice were maintained in the radioisotope facility with free access to drinking water containing ^{137}CsCl (0 and 100 Bq/ml).

Mice of the tenth generation (F_{10}) were assessed γ-H2AX foci in hepatic cell, micro nuclei test, chromosome aberration analysis, tumorigenicity of lung, oxidative stress, and metabolomic analysis in heart and liver.

Suppression of the tumor growth, increasing of the γ-H2AX focus number, and the metabolite of oxidative stress were observed in ^{137}Cs drinking mice, however, no significant difference was observed in the analyses of micro nuclei test, chromosome aberration, and tumor incidence.

* Correspondence to: Hiroo Nakajima, Tel: +81-6-879-3815, Fax: +81-6-6879-3819, E-mail: nakajima@radbio.med.osaka-u.ac.jp.

INTRODUCTION

The health effects of radioactive contamination on people living in contaminated areas such as Belarus, Russia, and Ukraine have been studied in various ways in the nearly 30 years since the Chernobyl Nuclear Power Plant accident. Among these studies have been reports on acute and late radiation sickness due to high-dose exposure in people involved in decontamination and fire control efforts immediately after the accident [1 ,2]. In addition, childhood exposure to radioactive iodine reportedly has a clear effect on the occurrence of pediatric thyroid cancer [3, 4]. However, various reports on the health effects on residents of radioactively contaminated areas, including other cancers, developmental deformities, and genetic effects, have yet to be confirmed.

The current absence of conclusive evidence is due to the time required for indices, such as cancer onset and effects on later generations, to occur. In addition, the influence of low-dose radiation exposure on genetic effects and carcinogenesis cannot be ignored, despite the difficulty of accurately determining dose, variable exposure doses, and the potentially confounding effects of chemical substances from environmental pollution and living conditions.

Amid unresolved concerns over the health effects of low-dose radiation exposure on residents of contaminated areas, the Fukushima No.1 Nuclear Power Plant accident, caused by the tsunami of the Great East Japan Earthquake, occurred in 2011, 25 years after the Chernobyl nuclear accident. Low-dose radiation exposure has become a real concern to many residents in Fukushima as well as Chernobyl, which has led to various studies in Japan on the effects of low-dose radiation exposure.

^{137}Cs, with a half-life of 30.2 years, is a major nuclide that persists in contaminated areas. Early reduction of radioactive contamination due to natural purification is expected, but the purification rate of ^{137}Cs in undisturbed soil is relatively slow. In addition, internal exposure by organisms living in such environments, including humans, can be expected to continue for multiple generations, much longer than the initial external exposure. However, various studies of Hiroshima and Nagasaki atomic bomb survivors have found no significant differences in genetic mutations between exposed and non-exposed groups [5-7]; studies on decontamination workers in Belarus and their children have also reported no significant next-generation effects thus far [8-10]. Because these investigations can only study the F_1 generation, the presence or absence of accumulated transgenerational effects is unclear. However, even if accumulated transgenerational effects occur at exposure doses lower than the average Hiroshima and Nagasaki doses, it will take several generations and potentially several hundred years until they become apparent. Therefore, new methods other than epidemiological studies must be developed for risk assessment of the health effects on descendants of those exposed to low-dose radiation caused by radioactive contamination from the Fukushima and Chernobyl accidents.

This study used inbred A/J strain mice, which have shorter generation times and more uniform genetic backgrounds than humans. A simulation experiment in a radioisotope facility that reproduced Chernobyl's ^{137}Cs environmental pollution was initiated in 2007 to examine the transgenerational health effects caused by low-dose radiation exposure.

This study has bred many generations of mice administered 100 Bq/mL ^{137}CsCl water ad libitum as drinking water and investigated the carcinogenic, genetic, and physiological effects in at least 10 generations of offspring to predict the health effects of low-dose radiation on humans.

1. EXPERIMENTAL METHODS

Two mouse pairs were selected from among A/J strain littermates bred in the radioisotope facility. One pair was assigned to the control group and bred while offered drinking water ad libitum; the other pair was administered a low-dose internal radiation exposure of ^{137}CsCl water (100 Bq/mL). Harvested offspring from both pairings were sibling-mated for more than 10 generations, corresponding to a human generation turnover of approximately 300 years. Both 10th generation groups were analyzed for the following:

1) ^{137}Cs distribution in each organ after long-term ad libitum intake of ^{137}CsCl water (10, 100 Bq/mL).
2) γ-H2AX focus number as an indicator of DNA double-strand break repair initiation in all cells due to radiation exposure.
3) Detection of chromosomal abnormalities common to all cells by multi-color FISH and micronucleus tests.
4) Whole genome sequencing to detect base sequence mutations in germ cells that could accumulate in non-coding regions in each generation.
5) Quantification of naturally occurring and urethane-induced lung tumor incidence and mean tumor volume at 10 months of age to investigate the effects of low-dose radiation on carcinogenesis.
6) The effects of low-dose radiation on myocardium and blood vessel cell structures.
7) Metabolomic analysis of heart and liver tissues in the ^{137}Cs water and control groups.
8) Balance of oxidative and anti-oxidative stress in the body.

2. RESULTS

2.1. *In Vivo* Dynamics and ^{137}Cs Dose

Figure 1a shows the accumulation of ^{137}Cs over time in each organ in the first generation of A/J strain male mice after continuous 8-month ad libitum intake of drinking water supplemented with an aqueous 100 Bq/mL cesium chloride solution. An average daily water consumption of 4.5 mL corresponded to 450 Bq ^{137}Cs intake per mouse per day and average intake of 16 Bq per 1 g body weight for a 28-g mouse. The accumulated level of ^{137}Cs increased slowly to approximately 10 times the average daily intake per 1 g body weight (16 Bq) in muscles by 3 weeks; levels in other organs reached an equilibrium concentration less than half that of muscles. As shown in Figure 2, a single oral dose of ^{137}Cs resulted in a peak muscle tissue concentration similar to intake concentration levels per 1 g body weight; however, chronic and continuous dosing resulted in an accumulated concentration approximately 10 times the intake concentration per 1 g body weight. Figure 1b compares accumulation in 10 and 100 Bq/mL water concentrations.

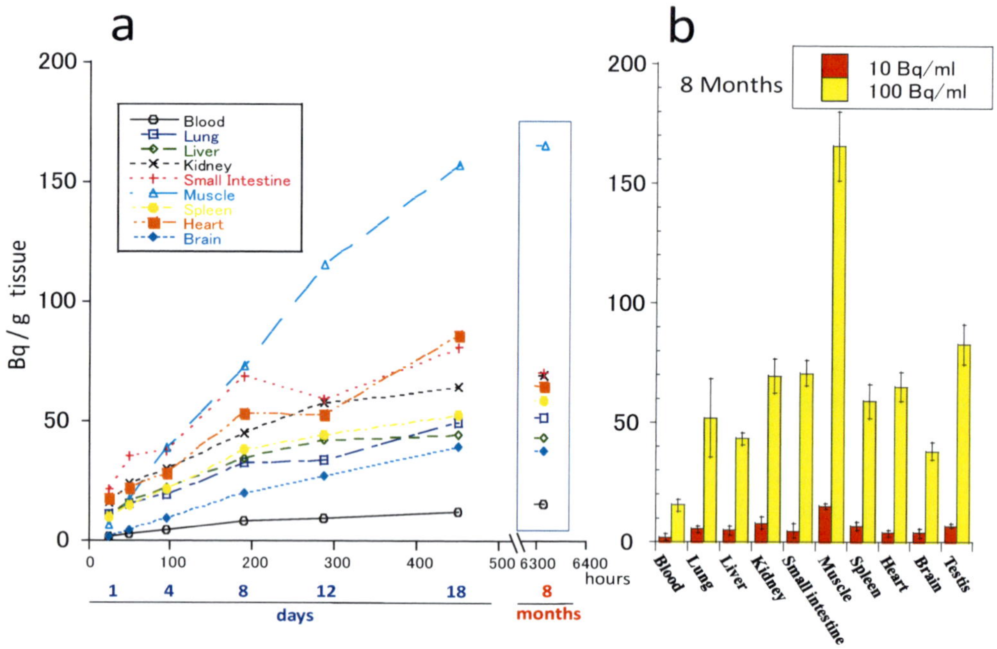

Figure 1. Accumulation of ^{137}Cs in mouse organs. a: Organ-specific temporal changes in accumulated ^{137}Cs levels with ad libitum ^{137}Cs water (100 Bq/mL). b: Body distribution of ^{137}Cs after 8 months of ad libitum ^{137}Cs water (10 Bq/mL or 100 Bq/mL).

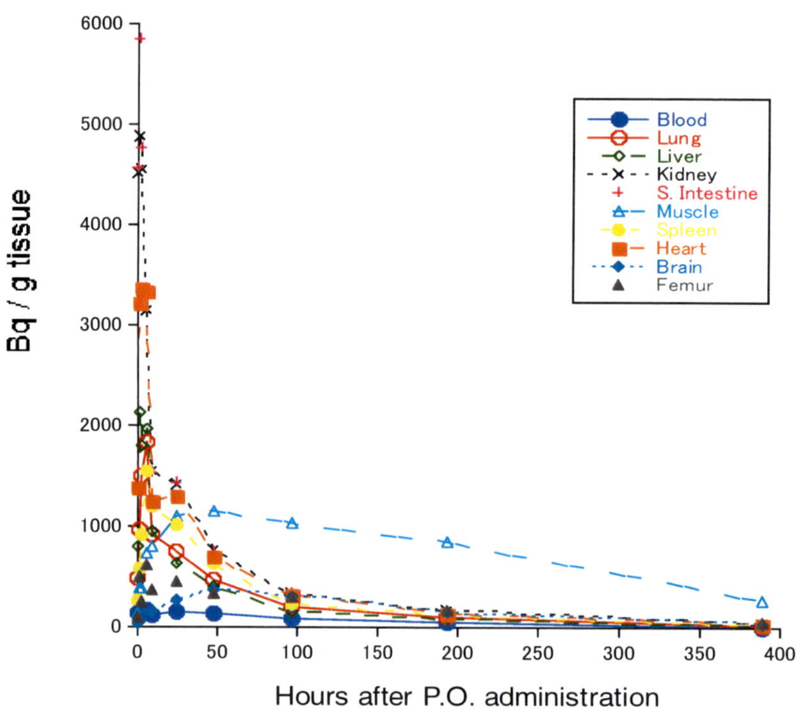

Figure 2. Organ-specific temporal changes in ^{137}Cs levels after administration of a single oral dose of ^{137}Cs (1,000 Bq/g body weight).

The equilibrium concentration was one-tenth for a corresponding one-tenth water concentration. Thus, if ^{137}Cs is ingested continuously, the accumulated levels will peak depending on the organ tissue and intake amount (Figure 2). Based on this observation, fixed internal exposure doses of 160 Bq/g and 80 Bq/g in the muscles and other organs, respectively, have been continuously administered for at least 15 mouse generations.

2.2. Genomic Effects

Increased levels of phosphorylated histone H2AX (γH2AX), a marker for initiation of DNA double-strand break repair, were observed in mice ingesting 100 Bq/mL water, suggesting chronic stress on the genome. Although DNA base sequence mutations were likely, there were no differences in micronucleus test results between the first and tenth generations. Similarly, multi-color FISH tests for chromosomal aberrations in bone marrow cells showed no differences between the control group and the tenth generation of mice. These results suggest that almost all damage was repaired or that limited genome damage occurred. Because small genomic mutations could occur, be inherited, and accumulate in each successive generation in non-coding genomic regions (introns and intergenic) and have no effect on survival, we compared levels of accumulated base mutations between the offspring exposed to continuous low-dose radiation and control mice of the same generation that started from a littermate mouse.

2.3. Carcinogenesis Effects

No difference in lung tumor incidence per mouse was observed between the two groups. However, the average tumor volume of naturally occurring or urethane-induced tumors at 10 months of age was significantly smaller in the ^{137}Cs water group compared to control group, suggesting an inhibition of tumor proliferation by ^{137}Cs. These findings suggest that low-dose internal ^{137}Cs exposure does not induce lung tumorigenesis in mice. Even more importantly, the result possibly suggests the expression of hormesis or the adaptive response.

2.4. Physiological Effects

A significant increase in oxidative stress was noted in the ^{137}Cs water group, but there was no significant difference in the total antioxidation capacity, and the oxidative stress of mice living in these experimental conditions fell within the expected range. We are currently creating a detailed report addressing points 2.1 to 2.4.

Conclusion

Until now, the low sensitivity of detection systems has prevented detailed studies on the biological effects of low-dose radiation exposure. There are also various barriers to this type of study, such as the complexity of animal experiments to determine the effects of long-term low-dose internal exposure, the challenge of obtaining clear results, and the relatively low impact of these findings in the research community.

However, animal experiments and individual evaluations are crucial for understanding the true effects of exposure to low-dose radiation. Regardless of the adverse or beneficial effects of low-dose radiation exposure on genes, cells, and tissues, it is ultimately most important to understand the potential range of effects on individual life expectancy. Studying the experimental effects on individual animals allows a full survey of all harmful and beneficial physical effects.

The results of these Fukushima and Chernobyl simulation experiments are informative whether or not biological effects are identified. While novel discoveries are generally considered positive research findings, negative results in this study can also contribute to a sense of security for people living in contaminated areas. Studies that offer useful results with either positive or negative results are less likely to have observational bias. However, researchers are reluctant to pursue this field of study, as evidenced by the low impact index of academic papers on this topic in recent years. Although this is a research area that lacks a research leader, increased interest in the effects of low-dose radiation exposure on individuals is anticipated. The author hopes that many scientists are interested in this field.

REFERENCES

[1] United Nations Scientific Committee on the Effects of Atomic Radiation (2011). "Annex D. Health effects due to radiation from the Chernobyl accident". Sources and Effects of Ionizing Radiation, UNSCEAR 2008 Report vol. II: Effects, Report to the General Assembly Scientific Annexes C, D and E. New York: United Nations. p. 58. ISBN 978-92-1-142280-1. "Of these 134 patients, 28 died within the first four months, their deaths being directly attributable to the high radiation doses (two other workers had died from injuries unrelated to radiation exposure in the immediate aftermath of the accident)."

[2] United Nations Scientific Committee on the Effects of Atomic Radiation (2011). "Annex D. Health effects due to radiation from the Chernobyl accident". Sources and Effects of Ionizing Radiation, UNSCEAR 2008 Report vol. II: Effects, Report to the General Assembly Scientific Annexes C, D and E. New York: United Nations. p. 189. ISBN 978-92-1-142280-1. "Table D7. Causes of death among Chernobyl ARS survivors in the later period [B10, B39, B41, B42, B44, G9, U3]"

[3] Demidchik YE, Saenko VA, Yamashita S (2007) Childhood thyroid cancer in Belarus, Russia, and Ukraine after Chernobyl and at present. *Arquivos Brasileiros de Endocrinologia & Metabologia* 51: 748-62.

[4] Shibata Y, Yamashita S, Masyakin VB, Panasyuk GD, Nagataki S (2001) 15 years after Chernobyl: new evidence of thyroid cancer. *Lancet* 358: 1965–66.

[5] Neel JV, Schull WJ, eds.(1991) The Children of Atomic-bomb Survivors: A Genetic Study. Washington DC: National Academy Press.

[6] Fujiwara S, Suyama A, Cologne JB, Akahoshi M, Yamada M, et al. (2008) Prevalence of adult-onset multifactorial disease among offspring of atomic bomb survivors. *Radiation Research* 170: 451-7

[7] Kodaira M, Roy H, Kamata N, Furukawa, K, Takahashi N, et al. (2010) No evidence of increased mutation rates at microsatellite loci in offspring of A-bomb survivors. *Radiation Research* 173: 205-213.

[8] Livshits LA, Malyarchuk SG, Kravchenko SA, Matsuka GH, Lukyanova EM, et al. (2001) Children of chernobyl cleanup workers do not show elevated rates of mutations in minisatellite alleles. *Radiation Research* 155: 74-80.

[9] Kiuru A, Auvinen A, Luokkamäki M, Makkonen K, Veidebaum T, et al. (2003) Hereditary minisatellite mutations among the offspring of Estonian Chernobyl cleanup workers. *Radiation Research* 159: 651-655.

[10] Furitsu K, Ryo H, Yeliseeva KG, Thuy LTT, Kawabata H, et al. (2005) Microsatellite mutations show no increases in the children of the Chernobyl liquidators. *Mutation Research* 581: 69-82.

In: Fukushima Nuclear Accident
Editor: Shizuyo Sutou
ISBN: 978-1-63463-748-0

Chapter 14

COMMENTARY 2: ABNORMALITIES OF BUTTERFLIES CAPTURED AROUND FUKUSHIMA AREAS MIGHT BE CAUSED BY NON-GENETIC BUT TOXIC EFFECTS

Shizuyo Sutou*

School of Pharmacy, Shujitsu University, Naka-ku,
Okayama-shi, Okayama, Japan

ABSTRACT

Hiyama et al. reported that high frequencies of abnormalities were found in pale grass blue butterflies *Zizeeria maha* that had been collected around the Fukushima area and concluded that the abnormalities were genetically induced. It appears that most abnormalities presented in the paper belong to smaller, shorter, or undeveloped changes, and might be explicable by the introduction of some toxicities. Also the transmission pattern that abnormalities were enhanced in F_1 generation violates the Mendelian law of inheritance. Actually, experiments designed to demonstrate their findings were biased and were premature to support that conclusion. Abnormality data of butterflies collected in 2012, 2013 or 2014, if available, are necessary to ascertain whether the abnormalities were genetic as the authors insist.

INTRODUCTION

When I attended the 17th International Congress on Biotechnology in Animal Reproduction on September 13, 2012, at the University of Leipzig, Germany, a speaker referred to the genetically hazardous effects of radioactive contaminants on butterflies that had been collected around the Fukushima area. I made a comment that the effects were not confirmed to be genetic, but the audience seemed to have received the message that the abnormalities had resulted from genetic damage. Because the same subject was sensationally reported on television without relevant commentary, I suspect that most Japanese people

*Correspondence to: Shizuyo Sutou, Tel & Fax: +81-86-271-8357, Email: sutou@shujitsu.ac.jp

believed that radionuclides discharged from the Fukushima Daiichi Nuclear Power Plant (FDNPP) are genetically very dangerous and dreadful, increasing fears of radiation unduly. As indicated clearly in this book, the linear no-threshold (LNT) is not valid anymore. Large-scale research into genetic effects of atomic bombings on survivors in Hiroshima and Nagasaki demonstrated that the bombs and their associated radiation had not exerted any inheritable genetic alteration. In fact, low-dose radiation might even be beneficial: accumulated data demonstrate that not only radiation but also chemical substances induce homeotic effects and adaptive responses. Therefore, let me examine critically the report by Hiyama et al. [1] who argue that high frequencies of abnormalities in butterflies are genetically induced. Actually, experiments designed to demonstrate their findings were biased and were premature to support that conclusion. The major issue of the LNT model is induction of cancers and that of inheritable genetic alterations is secondary, because the latter is almost negligible in humans. Of course humans and butterflies are widely different. However, it is incredible that genetic alterations are so high as to reach 20, 30, 40% or more in the insects.

1. Abnormalities do not Seem to Result from Genetic Damage

The authors investigated possible effects of radiation from FDNPP on the pale grass blue butterfly *Zizeeria maha*, and concluded that radionuclides from FDNPP caused physiological and genetic damage to the butterflies. Their conclusion is apparently premature if genetic damage is used to mean abnormalities resulted from gene mutations. M. Obinata shed new light on the findings (Chapter 15). His interpretation is that changes observed in the butterflies are not the manifestation of gene mutations, but that damage to some proteins such as Hsp90 could have otherwise masked mutations by making abnormal proteins normal. This is an attractive idea, but my interpretation differs: the abnormalities were not caused by gene mutations but by some toxicity including cell killing and cleavage of critical molecules such as miRNA. When these cells or molecules are sensitive to radiation, abnormalities can be explained more simply and directly.

Abnormalities include loss and weak expression of spots (Figure 1b in [1]). Neither the increase nor intensification of spots suggests a possible manifestation of genetic changes. Findings that dislocated spots are not situated at distal parts but at proximal parts of the wings, suggesting suppressed cell division or death of primordial cells at or close to the basal area of the wings. The enlarged spots (Figure 1b, right individual, ibid.) amount to five or so. They are fewer than those of the controls, which show around 10. Moreover, the widths of the affected wings are clearly narrower than those of the controls. Consequently, the enlarged spots might be fusion products of, say, two spots after in-between cells were killed or eliminated.

Figure 2d (ibid.) portrays a 'miniaturized' foreleg tarsus, an 'undeveloped' middle leg tarsus, and 'undeveloped' pulpi, suggesting manifestation of some toxicities. Shortened antennae, (Figure 2, ibid.) are also explicable by toxicity. Figure 2e (ibid.) shows a dented eye, which might be formed if sufficient cell numbers were not supplied to its full development. Figure 2f (ibid.) shows a butterfly with a smaller right wing than a left one.

This might be explained by inhibition of cell division or cell killing at the basal area of the right wing, or by disturbance of normal development. Figure 2f (ibid.) also shows folded and rumpled wings. These might be explained by unilaterally unbalanced cell divisions or by repression of normal differentiation. Refraining from further citations, it appears that most abnormalities presented in the paper belong to smaller, shorter, or undeveloped changes, and might be explicable by the introduction of some toxicities.

2. Observed Abnormalities do not Follow the Mendelian Law of Inheritance

Most cells of adult butterflies are terminally differentiated. The effects of radiation at this final stage must be minimal. Abnormalities that are apparent in adults must have resulted from damage in somatic cells at preceding stages such as the larval stage. Somatic cells are diploid, having two copies of a gene. If the abnormalities found in butterflies had resulted from gene mutations, then these must be dominant because it is quite difficult to expect that both allelic genes (ca. 14,000 genes in the fruitfly *Drosophila melanogaster*) were mutated or damaged simultaneously in the case of recessive genes. The ratio of dominant mutations to recessive ones, however, is very small in general. Consequently, the extremely high rates of abnormalities, reaching up to 60% (Table 1 in [1]), are difficult to understand based on the Mendelian law of dominance. A simple calculation demonstrates that when a female with a dominant mutation (A, a) is mated with a normal male (A, A), where A is the normal wild type and a is the mutated one, two AA (normal) and two Aa (abnormal) are produced, and abnormal offspring is expected to be reduced by half. When the gene is recessive, all F_1 offspring are phenotypically normal. In fact, abnormality rates in F_1 were 18.3%, 51.9%, and 60.2%, which are higher than that of parents (12.4%) (Table 1, ibid.). The rate of F_2 (33.5%) is also higher than that of the parents. These extremely high rates cannot be explained by gene mutations irrespective of whether they are dominant or recessive. The transmission pattern that abnormalities were enhanced in F_1 generation violates the Mendelian law of inheritance. In other words, the abnormalities must not be genetic. Fundamentally, gene mutation rates in various detection systems are a magnitude of $1/10^4$ at most. There must be other causes that induced the high rate of abnormalities.

3. Mischievous Experimental Design to Detect Gene Mutations

The authors mated affected female butterflies with normal males. Then the offspring were examined. Genes were transmitted from parents to offspring through germ cells this time. However, the experimental design of crossing between affected female and normal males is readily biased and quite mischievous because offspring from affected females are influenced not only by genetic changes but also by female-specifically associated conditions such as egg sizes, nutrients and ingredients in eggs, and hormonal and pheromonal effects. It is most regrettable that we cannot interpret the experimentally obtained results correctly because the experimentally obtained results from the reciprocal cross—between affected males and normal females—are not presented as though we are asked to interpret the results without control experiments. Genetic effects can be evaluated more directly by the cross between affected males and normal females because sperm is a kind of a pure DNA carrier and female-specific effects are negligible. Therefore, data from this cross are strongly required as well as a cross between normal males and normal females.

There are many mutagenicity testing methods. The dominant lethal test is one of them. To carry out this test, usually, male mice or rats are treated with a test compound. They are then crossed with virgin females. The reverse combination is possible, but not employed practically to avoid the possible non-genetic effects of the compound on females in which hormonal control, ovulation, fertilization, embryo development and so forth might be affected.

4. Possible Explanation of High Abnormality Rates: Defective Eggs

Eggs usually contain nutrients to enable larvae to grow for a certain period without eating, depending on species. Malnourished or toxically affected females usually lay small eggs containing fewer nutrients or components. I do not know if the authors measured the sizes of eggs, but they must have been smaller than normal ones. Larvae hatched from these radiation-affected eggs and pupae derived from these larvae would be smaller than normal ones and be defective. As a result, normal differentiation and development were hampered, leading to abnormalities. This idea is a simple plausible explanation for the widely diverse miniature-oriented abnormalities observed in the butterflies caught around the Fukushima area.

The authors examined abnormalities in F_2, which occurred more often than in F_1. This is explained simply by the fact that the defective egg cycle was repeated twice. In the F_2 experiments, abnormal females were crossed with normal males. This crossing is controversial, as discussed above. A reciprocal crossing between abnormal males and normal females is highly requested as a comparative control. Otherwise, correct evaluation of the results is difficult. Butterflies caught in September showed more abnormalities than those in May. This might also be similarly explicable by repeated defective egg cycles.

5. Small RNA Molecules Might be Involved in Defective Eggs to Lead Abnormalities

As a genetic material, DNA is vitally important: only two sets of genomic DNA are kept in the nucleus. Consequently, living organisms are equipped with effective DNA-protection systems including various repair mechanisms. I presume that the pale grass blue butterflies have effective DNA repair systems as usual, at least in the larval stage and probably in the pupal stage. If so, then DNA lesions induced by radiation would be repaired almost completely. This is one reason to consider that the abnormalities were not genetic. If they cannot repair DNA lesions, then most cells are killed before lesions are fixed as mutations. In this case, abnormalities are explained by cell killing.

Therefore, irrespective of the presence or absence of functional DNA repair systems, abnormalities by genetically inheritable mutations are excluded. Then, what are causes of the high abnormalities? Recently, Kikuchi et al. reported that the sex of silkworm *Bombyx mori* is determined by a short RNA [2].

In most moths and butterflies, sex is determined by the ZW system, in which a female has one Z and one W chromosome and a male has two Z chromosomes. Therefore, the W chromosome is responsible for female sex, but it codes no proteins. Instead, the *Fem* region in the W chromosome produces a short 29mer *Fem* piRNA, which cleaves *Musc* mRNA produced by the Z chromosome. Without the Musk protein, *Bmdsx*—*doublesex* of *Bombyx mori*—is female-specifically spliced. A female body is formed by default. Because males lacking the W chromosome lack *Fem* piRNA, Musk protein helps to produce male-specifically spliced *Bmdsx*, leading to a male body. This research is instructive for genetic control of butterflies. A small RNA is the master regulatory molecule of the cascade of sex determination. The initial number of *Fem* piRNA is probably minimal. miRNAs are known to play pivotal roles in early stages of differentiation and development in many organisms. Most abnormalities could be explained by degradation of such RNAs if small RNAs such as miRNA and others such as *Fem* piRNA are involved in the cascades of developmental processes in pale grass blue butterflies and if these are sensitive to radiation. RNA is more labile than DNA. Moreover, there are no RNA repair systems. Most radiation energy is used to produce reactive oxygen species, which might attack developmentally important RNA species. This is an example to explain the butterfly abnormalities.

In fact, microRNAs play critical roles in developmental wings of the neotropical butterfly *Heliconius melpomene* [3]. The doublesex/mab-3 related (DMRT) family is widely involved in sexual determination; *doublesex* is one of them [4]. For example, alternative splicing of a *doublesex* transcript determines the sex in *Bombyx mori*, as shown above. Very interestingly, the sex-limited mimicry in swallowtail butterflies—non-poisonous butterflies mimic wing patterns of poisonous ones—is shown to be controlled by a Mendelian supergene *doublesex* [5].

6. Relevant Information Related to Abnormalities Observed in Another Insect

Recently, Akimoto reported that gall-forming aphids collected around the Fukushima area showed high morphological abnormalities that reached 13.2% in samples collected in 2012. The ratio of the controls was 0.0–5.1% (3.8% on average) in the case of *Tetraneura sorini* [6]. Also, *T. nigroabdominalis* showed higher abnormalities. However, second-generation larvae of *T. sorini* produced only 0.37% abnormalities. Samples of both *T. sorini* and *T. nigroabdominalis* collected in 2013 were markedly normal, indicating that the abnormalities were not inherited and suggesting that they are not genetic consequences. Abnormality data of the pale grass blue butterflies collected in 2012 or 2013 or even 2014, if available, are necessary to ascertain whether the abnormalities were genetic as the authors strongly insist.

REFERENCES

[1] Hiyama A, Nohara C, Kinjo S, Taira W, Gima S, et al. (2012) The biological impacts of the Fukushima nuclear accident on the pale grass blue butterfly. *Scientific Reports* 2:570. doi: 10.1038/srep00570.

[2] Kiuchi T, Koga H, Kawamoto M, Shoji K, Sakai H, et al. (2014) A single female-specific piRNA is the primary determiner of sex in the silkworm. *Nature* 509: 633-636.

[3] Surridge AK1, Lopez-Gomollon S, Moxon S, Maroja LS, Rathjen T, Nadeau NJ, et al. (2011) Characterisation and expression of microRNAs in developing wings of the neotropical butterfly Heliconius melpomene. *BMC Genomics* 12: 62. doi: 10.1186 /1471-2164-12-62.

[4] Artyom K (2012) Dmrt genes in the development and evolution of sexual dimorphism. *Trends in Genetics* 28: 175-184.

[5] Kunte K, Zhang W, Tenger-Trolander A, Palmer DH, Martin A, et al. (2014) doublesex is a mimicry supergene. *Nature* 507: 229-232.

[6] Akimoto S (2014) Morphological abnormalities in gall-forming aphids in a radiation-contaminated area near Fukushima Daiichi: selective impact of fallout? *Ecology and Evolution 4*: 355-369.

In: Fukushima Nuclear Accident
Editor: Shizuyo Sutou

ISBN: 978-1-63463-748-0

Chapter 15

COMMENTARY 3: ABNORMAL BUTTERFLIES CAPTURED AROUND FUKUSHIMA AREAS MAY BE CAUSED BY HSP90-MEDIATED MECHANISM

Masuo Obinata*
Tohoku University, Professor emeritus, Ikeda-machi,
Kita-azumi-gun, Nagano, Japan

ABSTRACT

The collapse of the Fukushima Daiichi Nuclear Power Plant (FDNPP) caused a massive release of radioactive materials to the environment. Hiyama et al reported that the accident caused physiological and genetic • damage to the pale grass blue butterfly *Zizeeria maha*, a common lycaenid butterfly in Japan [1]. I would like to make a short comment on their report on the different view from the accompanying comment by Sutou who argues that the abnormalities of butterflies were caused by non-genetic but toxic effects [2].

INTRODUCTION

First, I briefly summarize their report. While some of the adult butterflies collected in the Fukushima area in May 2011 showed relatively mild abnormalities, the adult butterflies collected in September 2011 showed more severe abnormalities than those collected in May. They reported that the F_1 offspring from the females showed more severe abnormalities, which were inherited by the F_2 generation. Similar abnormalities were experimentally reproduced in individuals from a non-contaminated area by external and internal low-dose exposures.

They concluded that germ-line cells of the individual butterflies were genetically damaged by the massive external exposure from iodine and other artificial radionuclides that

* Correspondence to: Masuo Obinata, Tel: +81-261-62-9231, Email: akiobi8m@nifty.com

were released at one burst immediately after the accident from the FDNPP on 12 March, 2011. The subsequent damage may result primarily from low-level internal exposure caused by ingested leaves. The cumulative effects of the external and internal exposures could have resulted in the deterioration at the population level. They concluded that the present outbreak of abnormal individuals in the Fukushima area was caused by random genetic mutations in addition to physiological effects due to the artificial radionuclides from the FDNPP.

Their results were misinterpreted by BBC News as "The team concluded that this higher rate of mutation came from eating contaminated food, but also from mutations of the parents' genetic material that was passed on to the next generation, even though these mutations were not evident in the previous generations' adult butterflies." (13 August 2012 Last updated at 16:17 'Severe abnormalities' found in Fukushima butterflies' by Nick Crumpton)

Later, similar high morphological abnormalities were reported in gall-forming aphids collected around the Fukushima area in 2012 [3].

Although the statistical evaluation on the abnormalities found in their report is necessary, I agree that the higher abnormalities found in this butter fly were caused by a massive release of radioactive materials to the environment through the collapse of the FDNPP. However, I assume that the abnormality is not due to the genetic mutation [2], but is caused by Hsp90-mediated genetic capacitance [4].

Hsp90 and Buffering of Genetic Variation

In 1942, Waddington reported the expression of novel phenotypes under environmental stress in a variety of organisms, and such environmentally induced phenotypes, once exposed, can become independent of the inducing stress after several generations of selection. He proposed that traits are buffered against minor environmental insults and genetic polymorphisms of small effects. This phenomenon is called ''canalization''. Environmental stress can reveal genetic variants, presumably because it compromises buffering systems. If selected for, these uncovered phenotypes can lead to heritable changes and this phenomenon is called "assimilation" [5].

Both hypotheses are interesting to explain the apparent inheritance of environmentally induced characters, but the existence of and potential molecular mechanisms leading to canalization and genetic assimilation had not been elucidated for long time.

However, recent work by Lindquist and her group provided evidence that the chaperone Hsp90 can serve as such a buffer in the fruit fly *Drosophila melanogaster* and the mustard plant *Arabidopsis thaliana* [4]. Hsp90 is an essential protein in most eukaryotes, and abundant at normal temperatures, but induced by stress. Under physiological conditions, Hsp90 dynamically interacts with a diverse but highly select set of inherently unstable 'client' proteins, such as protein kinases, transcription factors. Hsp90 keeps these metastable proteins poised for activation until they are stabilized by conformational changes. The requirement of many principal regulatory proteins for Hsp90 renders entire pathways sensitive to decreases in its function.

Rutherford and Lindquist [6] reported in *Drosophila* that when removing half of Hsp90 protein, phenotypic variation affecting nearly any adult structure is produced. This variation was observed in specific variants depending on the genetic background and occurring both in laboratory strains and in wild populations. An interesting point of this finding is that multiple, previously silent, genetic determinants produced these variants and, when enriched by selection, they rapidly became independent of the Hsp90 mutation. This suggests that widespread variation affecting morphogenic pathway exists in nature, but is usually silent. Widespread morphological variation may accumulate under neutral conditions because Hsp90 buffers this variation.

When Hsp90 buffering is compromised, for example by temperature, cryptic variants are expressed and selection can lead to the continued expression of these traits, even when Hsp90 function is restored. This provides a plausible mechanism for promoting evolutionary change in otherwise entrenched developmental processes and Hsp90 appears to fulfill Waddington's concept of a developmental buffer and the effective molecule of canalization can lead to the assimilation of novel traits.

Furthermore, the use of Hsp90 as a capacitor for the conditional release of stores of hidden morphogenic variation may have been adaptive for particular lineages, perhaps allowing the rapid morphological radiations that are found in the fossil records.

ABNORMAL BUTTERFLIES MAY BE CAUSED BY HSP90-MEDIATED MECHANISM

Based on the buffering activity of Hsp90 on the widespread morphological variation, I assume that higher abnormality of butterflies captured around Fukushima areas may be caused by Hsp90-mediated mechanism.

Butterflies are generally considered useful environmental indicators and a species of lycaenid butterfly, *Zizeeria maha* is suitable because its wing colour patterns are sensitive to environmental changes. The same authors previously reported that the cold-shock-induced phenotype was heritable and partially assimilated genetically in the breeding line, on the northward range expansion and physiological adaptation of the butterfly [7,8]. Therefore, it is possible that morphological abnormality of the butterfly is produced by a variety of stresses including low dose radiation exposure. Under the low dose radiation exposure that may induce high ROS level in the cells, many proteins may exhibit the slightly destabilized state recognized by Hsp90.

Hsp90 typically has a dynamic interaction with its targets, undergoing cycles of binding and release [9]. Under stress, the presence of new substrates with higher binding affinities may cause Hsp90 activity to become limiting for certain client proteins, even though Hsp90 is highly abundant. In addition, Hsp90-dependent, natural metastable client proteins are likely to be exquisitely sensitive to conditions that promote protein unfolding, thus, limiting Hsp90's ability to keep clients in an activatable state. This hypersensitivity condition may expose cryptic polymorphisms in connected pathways. Under this condition, some destabilized proteins may irreversibly aggregate or be degraded, reducing the function of entire pathways and such narrow developmental windows may allow persistence of the altered phenotype even with stress-induced expression of Hsp90 or a return to ambient conditions. Beyond cytoplasmic protein folding, titration and hypersensitivity may exert effects at all levels of organismic function including alterations in protein trafficking or degradation, facilitating intracellular protein transport across mitochondrial membranes, partitioning proteins into folding or degradation pathways. Limited Hsp90 capacity may cause proteins to be present in an inappropriate setting and increasing the chances for aggregation or improper interactions and may induce a variety of developmental and morphological abnormalities. It is probable that increase of abnormal butterflies after the collapse of the FDNPP in Fukushima area may due to the manifestation of the hidden mutations which had been accumulated previously.

Thus, abnormalities in the butterflies captured around Fukushima areas can be explained by the Hsp90 mediated mechanism without any genetic mutation.

Reliable Evaluation System for the Biological Effects of Low-Dose Radiation Exposure on Animals

Reliable evaluation of the biological influences of the artificial radionuclides from the FDNPP is lacking and the lack of information raises serious concerns about biological influences on humans, because the effects of low-dose radiation exposure on animals, including humans, are still a matter of debate despite the relatively rigorous documentation of physiological damage to animals from external high-dose radiation exposure [10-12].

In terms of evaluation of the biological effects of low-dose radiation exposure on animals, the demonstration that abnormalities were experimentally reproduced in individuals from a non-contaminated area by external and internal low-dose exposures in the laboratory [1] may be a promising, if we consider the mechanism of Hsp90-mediated genetic capacitance. This assay system is different from the genetic mutability assay system which could not get significant response at low-dose radiation levels, but may provide new evaluation method for low-dose radiation effects.

References

[1] Hiyama A, Nohara C, Kinjo S, Taira W, Gima S, et al. (2012) The biological impacts of the Fukushima nuclear accident on the pale grass blue butterfly. *Scientific Reports* 2:570. doi: 10.1038/srep00570.

[2] Sutou S (2014) Chapter 14, Commentary 2: abnormal butterflies captured around Fukushima areas may be due to non-genetic but toxic effects. In: Sutou S, Tanooka H, Doss M. editors. *Fukushima Nuclear Accident: Global Implications, Long-Term Health Effects, and Ecological Consequences.* New York: Nova Science Publishers, Inc. 225-230.

[3] Akimoto S (2014) Morphological abnormalities in gall-forming aphids in a radiation-contaminated area near Fukushima Daiichi: selective impact of fallout? *Ecology and Evolution 4*: 355-369.

[4] Queitsch C, Sangster TA, Lindquist S. (2002) Hsp90 as a capacitor of phenotypic variation. *Nature* 417:618–624.

[5] Waddington CH (1942) Canalization of development and the inheritance of acquired characters. *Nature* 150:563–565.

[6] Rutherford SL, Lindquist S (1998) Hsp90 as a capacitor for morphological evolution. *Nature* 396:336–342.

[7] Hiyama, A., Taira, W. & Otaki, J. M. (2012) Color-pattern evolution in response to environmental stress in butterflies. *Frontiers in Genetics* 3:15.

[8] Otaki J.M, Hiyama A, Iwata M, Kudo T (2010) Phenotypic plasticity in the range-margin population of the lycaenid butterfly Zizeeria maha. *BMC Evolutionary Biology,* 10:252-265

[9] Sangster TA, Lindquist S, Queitsch C (2004) Under cover: causes, effects and implications of Hsp90-mediated genetic capacitance. *BioEssays* 26: 348–362.

[10] Brenner, DJ (2009) Extrapolating radiation-induced cancer risks from low doses to very low doses. *Health Physics* 97: 505–509.

[11] Sazykina, T. G., Kryshev, A. I. & Sanina, K. D. (2009) Non-parametric estimation of thresholds for radiation effects in vertebrate species under chronic low-LET exposures. *Radiation and Environmental Biophysics* 48: 391–404.

[12] Moller, A. P.& Mousseau, T. A. (2011) Efficiency of bio-indicators for low-level radiation under field conditions. *Ecological Indicators* 11: 424–430.

In: Fukushima Nuclear Accident
Editor: Shizuyo Sutou
ISBN: 978-1-63463-748-0

Chapter 16

SUMMARY OF THE FUKUSHIMA DISASTER

Shizuyo Sutou*
School of Pharmacy, Shujitsu University, Okayama, Japan

DESTRUCTION WROUGHT BY THE EARTHQUAKE AND TSUNAMI

More than three and half years have passed since the magnitude 9.0 earthquake, the Great East Japan Earthquake (GEJE), occurred at 14:46 on March 11, 2011. More destructive than the earthquake itself were the accompanying tsunami waves, which devastated long coastal areas of eastern Japan. The disaster claimed 15,889 lives. According to the National Police Agency, 2,601 residents were still missing as of September 10, 2014. Disaster-associated deaths reached 3,089 as of March 31, 2014, according to the Reconstruction Agency. This number includes those who have died of physical and mental fatigue attributable to refuge life, those who died during transfer from homes to refuges, and those who died from lack of primary care because of impaired function of hospitals. More than half (55%) of those disaster-associated deaths were of Fukushima residents. Consequently, a total of more than 20,000 have died as a direct result of this disaster. Injured people were 6,152. Evacuees have decreased in number from 328,903 on November 17, 2011, to 263,392 as of February 13, 2014.

The number of houses that had been completely destroyed, half destroyed, and partially destroyed were 127,367, 273,335, and 744,539, respectively. Many houses were flooded above the floor level (3,352) and below the floor level (10,218). Non-residential houses or buildings destroyed were 57,880. Other damage includes destruction of roads (4,198 sites), bridges (116), embankments (45), and railways (29). Landslides have occurred at 45 sites. Additional damage includes destruction of port facilities and vessels including fishing boats. The Ministry of the Environment announced on March 31, 2014 that tsunami wreckage of 20 million tons and tsunami deposits of 10 million tons had been disposed of in 12 prefectures aside from Fukushima Prefecture, where 71% of wreckage and 46% of deposit had been treated. Of course radioactive contamination is responsible for the delay in the disposal. People who lived in tsunami-devastated areas were once in desperate conditions, but they

* Correspondence to: Shizuyo Sutou, Tel & Fax: +81-86-271-8357, E-mail: sutou@shujitsu.ac.jp.

have been able to find hope of restoration and vision for a better life in the near future. Now, the most disaffected are those who lived around the Fukushima Daiichi Nuclear Power Plant (FDNPP), which was damaged heavily by the tsunami and its aftermath.

Survey of Refuge Life

For the first time since the earthquake, the Fukushima prefectural office conducted a questionnaire survey of 58,627 households during January and February, 2014. Of those questionnaire forms, 26,080 yielded responses. The results were publicized on April 28. The numbers of evacuees in and out of Fukushima Prefecture were 132,500. More than half of the families (14,429) of 26,080 live an evacuee lifestyle. Almost half of the families (48.9%) are separated to 2 (33.3%), 3 (12.1%), 4 (2.9%), or >5 (0.6%) places. Approximately half of the evacuees (44.7%) in Fukushima Prefecture want to return to their former places, while only a small fraction (17.5%) of those who live outside Fukushima Prefecture want to return to their homes. A quarter of those people (26.4%) chose to stay at the present refuge locations. Most importantly, 67% people expressed mental or physical disorders, indicating that life as an evacuee is extremely stressful.

Refuge life was forced upon people by government in response to radio-contamination. Refuge life with fear of radiation is so difficult that many people have been affected by suicide, divorce, separation of family members, migration and settlement to another places, mental illness, and so forth. The Cabinet Office reported that 129 (56 in Fukushima Prefecture) people had committed suicide until July, 2014, in association with GEJE—55 (10, ibid.) in 2011, 24 (13) in 2012, 37 (23) in 2013, and 13 (10) in 2014. Some numerals listed above demonstrate that GEJE caused tremendous material and human damage. Human damage includes not only physical but also mental damage. The major influence of the Fukushima tragedy derives from fear of radiation. Therefore, it is important to ascertain whether such fear is justified.

Scrutiny of the LNT Model

Japan is the only nation in the world to have experienced the devastation of a nuclear attack, the distribution of energy from which was thermal radiation (including light, 35%), blast (pressure shock wave, 50%), and nuclear radiation (immediate and delayed, 15%) [1]. Thermal and blast energy, not radiation, were the major factors contributing to atomic bomb damage and death. Nevertheless, Japanese people may have some misunderstanding, at least in part, that a major part of the overall death toll was attributable to radiation. Such misunderstandings have helped formulate the current Japanese mental attitude of radiation phobia, which seems to constitute a background for their ready acceptance of the idea that radiation is limitlessly hazardous independent of the dose, as implied by the linear no-threshold (LNT) model.

In addition to the atomic bomb experience, fear of radiation is generated by the low radiation dose limit of 1 mSv/y for the general public set by the Japanese government and the recommendation to use the LNT model by the International Commission on Radiological

Protection (ICRP), which insists that the cancer risk is linearly proportional to the radiation dose, with no threshold [2]. When theories or models fail to explain data, they become invalid: so the LNT model is not valid. The LNT model torments Fukushima people and leaves them spell-bound. Not only residents but also farmers and fishery workers have been affected by harmful rumors that their products or catches are radio-contaminated. Consequently, they are rejected by markets.

The LNT model is based on analyses of cancers among atomic bomb survivors in Hiroshima and Nagasaki [3]. Solid cancers and tumors of the esophagus, stomach, colon, liver, lung, bladder, and breast increased as the radiation dose increases. No dose–effect relation was found in those of the pancreas, kidney, or prostate. No increase was detected in cancers or tumors of the rectum, gall bladder, or uterus. Cases of bile-duct cancers were, contrarily, reduced by irradiation. Combining all the cases, an upward-sloping line was obtained. The linear-quadratic model fit to the line best, but the linear model was not deviated significantly from the linear-quadratic line. Therefore, ICRP adopted the linear model. Cancer and tumor cases below 200 mSv were not significantly different from those of the comparative controls. This 200 mSv might therefore be set as the practical threshold. Standing on the side of safety, however, linearity was adopted, even though the originally compiled data could set the threshold at 40 mSv. Fundamentally, solid cancers and tumors are problematic. Radiation doses are not perfectly estimated. Identification of cancers or tumors is not always easy. Sensitivity of tissues or organs to radiation varies to a great degree. The onset of cancers and tumors requires varied lengths of periods of 10–40 years or more. Estimation of background levels of cancers and tumors as the basis for comparison is also extremely difficult because the levels differ depending on the age of exposure, sex, historical period, genetic background and so forth. Considering these factors, assumption of linearity as the LNT model based on solid cancers and tumors appears to be overly simplified and forced. In this regard, the cases of leukemia are much more reliable.

Leukemia is a tumor of blood cells. Its incidence reaches a peak several years after exposure to radiation. Atomic bomb-induced leukemia peaked a long time ago. The result was fixed, although analyses of solid cancers and tumors are still ongoing. The linear-quadratic model fits the incidence of leukemia [4] as for thyroid tumors [3]. In the linear-quadratic model, cancers and tumors increase linearly at first and then increase exponentially. As doses increase further, however, the cancer and tumor incidence reaches a peak and then decreases gradually because people exposed to higher doses die before cancers and tumors develop. Linearity of the LNT model must come into existence in a limited range of a linear-quadratic curve. The linear-quadratic model itself seems to come into existence in a limited, moderate dose range excluding lower and higher doses. Because many biological responses follow the sigmoidal curve, where responses do not appear at first, arise to some degree, gradually increase, and reach a peak. Then they fade out. For example, a bacterial growth curve and the dose–effect curve of a drug follow a sigmoidal curve. Radiation-induced cancers and tumors are likely to follow this curve closely as biological processes.

Irrespective of dose-response shape being the linear-quadratic curve or the sigmoidal curve, the effect of low-dose radiation below 200 mSv on cancer and tumor induction has not been demonstrated clearly. No statistically significant difference has been detected between the exposed and control groups in Hiroshima and Nagasaki. Therefore, the effect, if any, must be negligibly small. Moreover, radiation effect estimates from data in Hiroshima and Nagasaki are exaggerated because the radiation exposure occurred in a short period of time at a high dose-rate that induced 10–100 times higher biological effects than lower dose-rates [5]. In fact, ICRP acknowledges the dose-rate effect and adopts 2 as the dose and dose-rate effectiveness factor (DDREF), but this value is apparently too small and therefore exaggerates low-dose radiation effects.

Some Data Contradictory to the LNT Model

Because several contributors in this book refer to data that are not explained by the LNT model, a few can be listed here. More than 2,500 nuclear bomb tests do not seem to have caused severe health hazards. People living in areas with high background radiation levels appear to be healthy. Japanese people are exposed to 4 mSv radiation on average for medical examinations or treatments. The LNT model predicted that 25,000 would die of cancers and tumors annually from 4 mSv annual dose. In fact, Japanese people have extremely high longevity among the world's people. These few examples support the idea that low-dose radiation is not hazardous to our health and that the LNT model is not relevant. The French Academy of Sciences – French National Academy of Medicine clearly discounts the LNT model [6]. It is of great interest to learn that the higher the radon levels in residences are, the lower the observed lung cancer mortalities. The dose-response relationship fits the linear-quadratic curve and deviates greatly from the LNT model [7]. Induced mutation frequencies in both somatic and germ cells were also found to follow the linear-quadratic curve [8]. These findings [7,8] not only negate the LNT model but also suggest the existence of special biological responses to low dose radiation. Indeed, several contributors to this book describe that low-dose radiation is not harmful, and that it is even beneficial. The idea of net benefit from radiation-stimulated beneficial changes is called hormesis. Some people prefer to call it an adaptive response. Hormesis, or adaptive response, is quite plausible when one examines the evolutional history of living organisms on the earth and the mechanisms of action of radiation on biological systems.

Evolutional View of Radiation Effects

When primitive life appeared on the earth around 3.8 billion years ago (Ba), the environment was anaerobic. Oxygen was toxic. When limited to K-40, the half life of which is 1.2 billion years, concentrations of K-40 on the earth were 2, 4, and 8 times higher, respectively, than those of the present day at 1.2, 2.4, and 3.6 Ba. Considering radiation from other sources, radiation would be much higher. Therefore, organisms without mechanisms to neutralize radiation effects would not have evolved. Photosynthetic organisms started to produce oxygen. The contents of oxygen in the atmosphere began to increase from around 2.5 Ba. Since then, aerobic organisms started to appear. Organisms without mechanisms to dispose of reactive oxygen species (ROS) were unlikely to have evolved. Considering the fact that radiation exerts toxic effects on organisms mostly by producing ROS, common mechanisms to address ROS produced by both radiation and aspiration are expected to have evolved. The Nrf2 system described in this book (chapter 7) is very likely to play a pivotal role in this process.

The essential task of organisms on the earth is to maintain homeostasis on their own. Our body temperature is maintained in a narrow range close to 37°C. Our blood pH is maintained at pH 7.4. Our heart beats, respiration rates, blood sugar levels, menstrual cycles, and many other biological indices are maintained at certain levels with marginal fluctuations by feedback control mechanisms. ROS levels are quite likely to be under the control of homeostasis. A television program reported that, when rats were captured in the Chernobyl Exclusion Zone and examined for hazardous effects, neither DNA damage nor elevation of DNA repair enzymes was detected. Only ROS scavenger levels were heightened. A recent report describes that when 16 species of birds were examined near Chernobyl, the glutathione, an ROS scavenger, level was found to be increased. The authors insist that hormesis is functioning under intense radiation levels [9]. Animals under strong radiation are expected to benefit by erasing ROS before DNA damage occurs.

Quantitative Consideration of ROS Levels

Granted that quenching of ROS produced by respiration and radiation is under the control of homeostasis, what is the controllable range for the ROS-quenching mechanism? The following calculation is made mainly according to references [10, 11]. Numerous ROS (10^9/cell/day) are produced by respiration and cause 10^6 base damage. Moreover, single strand breaks/cell/day are formed, but these are repaired efficiently to leave behind 10^2 lesions/cell/day. The ROS produces 0.1 double strand breaks/cell/day that are less efficiently repaired to leave behind 0.01 lesions/cell/day. However, 0.005 base damage and single strand breaks/cell/day are formed by 1 mSv radiation and are repaired as efficiently as for those produced by respiration to leave behind 0.0000005 lesions/cell/day. Therefore, the ratio of lesions made by respiratory ROS to 1 mSv radiation is 2×10^7. Radiation produces more double strand breaks than respiratory ROS and forms 0.001 breaks/cell/day that are less efficiently repaired to leave behind 0.0001 lesions/cell/day. Consequently, the ratio of lesions produced by respiratory ROS to 1 mSv radiation is 1×10^3. Therefore, irrespective of base damage, single strand breaks, or double strand breaks, daily respiration produces at least 1×10^3 times more DNA damage than 1 mSv radiation does. In other words, 1,000 mSv is equivalent to daily respiration. Provided that homeostasis can control additional ROS equivalent to daily respiratory ROS, an acceptable or tolerable radiation dose would be 1,000 mSv or 1 Sv.

Beyond the Spell of LNT

This book was compiled for all people who are concerned about radiation. Some major targets are the following: Those people working at nuclear power plants must feel relieved by reading this. Governmental authorities must assume heavy responsibility. The Ministry of Health, Labour, and Welfare (Japan) set a new standard on April 1, 2012, that radioactivity in food and water/milk be less than 100 and 50 Bq/kg, respectively, while the corresponding US standards are 1,000 and 1,200 Bq/kg. This strict standard only serves to stir up feelings that

Table 1. Critical events in the Fukushima Daiichi disaster are chronologically tabulated

Date	Agent and/ or time[1]	Event
1971.03.26	FDNPP[2]	Unit 1 (0.46 MW) of FDNPP went into operation.
1974.07.18		Unit 2 (0.784 MW) went into operation.
1976.03.27		Unit 3 (0.784 MW) went into operation.
1978.10.12		Unit 4 (0.784 MW) went into operation.
1978.04.18		Unit 5 (0.784 MW) went into operation.
1979.10.24		Unit 6 (1.1 MW) went into operation.
2011.03.11	14:46	The 2011 earthquake of a magnitude 9 occurred.
	14:46	Unit 1 automatically scrammed.
	14:47	The main turbine of the Unit 1 was automatically shut down.
	14:47	Unit 2 automatically scrammed, the main turbine was automatically shut down. The emergency diesel generator automatically started up.
	14:47	Unit 3 automatically scrammed. The main turbine was shut down manually.
	14:47	The emergency diesel generator of the Unit 5 automatically started up.
	14:47	The emergency diesel generator of the Unit 6 automatically started up.
	14:48	The emergency diesel generator of the Unit 3 automatically started up.
	14:52	The isolation condenser of the Unit 3 automatically started up.
	TOPCO[3] 15:06	The Headquarters for emergency Disaster Control was set up.
	Jp. Gov.[4] 15:14	The Headquarters for Extreme Disaster Management was set up in the Cabinet Office.
	15:27	The first wave hit the FDNPP; there were no hazards.
	15:35	The second wave hit the FDNPP; major buildings of Units 1 to 6 were flooded.
	15:37	Unit 1 lost all electric supply.
	15:38	Unit 3 lost all electric supply, but DC electricity was active.
	15:38	Unit 4 lost all electric supply.
2011.03.11	15:40	Unit 5 lost AC-electric supply, but DC electricity was active.
	15:41	Unit 2 lost all electric supply.
	17:50	High radioactivity was detected at near the double door of Unit 1.
	Fukushima P.[5], 20:50	Fukushima P. ordered the inhabitants within a 2 km radius from FDNPP to evacuate.
	P.M.[6], 21:23	P.M. ordered the inhabitants within a 3 km radius from FDNPP to evacuate. P.M. ordered the inhabitants within a 3-10 km radius from FDNPP to evacuate indoors.
	21:51	Entrance into Unit 1 was prohibited because of high radioactivity in it.
	22:00～	Meltdown seemed to start in Unit 1.
2011.03.12	6:00	M. Yoshida, General Manager of FDNPP, ordered to prepare for ventilation in Unit 1.
	2:45	Pressures of inside and outside of the reactor containment vessel of Unit 1 were equal.
	2:55	It was confirmed that the reactor core isolation cooling system (RCIC) of Unit 2 was working.
	4:00～	Fresh water was started to pour into the Unit 1 reactor.
	P.M., 5:44	Prime minister ordered the inhabitants within a 10 km radius from FDNPP to evacuate.
	P.M., 7:11	Prime Minister arrived at FDNPP.
	P.M., 8:04	Prime Minister left FDNPP.
	9:15	An operator manually opened the ventilation valve of Unit 1.
	11:36	The reactor core isolation cooling system (RCIC) of Unit 3 stopped.
	12:35	The high pressure coolant injection (HPCI) of Unit 3 started automatically.
	14:30	The pressure in the drywell of Unit 1 reactor vessel was low; ventilation seemed successful.
	15:36	Explosion occurred in Unit 1.

Date	Agent and/ or time[1]	Event
	17:30	M. Yoshida ordered to prepare for ventilation in Unit 2 and 3.
	18:25	Prime minister ordered the inhabitants within a 20 km radius from FDNPP to evacuate.
	19:04	Sea water was started to pour into the Unit 1 reactor.
	20:45	Sea water containing borate was started to pour into the Unit 1 reactor.
2011.03.13	2:42	The high pressure coolant injection (HPCI, the first defense line) of Unit 3 was manually stopped.
	9:20	Ventilation was started in Unit 3.
	9:25	Fresh water containing borate was started to pour into the Unit 3 reactor.
	10:00~	Meltdown seemed to start in Unit 3.
	10:15	M. Yoshida ordered to start ventilation in Unit 2.
	12:20	Pouring of fresh water containing borate was finished in the Unit 3 reactor.
	13:12	Sea water was started to pour into the Unit 3 reactor.
	14:45	High radioactivity was detected at the double door of Unit 3.
2011.03.14	3:20	Sea water was started again to pour into the Unit 3 reactor.
	4:08	The temperature of the pool for spent nuclear fuel of Unit 4 was 84℃.
	6:30	The inner pressure of the containment building of Unit 3 became high.
	11:01	Explosion occurred in Unit 3.
	13:25	The high pressure coolant injection (HPCI) of Unit 2 stopped.
	15:30~	Sea water was started to pour into the Unit 3 reactor.
	19:57	Sea water was started to pour into the Unit 2 reactor.
2011.03.14	22:00~	Meltdown seemed to start in Unit 2.
2011.03.15	5:35	Jp. Gov. and Fukushima P. established a combined headquarters for emergency disaster control.
	6:55	Explosion occurred in Unit 4.
	9:38	A fire broke in Unit 4.
	P.M. 11:00	Prime minister ordered the inhabitants within a 20-30 km radius from FDNPP to evacuate indoors.
	11:00~	The fire in Unit 4 died down automatically.
2011.03.11-03.14		An east wind from March 11–14 carried large quantities of the radioactive substances over the Pacific.
2011.03.15-03.16		A west wind on March 15 blew back the substances toward the Japanese archipelago.
2011.03.15 — 03.16		Rain or snow from 5:00 p.m., March 15, to 4:00 a.m., March 16, was supposed to fix most contamination in Fukushima P.
2011.03.17	from 9:48	Sea water was poured from choppers onto the pool for spent nuclear fuel of Unit 3 (totally 4 times).
	TOPCO	TOPCO published the Decontamination plan of Fukushima Daiichi Nuclear Power Plant.
2011.03.18	13:30	Three openings were made on the roof of the Unit 5 building.
	17:00	Three openings were made on the roof of the Unit 6 building.
2011.03.20	14:30	The Unit 5 reactor was in cold shutdown (its coolant system was below 100℃ at atmospheric pressure).
	19:27	The Unit 6 reactor was in cold shutdown.
2011.03.22	10:35	Units 3 and 4 were supplied with electricity from outside.
2011.03.24		The feet of thee workers were soaked in contaminated water 15 cm in depth (no hazardous effects).
2011.03.25	15:37	Sea water was replaced with fresh water as a coolant in Unit 1.
	18:02	Sea water was replaced with fresh water as a coolant in Unit 3.
2011.03.26	10:10	Sea water was replaced with fresh water as a coolant in Unit 4.
2011.03.28		Trenches outside the turbine facilities were found to be full of contaminated water.
2011.06.24		Water treatment equipment was introduced to remove cesium-137 from contaminated water.

Table 1. (Continued)

Date	Agent and/ or time[1]	Event
2011.07.02	18:00	The recirculation cooling system using decontaminated water was adopted to the Units 1 and 2 reactors.
2011.12.16	17:00	Jp. Gov. and TOPCO declared that the 6 FDNPP reactors were in cold shutdown.
2011.12.21		TOPCO published the "Mid-and-Long-Term Roadmap towards the Decommissioning of TEPCO's FDNPP."
2012.02.10	Cabinet Office	The Reconstruction Agency was set up (closing: 2021.03.31).
2012.04.19		The FDNPP Units 2-4 were legally decommissioned.
2012.04.20		The FDNPP Units 1 was legally decommissioned.
2012.09.19	Min. Environ[7]	The Nuclear Regulation Authorities was set up.
2013.03.30	TEPCO	Trial runs of the ALPS multi-nuclide removal equipment started at FDNPP.
2013.07.08	NRA[8]	Enforcement of the new regulatory requirements for commercial nuclear power plants.
2013.09.03	Jp. Gov.	Jp. Gov. declared the Basic Policy for the Contaminated Water Issue at the TEPCO's FDNPP.
2013.09.15	KEPCO[9]	KEPCO's Unit 4 at the Ohi nuclear power plant was shut down for regular inspections. This left Japan without any nuclear power for the third time in 40 years.
2014.01.31	TOPCO	The FDNPP Units 5 and 6 were legally decommissioned, leaving 48 reactors in Japan as of February 1, 2014.
2014.03.31	Min. Environ.	Debris ($2x10^7$ tones) and earth ($1x10^7$ tons) in Kanto and Tohoku districts excepting Fukushima were processed.
2014.03.31	Min. Environ.	Decontamination operation was over for a village and two towns out of 11 areas in Fukushima P.
2014.04.01	Jp. Gov.	Evacuation order was cancelled partially in Miyakoji district of Tamura City, Fukushima P.
2014.04.02	UNSCEAR[10]	UNSCEAR stated that cancer levels are likely to remain stable in the wake of the 2011 FDNPP accident.
2014.06.02	TEPCO	Construction of an ice wall 30 m deep and 1,500 m long started to stop inflow of underground water into FDNPP.
2014.07.16	NRA	The Sendai Nuclear Power Plant of KEPCO passed the review of the new regulatory requirements.
2014.08.17	Jp. Gov.	Evacuation order was cancelled from October 1, 2014, fully in Kawauchi Village, Fukushima P.
2014.08.26	Electric Co's	Nine electric companies had filed 20 applications (12 facilities) for a safety review to resume operations.
2014.09.01	Fukushima P.	The Governor of Fukushima accepted the construction of interim storage facility of contaminated soil (from 2015.01).
2014.09.10	NRA	NRA decided to issue a certificate to resume operations of KEPCO's two nuclear reactors.
2014.09.11	Cabinet Sec.[11]	The Accident Investigation Board of the TOPCO's FDNPP Accident publicized testimonies from 19 persons.
2021		Removal of fuel debris will be started.
2051		Full commissioning of nuclear reactors will be over.

[1] Unless otherwise specified, events occurred at FDNPP.
[2] FDNPP: Fukushima Daiichi Nuclear Power Plant.
[3] TOPCO: Tokyo Electric Power Co., Ltd.
[4] Jp. Gov.: The Japanese Government.
[5] Fukushima P.: Fukushima Prefecture.
[6] P.M.: Prime Minister.
[7] Min. Environ.: The Ministry of Environment.
[8] NRA: The Nuclear Regulation Authorities.
[9] KEPCO: The Kansai Electric Power Co. Ltd.
[10] UNSCEAR: The United Nations Scientific Committee on the Effects of Atomic Radiation.
[11] Cabinet Sec.: Cabinet Secretariat.

low-dose radiation is dangerous. Most radiation-associated articles in magazines and newspapers take a pessimistic tone that low-dose radiation is dangerous. Journalists must be expected to learn the hormetic or adaptive response of radiation. Most activists who advocate that they protect the public from hazardous radiation usually emphasize the dangerous aspects of radiation, which tortures and torments the public by evoking undue fear and anxiety, which in turn might induce physical and mental disorders, familial separation, suicide, and so forth as described above. It is a brutal irony that the fear of radiation has been more dangerous than the radiation itself to the people in Fukushima. No one has been killed by radiation there, but more than 1,000 people have died as a direct result of fear and stress. The prime cause seems to reside in the traditional LNT model, on which new scientific light should be shed. Quantity changes quality. Strong instantaneous radiation (high dose rate) from atomic blasts is fatal just as drinking a large amount of alcohol in one gulp is fatal. Low-dose radiation over a prolonged period (low dose rate) is not only non-hazardous, but beneficial, as shown by hormesis or the adaptive response, just as a daily sip of alcohol might be helpful for longevity. The United Nations Scientific Committee on the Effects of Atomic Radiation (UNSCEAR) reported on April 2, 2013 that radiation had not affected Fukushima people's health and predicted that cancer incidence would not increase in the future. Critical events in the Fukushima Daiichi disaster are chronologically tabulated in Table 1.

REFERENCES

[1] Glaser A (2007). Effects of nuclear weapons, Princeton University. (http://www.princeton.edu/~aglaser/lecture2007_weaponeffects.pdf)

[2] The International Commission on Radiological Protection (2007) The 2007 Recommendation of the International Commission on Radiological Protection. ICRP Publication 103, *Elsevier Ltd,* Amsterdam.

[3] Preston DL, Ron E, Tokuoka S, Funamoto S, Nishi N, et al. (2007) Solid cancer incidence in atomic bomb survivors: 1958-1998. *Radiation Research* 168:1-64.

[4] Committee to Assess Health Risks from Exposure to Low Levels of Ionizing Radiation (2006) Free executive summary. Health risks from exposure to low levels of ionizing radiation: BEIR VII – Phase 2. P. 16.

[5] Tanooka H (2011) Meta-analysis of non-tumour doses for radiation-induced cancer on the basis of dose-rate. *International Journal of Radiational Biology* 87: 645–652.

[6] Aurengo A, Averbeck D, Bonnin A, Le_Guen B, Roland_Masse R et al. (2005)Dose-effect relationships and estimation of the carcinogenic effects of low doses of ionizing radiation. http://lowrad.wonuc.org/lowrad/lowrad-bulletin.htm

[7] Cohen BL (1997) Lung cancer rate vs. Mean radon level in U.S. counties of various characteristics. *Health Physics* 72: 114–119.

[8] Vilenchik MM, AG Knudson, Jr AG (2000) Inverse radiation dose-rate effects on somatic and germ-line mutations and DNA damage rates. *Proceedings of National Academy of Sciences of the United States of America USA* 97: 5381-5386.

[9] Galvan I, Bonisoli-Alquqti A, Jenkinson S, Ghanem G, Wakamatsu K, et al. (2014) Chronic exposure to low-dose radiation at Chernobyl favors adaptation to oxidative stress in birds. *Animal Physiological Ecology*. Doi: 10.1111/1365-2435.12283

[10] Jerry M. Cuttler JM, Pollycove M (2009) Nuclear energy and health and the benefits of low-dose radiation hormesis. *Dose-Response* 7: 52–89.
[11] Tateno Y (2001) *Radiation and health*. Iwanami Shoten, Tokyo (in Japanese).

INDEX

C

E

F

I

J

K

L

M

N

O

P

Q

R

S

T

U

V

W

X

Y

Z

β

γ